AF517715

21st Century Biology and Agriculture

Wildlife Ecology and Conservation

- Mundanthra Balakrishnan
Professor, Department of Zoological Sciences, College of Natural Sciences, Addis Ababa University, Addis Ababa, Ethiopia

Published by:

SCIENTIFIC PUBLISHERS (INDIA)

Jodhpur
–

5 A, New Pali Road
P.O. Box 91
Jodhpur - 342 001
INDIA

Delhi
–

4806/24, Ansari Road
Daryaganj
New Delhi - 110 002
INDIA

E-mail: info@scientificpub.com
Website: www.scientificpub.com

Print 2025

Cover illustration: This textbook series *21st Century Biology and Agriculture* launched by Scientific Publications, Jodhpur (India), celebrates the native, but critically-threatened bird of the western semi-arid grasslands and scrubs of India, the great Indian bustard *Ardeotis nigriceps* (Gruiformes: Otididae). The great Indian bustard has been a part of Indian culture and tradition, known as *gonãdh* (*Sanskrit*), because the male call closely resembles the mooing of cows. Artist: Urvashi Sharma.

ISBN: 978-81-7233-974-6

eISBN: 978-93-87307-70-4

Printed in India

Foreword

The year 2015 is a landmark year in international development. It marks the transition from Millennium Development Goals to a new post-2015 agenda for sustainable development, with new Sustainable Development Goals as pillars of this new development agenda, and for advancing international efforts to cope with global climate change.

Sustainable development and wildlife as defined by Mundanthra Balakrishnan in *Wildlife Ecology and Conservation* are closely interlinked. At the same time, as we address our development challenges, we are at the mid-way mark in the process of trying to achieve the goals and targets of the Strategic Plan for Biodiversity 2011–2020. All these efforts are aimed at meeting the challenges of alleviating poverty, improving human health and providing clean water, food and energy for all. It has become ever clearer that conserving our biological diversity is *sine qua non* not only for making sustainable development become reality but for the well-being of humankind itself.

How does all this relate to Prof. Balakrishnan's book *Wildlife Ecology and Conservation*?

Education for all and capacity building are key elements in the processes discussed above. This is where the book by Prof. Balakrishnan comes into play. It serves as an element in these contexts and, more specifically, at its target clientele, the college and university students and teachers in the Indian subcontinent. The book uses the right approach, viz. a connotation of wildlife as encompassing all organisms not directly under human control and which build the fabric of our "natural ecosystems". The latter term is however discussed controversially: there are virtually no ecosystems, which have not been influenced by humans to various extents. Again, wildlife traditionally refers to non-domesticated animal species, but has come to include all animals, plants, fungi and other organisms, which grow or live wild in an area and have not been introduced by humans. As a matter of fact, the distinction between wildlife and biodiversity, not even in contexts such as agro-ecosystems or urban ecosystems ("urban wildlife"), may not be as straight forward as commonly thought.

Loss of wildlife and biodiversity has been a serious issue and challenge at all levels, from local to international. Bringing biodiversity loss to a complete halt, as repeatedly requested and currently aimed at in the international "Vision 2050" for biodiversity, may be too lofty a goal to be achievable in the coming decades, or it may be even impossible to be achieved in its entirety, lastly a result of continued growth of the human population and concomitant rising pressure on natural resources.

Wildlife has experienced enormous losses. According to a recent report of the World Wide Fund for Nature (WWF), between 1970 and 2014 wildlife has been globally reduced by over 50%. Severely affected have been the tropical realms, in particular the Neotropical and Indo-Pacific Realms. Global species extinction rates are generally estimated 1,000 times or even more than the normal or background extinction rate.

India, a mega diverse country, contains two of the 34 global hotspots of biodiversity. Neighbouring countries in South Asia have two more biological hotspots, which are partially contiguous to India. Special efforts have been taken to conserve biodiversity and wildlife in India. For instance, in line with the Aichi Targets, India has defined its specific targets to address the different facets of biodiversity. In international debates, India has been a forerunner in stressing the need for a global change in our production and consumption patterns. In view of the continued human population growth and increasing pressure on natural resources and ecosystems, this change may be needed ever more for the conservation of biodiversity for our future generations.

I have known Mundanthra Balakrishnan for a long time. We first met when he still worked in southern India, at the Department of Zoology, University of Kerala, and I carried out research on the biogeography of India and Sri Lanka. Since then we have been in regular contact.

Wildlife Ecology and Conservation is not just another book on wildlife ecology and conservation. Its primary focus is on education. Accordingly, its design reflects the series focus on "self-directed learning", as pointed out in the series preface at the beginning of the book. *Wildlife Ecology and Conservation* is targeted at college and university students and teachers in the Indian subcontinent. A major add-on is the 13 text boxes in which specific aspects are highlighted. These include species-specific situations, examples of ecosystems, major institutions working on wildlife issues in India, butterfly farming, and specific methodology used in wildlife research.

Depth and breadth of discussing wildlife and conservation issues is a major feature and strength of Prof. Balakrishnan's book. This holistic approach is reflected *inter alia* in the introduction which sets the scene, an overview of the biogeography of India and major ecological aspects of wildlife, followed by chapters on habitats, predator–prey interactions, ecological principles, conservation with emphasis on India and the South

Asian subregion, human–wildlife conflicts, economic aspects related to wildlife, protected areas and wildlife conservation, conservation biology, and international instruments. The book concludes with two chapters on methods used in ecological studies of wildlife. Each chapter ends with content-related questions posed to the reader. Both theoretical and practical aspects of wildlife ecology and conservation are, therefore, presented.

With rapid population growth and development in India, conservation of biodiversity and wildlife and related education efforts have become ever more pressing an issue. Specifically with regard to this situation, I hope that this book will make a difference. Moreover, I do hope that *Wildlife Ecology and Conservation* will become a standard textbook for education about wildlife and its conservation, not only for current and future generations of students and teachers of India, but also at regional and international levels.

Mechernich,
June 11, 2015

Prof. Dr. Walter R. Erdelen
Former Assistant Director-General for Natural Sciences
UNESCO

Series Preface

With great pleasure, the Scientific Publishers (Jodhpur) and I launch the third volume of the postgraduate textbook series under the general title *21st Century Biology and Agriculture* for use in the Indian subcontinent.

This series aims at fulfilling the knowledge needs of postgraduate learners in agriculture and biology, focussing on self-directed learning. Keeping this point in full view, these volumes would dilate on contemporary information in chosen themes in a pertinent, but brief backdrop of historical knowledge, with appropriate textual information laced with relevant illustrations. Most importantly, these books aim to cater to the self-learning needs in passionate and committed learners. By self-learning, the teacher's role turns into mentoring rather than tutoring. Vital details have been 'box'ed so that learners can internalize them easily, swiftly, and forever. Care has been exercised to integrate examples from the Indian subcontinent, so that the learners can relate to concepts and principles quickly. The most critical aspect is the inclusion of specific case studies and interactive-mode of learning, so that learners can learn about the day-to-day issues and application of theory that surround the nominated theme effectively. Every effort has been meticulously made to see that the books launched under this series are easily readable and user friendly.

Tanay Sharma of Scientific Publishers (Jodhpur) readily agreed to my demand to identify and recruit an efficient copy editor, who has done a neat job.

This book entitled *Wildlife Ecology and Conservation* executed by M. Balakrishnan is the third book of this series. Balakrishnan is a well-known wildlife ecologist of India. Presently he is working as professor of Zoology, Addis Ababa University in Ethiopia. He has contributed to diverse aspects of wildlife biology, ethology and ecology. Scientific Publishers (Jodhpur) and I consider his acceptance to write this book a great honour.

Many more similar titles have been enlisted and will be appearing in the near and far future. Scientific Publishers (Jodhpur) and I thank Balakrishnan for readily and willingly accepting to our request by executing this task, and importantly doing it speedily.

India is a land of varied landscapes. One of the remarkable natural icons of India is the Great Indian Bustard, the populations of which are restricted to Western India. Because this series is being published from Jodhpur, I thought it will be appropriate to remind ourselves of the uniqueness of this splendid bird and make every effort of conserve its diminishing population.

All books slated to appear in the series *21st Century Biology and Agriculture* will celebrate this Indian icon.

I am confident that learners will benefit from the information and knowledge shared by M. Balakrishnan; I am also confident that post-graduate teachers of biology and agriculture in the Indian subcontinent would find this material appropriate to prescribe this book as a learning resource.

Anantanarayanan Raman
Editor-in-Chief, Textbook Series
21st Century Biology and Agriculture
School of Agricultural & Wine Sciences
Charles Sturt University
Orange, NSW 2800, Australia

Preface

This textbook on Wildlife Ecology and Conservation is prepared as requested by Scientific Publishers, Jodhpur, for the use of students of College and University students and teachers in India. My long-term experience of over 40 years at the University of Kerala, Agricultural University of Norway and Addis Ababa University, Ethiopia, where I have been dealing with this subject area in teaching and research have helped me to put in all essential information with suitable examples to introduce the subject matter in a simple way, and I hope our students can easily grasp and understand the basic concepts of wildlife ecology and conservation biology using this book. For a better understanding of the subject, I have drawn more examples from India with additional examples from Africa and elsewhere in the tropics.

As a deviation from routine textbooks, I have included 13 special boxes in this book. Some of these boxes are case studies, some are typical examples and others are details of techniques. These boxes are prepared by eminent scientists having long-term experience in the respective subject areas. I expect that these boxes will serve for details of the subject with first hand elaborations by well-known scientists, which will be of high use for the student community to know at least some of the scientists in their own fields of interest. Contributors of these boxes have also agreed that anyone who may read these boxes can contact them for further information and ideas, if required, and all of them have given their full postal address and e mail IDs for this purpose. I am extremely thankful to each of these contributors for devoting their time in response to my request and providing me their text and figures on time.

I thank Dr. S. Anilkumar (one of my former post-graduate and doctoral students at the University of Kerala, India), now at Jawaharlal Nehru Tropical Botanic Garden and Research Institute, Thiruvananthapuram, Kerala for helping me in the preparation of the drawings used in this book. My son, Sharon M. Balakrishnan is thanked for his help in final editing of figures and photos included in this book. He has also supported me on computer technical problems during the final editing of the text, figures and photos. I am also thankful to a number of my post-graduate and doctoral students whom not only I taught, but from whom I also learned various issues of wildlife ecology, conservation and management I have presented in this book. I am also thankful to them for their valuable feedback on my ecology, biodiversity and other wildlife related courses.

The text of this book was prepared using my spare time in Addis Ababa. In this context, I am indebted to Addis Ababa University, particularly to the Department of Zoological Sciences for giving me excellent working facilities and comfortable working atmosphere. Professor Afework Bekele, former Dean, Faculty of Science has been supporting me throughout my over 13 years of career with Addis Ababa University, which made me to continue here for such a long time even beyond my expectations. I sincerely thank my other Ethiopian colleagues, Dr. Kifle Dagne, Dr. Dawit Abate, Dr. Gurja Belay, Dr. Ababe Getahun and Dr. Tilaye Wube, under whom I have worked from time to time, for their support.

I take this opportunity to thank Professor Anantanarayanan Raman, Charles Sturt University, Australia, for inviting me to write this book. Mr. Tanay Sharma, Scientific Publishers (India) and his team of officials are thanked for their immediate and efficient responses on all my clarifications, particularly during the final stage of the preparation of this book.

I have reproduced parts of my publications appeared in few of the scientific journals as part of student review, exercise and case study for which I thank the Society of Photo–Optical Instrumentation Engineers (SPIE); National Institute of Ecology, New Delhi; International Scientific Publications, New Delhi; International Society for Tropical Ecology; Blackwell Publishing Ltd; Indian Academy of Sciences; Elsevier; Association for Tropical Biology and Conservation; Bombay Natural History Society; and the International Union for Conservation of Nature and Natural Resources.

I will fail in my attempt if I do not thank my parent institution, the Department of Zoology, University of Kerala, Thiruvananthapuram, where I started my academic career as a doctoral student and retired as a full Professor. This institution was responsible for my academic advancements and success, with its limited facilities and funding just like any of the other Indian State Universities. The earlier experience I had with the Kerala Forest Research Institute, Peechi as a staff of its Division of Wildlife made me to advance as a wildlife ecologist, for which I thank my then senior colleague, Dr VS Vijayan, who introduced me to the field of wildlife ecology. It was with him that I first made wildlife field observations in the famous Silent Valley forests of Kerala.

I have made my wife, Mrs. P.C. Roopa to suffer for a long time during the preparation of this text, as I have been spending much of my time at home to complete this task. I take this opportunity to thank her for her patience.

M. Balakrishnan

Addis Ababa,
March 30, 2015.

List of Boxes

Contents

Introduction 1

Wildlife – definition and scope, wildlife ecology and management, biomes of the world, biogeographic zones of India, ecology of natural communities, niche in wildlife ecology, ecological succession, diversity and stability, association between species, food specificity and habitat association of wildlife

---

1.1. WILDLIFE - DEFINITION AND SCOPE

The term 'wildlife' implies all the organisms living in their natural habitats. This means that any of the organisms living outside the direct human control are wildlife. Domestic organisms, laboratory organisms and those in farmlands are not wildlife forms. Wildlife denotes not only large animals, but it also encompasses all wild organisms — large and small, plants and animals (fungi, wild orchids, flagellates, insects, fish, frogs, snakes, tortoise, birds, deer, tiger, elephant, whale and so on). It differs from biodiversity as biodiversity encompasses all organisms on the planet Earth — wildlife, domestic organisms and organisms in farmlands and in laboratories (Narendran and Balakrishnan 2008). Wildlife is the basis of natural ecosystems where specific combination of plants and their depending animals coexist and sustain naturally. Wildlife contributes essential resources such as oxygen, and also absorbs carbon dioxide and regulates climate. The greenhouse effect is brought under control by natural regulation of wildlife of various forms. When wildlife is disturbed (e.g., clearing of forest habitat), wild habitats loose scope for carbon sink, which leads to the process of global warming (IPCC 2007, Meinshausen et al 2009).

In general, it is assumed that wildlife denotes only mammals such as elephant, lion, deer, rhinoceros and other large animals. This is not true. All organisms, both plants and animals, whether small or large, seen in natural habitats are wildlife (Boxes 1.1 – Figs 1 to 4, 1.2 – Figs 5 to 9, 1.3 – Fig 10). Larger herbivores have a major role in controlling natural populations of a variety of herbaceous species. Similarly, larger carnivores such as the lion (*Panthera leo* Linnaeus, 1758) or the tiger (*Panthera tigris* Linnaeus, 1758), being placed at the apex of food web, have major roles in controlling populations of artiodactyls. Larger artiodactyls have control over the populations of a variety of fodder-producing species. Similarly, small mammals such as rodents are also important in natural habitats as regulators of ecosystems because they are ecological engineers. Insects form the largest group of animals in any type of habitat. They form important part of diverse ecosystems as many of the insects act as agents of pollination. They form food of many species of animals (e.g., amphibians). They feed upon vegetation and regulate the growth of such species. In general, we consider insects as vermin due to their role as vectors, pests and parasites, but they are important in regulation and maintenance of the habitat where they live. This shows that any species of animals, whether large or small, has its own role in nature. Similarly, plants also play a major role in regulating habitats by directly providing food, and hiding and nesting areas for a variety of animals. Thus, a combination of plants and animals in any area forms the strength of the habitat characteristics and the wildlife of the area.

Box 1.1

Caecilians – The Elusive Limbless Amphibians

Ramachandran Kotharambath[1] *and Oommen V Oommen*[2]

[1]Department of Zoology, Government College, Chittur, Kerala, India
(E-mail: ramachandrank6@gmail.com)
[2]Kerala State Biodiversity Board, Thiruvanthapuram, Kerala, India
(E-mail: oommenvo@gmail.com)

Amphibians are fascinating animals living around us. They are present in almost all terrestrial and freshwater habitats barring the coldest and the driest regions, and some of the most remote oceanic islands (Stuart et al 2008). Though there are three major types of living amphibians, frogs and toads (Order Anura) are more familiar, and to a lesser extent newts and salamanders (Order Caudata). The third group of amphibians, caecilians (Order Gymnophiona) is almost unfamiliar to not only the general public, but also to most of the biologists. The following is a quick look at the interesting aspects of one of the most secretive vertebrates on the Earth.

Caecilians are elongated animals with superficial resemblance to worms and snakes. Smaller species bear a resemblance to earthworms, while the larger ones look like snakes. The aquatic larvae of some species could easily be mistaken for eels. Unlike frogs, caecilians are found only in the tropics of Americas, Africa, Asia and Seychelles Islands off the coast of Africa. They are not reported from the Madagascar, New Guinea and Australasia. Caecilians are among the least studied vertebrate groups on the Earth. Of the >7,000 amphibian species known today, about 200 are caecilians. A caecilian species can be as short as 11 cm (*Grandisonia brevis* from Seychelles Islands) to as long as 150 cm (*Caecilia thompsoni* from Columbia).

Fig 1. *Uraeotyphlus* species

Caecilians are usually seen in leaf-litter or under the soil by the side of streams, rivers or other water bodies. They prefer a secretive life, mostly inside decaying vegetable matters or in wet soil. As the eyes are covered with skin, they have poor vision. In some species, the eyes are placed under the bone. Morphologically, caecilians are very different from other amphibian groups. Apart from the shape of snake, their entire skin is divided into annuli or folds, and most of the species possesses scales under the skin folds. All the caecilian species have a pair of sensory tentacles placed between eyes and nostrils. Most of the caecilian species are dark in colour, but some are remarkably colourful with shades of pink, yellow and brown. All the species show internal fertilization, and have diverse reproductive modes. Caecilians are normally perceived as rare due to their subterranean lifestyle. Some species of caecilians, like those of the Family Ichthyophiidae, can be seen slithering like snakes in the courtyard, garden or in the forest paths during or just after heavy monsoon showers.

Being a tropical country, India is home to diverse groups of caecilians. Currently, 35 species are described from India, of which 25 are from the Western Ghats and 10 are from the north-east. Caecilians of three families are known from India. There have been many new species discoveries in the last decade, especially from the Western Ghats. Some of the caecilian species have dense populations in their habitats.

Gegeneophis primus Kotharambath et al

Gegeneophis primus is one of the latest caecilian species (Family Indotyphlidae) discovered from the Western Ghats. It was found under the soil by the side of a mountain stream inside Sugandhagiri Cardamom Estate near Vythiri in Wayanad district of Kerala. A small species with maximum known length of 180 mm, *G. primus* could well be mistaken for earthworms.

Fig 2. *Gegeneophis primus*

Myths

The elusiveness of an animal species contributes to the formation of myths around it. Rural people believe in thousands of myths about the animals living around. For an animal with the underground life and snake-like body, it is not surprising to be associated with myths. In many parts of the Western Ghats, caecilians are associated with a belief that they are venomous. In some areas, they are considered as deadly venomous. In fact, caecilians generally do not bite humans, and are harmless. They produce secretions from the skin glands to keep away some of the natural predators. A tribal man in the Agasthyavanam Biological Reserve opined that *Ichthyophis tricolour* was more venomous than a King Cobra! But he had no knowledge of any case of caecilian bite or any casualty. There are people who know that caecilians are not venomous, but just unappealing. During field works, we have seen people who are ready to hold them in hands. However, most of the people do not even dare to touch a caecilian believing that they can be dangerous.

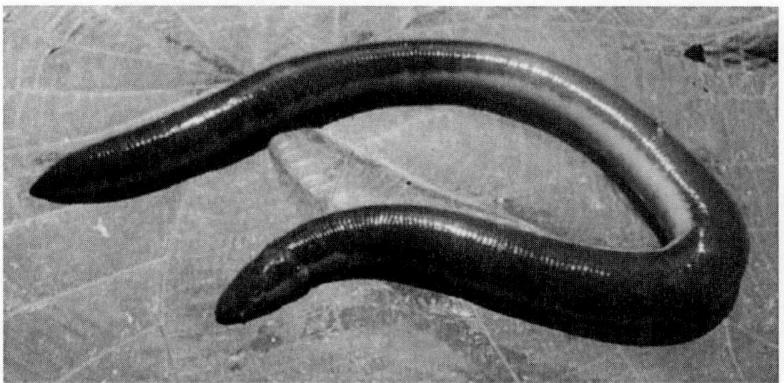

Fig 3. *Ichthyophis tricolor*

Fig 4. *Ichthyophis bombayensis*

Additional Readings

AmphibiaWeb: *Information on Amphibian Biology and Conservation* [web application] 2012 Berkeley, California: AmphibiaWeb. Available: http://amphibiaweb.org/

Kotharambath R, Gower DJ, Oommen OV and Wilkinson M 2012 A third species of *Gegeneophis* Peters (Amphibia: Gymnophiona: Indotyphlidae) lacking secondary annular grooves, *Zootaxa* **3272** 26–34

Oommen OV, Measey GJ, Gower DJ and Wilkinson M 2000 Distribution and abundance of the caecilian *Gegeneophis ramaswamii* (Amphibia: Gymnophiona) in southern Kerala, *Current Science* **79** 1385–1389

Stuart S, Hoffmann M, Chanson J, Cox N, Berridge R, Ramani P and Young B (eds) 2008 *Threatened Amphibians of the World*, Barcelona: IUCN and Conservation International

Box 1.2

Glimpses of the Indian Barbets

HSA Yahya

Department of Wildlife Sciences, Aligarh Muslim University,
Aligarh, India
(E-mail: hafiz_yahya02@yahoo.com)

Closely related to the Old World Honey guides (Indicatoridae) and the New World Puff birds (Bucconidae), barbets (Capitonidae) have a pan tropical distribution occurring in the three continents; Africa, Asia and South America. Most likely, they originated in Asia and were pushed across North America as well as across Europe when there was a wide land bridge connecting Siberia and Alaska. According to various authorities, the number of genera and species vary from 13 to 16 and 72 to 80, respectively. All the ten species (nine on the mainland and one endemic to Sri Lanka) occurring in the Indian subcontinent belong to the genus *Megalaima.* This genus comprises of 23 species and is the largest genus in the Family. Recent classification has grouped all *Megalaima* species under the Family Megalaimidae.

Paucity of information on Asiatic barbets prompted me to undertake a study on their ecology and biology mainly during 1977–1980 at Periyar Tiger Reserve, Kerala where two endemic species to Western Ghats, *M.*

viridis and *M. rubricapilla* occur, and casual observations have been made till now. The Indian barbets are gaudily-coloured birds, largely green with prominent markings of other body parts with crimson, yellow, golden, blue and brown colour. They are average-sized birds; the Himalayan great barbet (*M. marshalloram*) is the largest (33 cm) and the Coppersmith or Crimson-breasted barbet (*M. haemacephala*) is the smallest (17 cm). They are widely distributed in India. In most places, both the species occur. They show ecological isolation, which allows two congeneric sympatric species to coexist successfully.

Fig 5. Brown-headed barbet (*Megalaima zeylanica*)

Fig 6. A Crimson-breasted barbet (*Megalaima haemacephala*)

Barbets are among the most vociferous birds. Their loud duets, trios or quartets are characteristic features of each species. Among the eight species observed, only *M. rubricapilla* and *M. haemacephala* have apparently similar vocalization pattern; others have distinctly different calls. Owing to their cryptic colouration, their calls are of great help to them. Calls are effectively used for proclaiming and defending the roosting and breeding territories, for courtship and pair formation, for shift change during nest excavation and incubation, to lure out fledglings from the nest and lure them back to roost holes.

Barbets maintain regular roosting periods evidently governed by sunrise and sunset. On dark cloudy evenings and mornings, they settle earlier and emerge late. On one afternoon when total solar eclipse occurred in 1979 and it became dark an hour prior to its settling time, one *M. viridis* entered the hole for roosting, but came out immediately when the eclipse was over and the environs was bright again. Then, it settled finally for roosting at the regular time. Loud and shrill pre- and post-roosting calls are regular features of barbets. Roosting in self-excavated holes is advantageous for them in several ways. It protects them from rain and from predators such as owl. At times, the roost holes are renovated for nesting. Indian barbets roost singly; each in its own hole, and not in dormitories like some of their counterparts in South America or Africa. The different roosting hours in sympatric *M. viridis* and *M. rubricapilla* are suggested to be an indirect mean of ecological isolation as it affects their feeding cycle.

Barbets preen a great deal and in a variety of situations. The amount of preening while resting depends on the period of rest, the larger the rest the greater the amount of preening. Among the coexisting *M. viridis* and *M. rubricapilla,* only the former was found drinking and bathing from open natural rain-filled tree holes. A thorough preening is an essential part of a complete bath. While bathing, barbets do not tolerate other birds in the vicinity of its pool.

The Indian barbets are largely frugivorous, but consume considerable amount of insects, especially during the breeding season. Some of them sip nectar too. Food competition between coexisting species is not severe as normally they collect food from different feeding zones. Intraspecific competition and aggression among *M. viridis* is markedly more while feeding either on a fruit tree or with a mixed hunting party. Although *M. viridis* and *M. zeylanica* are, at times, alleged as minor pests on coffee cherries, during my long observation, I never found them damaging the crop like a pest. On the contrary, barbets may help in controlling serious Coffee Stem Borer (*Xylotrechus quadeipes*) and Swarm of Teak Defoliator (*Hyblaea purea*), and serve as a pest-control agent.

Barbets select dead and dry branches for excavating nests. Only barbets and woodpeckers are the primary tree cavity nesters among Indian birds. It is quite a laborious work to drill out a neat and curved chamber for laying eggs. The colour banding of several pairs of barbets during the study revealed many interesting insights of their hitherto unknown stories.

The clutch size of barbets varies between 2–3 and the incubation period between 13–15 days. During the daytime, both parents incubate and brood, but during nights, only females do so. Being a hole-nesting bird, the nestling period of barbets is quite long (34–38 days), while the fledgling period is short (4–5 days). Sometimes, *M. viridis* raise two broods in a season. Both the parents feed and clean the nest. It is interesting to watch barbets during the shift changing for incubation and for luring nestlings out of nest holes for the first time. The young ones are quite reluctant to leave the safe environs of the hole, while parents remain determined to bring them out. Similarly, the parents struggle a lot to take the young ones back into the roost hole in the evening. After initial training of 2–3 days, the young ones become quite confident and independent. On an average, their nesting success is about 66%.

Courtship display is mostly performed vocally. Males feed females after copulation. Barbets defend their nesting branches from other hole-nesting birds and from squirrels. The three-stripped palm squirrel often robs the nest of barbets. While several pairs of *M. viridis* may nest in close vicinity, the nests of *M. rubricapilla* are scattered. The scattering of nests of the latter may serve as an interspecific defence mechanism.

Barbets are ecologically and economically important as they help in seed dispersal and cross pollination of plants. They also control several harmful insects including termites and teak defoliator caterpillars. Shrinkage of forest areas, removal of road-side trees, reduction of gardens and large-scale urbanization are detrimental for these useful birds. In some parts of India (e.g., Arunachal Pradesh), people indulge in large-scale killing of barbets. Therefore, like several other species of birds, barbets are becoming vulnerable. They deserve due consideration in country's biodiversity conservation strategies and planning.

Fig 7. Barbets of the Indian Subcontinent: 1. Great barbet (*Megalaima virens*), size: 33 cm; 2a & 2b. Large green barbet (*M. zeylanca*), size: 27 cm, adult *M.z. caniceps* (2a) and adult *M.z. inornata* (2b); 3. Lineated barbet (*M. lineate*), size: 28 cm, adult (3a) and paler variant (3b); 4. Small green barbet (*M. viridis*), size: 23 cm, adult (4a) and juvenile (4b); 5. Yellow-fronted barbet (*M. flavifrons*) (endemic to Sri Lanka), size: 21 cm; 6. Golden-throated barbet (*M. franklinii*), size: 23 cm; 7. Blue-throated barbet (*M. asiatica*), size: 23 cm, adult (7a) and juvenile (7b); 8. Blue-eared barbet (*M. australis*), size: 17 cm, adult (8a) and juvenile (8b); 9. Crimson-throated barbet (*M. rubricapilla*), size: 17 cm, adult *M.r.mato* (9a) and juvenile (9b); 10. Crimson-breasted barbet (*M. haemacephala*), size: 17 cm, adult (10a) and juvenile (10b).

Fig 8. 24-day old, colour-banded *M. viridis*

Fig 9. Crimson-throated Barbet *(M. rubricapilla)* feeding the nestling

Additional Readings

Yahya HSA 1987 Roosting behaviour of barbets, *Megalaima* spp. *i Recent Trends in Ethology* (eds) M Balakrishnan and KM Alexander, Bangalore: Ethological Society of India 101–106 pages

Yahya HSA 1988 Breeding biology of barbets, *Megalaima* spp. with special reference to *M. viridis and M. rubricapilla malabarica* at Periyar Tiger Reserve, Kerala, *Journal of Bombay Natural History Society* **85** 493–511

Yahya HSA 1989 Habitat Preference of birds in the Periyar Tiger Reserve, Kerala, *Indian Journal of Forestry* **12** 288–295

Box 1.3

Mountain Nyala

B Afework

Department of Zoological Sciences, Addis Ababa University, Addis Ababa, P.O. Box 1176, Ethiopia
(E-mail: afeworkbekele@hotmail.com)

Mountain nyala (*Tragelaphus buxtoni*, Lydekker, 1910) is an endemic and endangered mammal of Ethiopia. It occurs only in the higher forests and moorlands of Arisi, Bale and Chercher mountains of south and south-eastern parts of Ethiopia from 1,800–3,400 m asl. It frequently occurs in dense cover at the edge of *Erica arborea* and *Phillipia abyssinica* heath zone at 3,300–3,500 m asl, but extends down into dense *Hagenia-Podocarpus* forests. The species has also been observed at the lowest range as low as 1,800 m asl at Mt. Abaro and as high as 3,800 m asl or more in a more open Afro-alpine moorlands above the heath zone.

Fig 10. A mountain nyala stag in Bale Mountains National Park, Ethiopia

It is a graceful big antelope with weight ranging between 150–300 kg. Males possess one-turn, strongly-spiralled horns. They lack manes on the throat. Markings are less distinct with a few faint stripes and spots on the flanks. Within its range, their abundance has been decreasing from time to time. The total population once was estimated at about 8,000 individuals. At present, this is estimated to be lower than 4,000 (Anagaw et al 2011). Increasing agricultural pressure and military strife were the leading causes of its decline. In the Bale Mountains National Park, at the head quarter, it is possible to see several hundreds of them within a short time (Befekadu and Bekele 2004). Indiscriminate hunting during the change over of government in 1990 reduced the numbers drastically. However, the population has slightly recovered recently. A population survey conducted in the head quarter area between 2003 and 2005 showed 830–908 individuals where 58% was adults, 25% sub-adults, 9% juveniles and 5% calves with female:male ratio of 3:1 (Mamo et al 2010). However, encroachment by livestock had damaged the habitat even near the head quarter area resulting in a change of the vegetation structure (*Hypericum revolutum*) by less preferred open grassland (Mamo and Bekele 2011). Agricultural expansion and further settlement have negatively affected the habitat availability of mountain nyala. Studies have shown that the vegetation at the edge of the Park have been reduced by 47%. The grazing pressure has resulted in a change in both vegetation type and structure between grazed and ungrazed areas. The once common *Hagenia* and *Juniperus* woodlands have been converted to open agricultural fields affecting the distribution of mountain nyala. Recent studies in Bale have identified 11 pockets of nyalas surviving along forest habitats most of which lie outside the Bale Mountains National Park. Using line transect method by counting droppings, the population was estimated at about 3,756 (Atickem et al 2011). The species is still an important source of trophy hunting and more than 50% of the estimated population occurs in controlled hunting areas occupying woodlands and avoiding settlement areas. The total suitable habitat for mountain nyala in the Bale Mountains was estimated at 8333 km^2 suggesting that there is a need to enlarge the size of the protected area within the suitable habitat by applying effective law enforcement to reduce human settlements and livestock grazing. Recently studies on foraging ecology and habitat use were carried out by radio-collaring 19 animals. Feeding records in 5,875 scan samples showed 73.4% grazing and 26.6% browsing. The home range size of solitary adult females and males was 22.37 ± 9.88 and 51.85 ± 22.56 km^2. Human activities have been known to affect the ranging pattern of the nyala. In areas where there is minimal human disturbance, home range extensions beyond GPS fixes have been recorded. Females are known to form small groups during the day and large groups during the night. Males do not show a stable grouping pattern, but can form large groups at night. Males are not territorial and do not defend females.

References

Atickem A, Loe LE, Langangen O, Rueness EK, Bekele A and Stenseth NC 2011, Estimating population size and habitat suitability for mountain nyala in areas with different population status, Animal Conservation **14** 409–418

Refera B and Bekele A 2004 Population status and structure of mountain nyala in the Bale Mountains National Park, Ethiopia, African Journal of Ecology **42** 1–7

Mamo T and Bekele A 2011 Human and livestock encroachment on the habitats of mountain nyala in the Bale Mountains National Park, Ethiopia, Tropical Ecology **52** 265–273

Mamo Y, Pinard MA and Bekele A 2010 Demography and dynamics of mountain nyala in the Bale Mountains National Park, Ethiopia, Current Zoology **56** 660–669

1.2. WILDLIFE ECOLOGY AND MANAGEMENT

Ecology has been in the forefront of biology since the beginning of biological studies. It emphasizes the life of organisms in relation to their surroundings. Environmental changes are prominent factors in the study of ecology as organisms have to adjust themselves with the environmental changes from time to time. Wildlife ecology deals only with natural habitats and the organisms in the wild, where human interference is only marginal. However, in the present day context, most of the natural habitats are frequently visited by humans for various purposes such as collection of various forest produces and for poaching. Clear-felling and fragmentation of natural forest are widespread all over the world. This affects natural populations of most of the organisms. Both aquatic and terrestrial fauna and flora are affected by human activities and pollution. It is in this context that wildlife ecology and management has developed as a separate entity from general ecology to deal specifically with ecology, conservation and management of natural organisms and their habitats.

Wildlife biologists and managers should know how and when to apply basic biological and ecological principles to manage wildlife populations and their habitats. Ecologically-based wildlife management can strengthen the fragile web of nature, and in doing so, it can enhance the quality of all organisms including human beings, and can ensure constant existence of biodiversity on the planet Earth. There is a need for trained wildlife biologists, wildlife ecologists, wildlife managers and conservationists to:

- census wildlife populations
- study various aspects of wildlife such as the reproductive potential of wildlife, harvesting potential and rates of harvesting of wildlife in managed areas and production of more game animals (birds and mammals hunted for sport)
- maintain and manage wildlife habitats so as to stock optimum wildlife populations
- survey and find out the organisms under threat and risk of extinction in various parts of the world
- restore wildlife populations where wildlife had been extirpated
- develop high quality zoos that can be instrumental in wildlife education and conservation

1.3. BIOMES OF THE WORLD

Regions with similar biological characteristics are called biomes. Biomes are groupings of ecosystems with similar environment and vegetation structure (Physiognomy). In addition to the marine biome, there are six major terrestrial biomes (forests, woodlands, shrublands, grasslands, semi-desert scrub and deserts). Forest biomes are subdivided into boreal, temperate, tropical, woodland and shrubland biomes. Grassland biomes include tropical savannah, temperate grassland, alpine grassland, and tundra (occur north of the tree line in North America and Eurasia, where the growing season spans the four summer months when snow melts). The biomes are further divided into units of greater similarity called ecosystems, based on environment and groupings of plants and animals. Thus, the world can be divided into broad ecological divisions, each of which has a characteristic vegetation and animal life. Each wildlife species selects its habitat from within these larger groupings. Hence, it is important to maintain all these biomes to represent the global wildlife populations.

Though the presence of biodiversity may vary from biome to biome, each of the biomes is equally important in wildlife ecology. For example, the species composition in a forest biome is much different from that of shrublands or deserts, but as shrublands and deserts are natural habitats, the species composition in such a habitat is specific and not seen elsewhere.

Hence, diverse natural habitats are required to be maintained to accommodate diversity of species. Most of the large herds of ungulates occur in grassland biomes (e.g., various antelopes in the African savannas). They have the ability to migrate in response to seasonal climatic and changing vegetation patterns, and hence, they place themselves in areas of highest food production from time to time. They avoid many of their predators, who cannot migrate to the same extent. These two abilities; to find temporary food patches and to avoid predators, allow higher density if they range in wider areas, than if the populations are restricted in their movements. Thus, large herds of herbivores are not just a consequence of the extensive areas of their selective biomes, but also of their ability to seek food in extensive areas and to avoid predators from time to time. A typical example is the African wildebeest (*Connochaetes taurinus* Lichtenstein, 1812), which migrates from area to area in large herds, numbering even up to 10,000 individuals, depending on the availability of water and fodder (Estes 1991).

1.4. BIOGEOGRAPHIC ZONES OF INDIA

Biogeography deals with the distribution of plants and animals in different geographical regions of the world through geological time. It gives us information on the dynamics of temporal and spatial distribution of organisms and ecosystems through the gradients of latitude, longitude and elevation. As physical, biological and edaphic features act as limiting factors in the distribution of organisms, communities of plants and animals in different geographical areas of the world differ widely from each other. They also show dynamic changes from time to time, particularly in relation with phenological factors. Phytogeography deals with the origin, distribution and environmental interrelationships of plants, whereas zoogeography deals with the origin, distribution and migration of animals. Modern biogeography often employs the use of Geographic Information System (GIS) to understand factors affecting distribution of organisms and to predict future trends in their distribution patterns.

A good account of the biogeography of the Indian sub-continent with reference to the habitats and distribution of animals is available in Prater (1980). Phytogeographically, India has nine zones, whereas zoogeographically, India has the following ten zones as described by Rodgers and Panwar (1988) (Fig 11):

1. Trans-Himalayan zone
2. Himalayan zone
3. Desert zone

4. Semi-arid zone
5. Western Ghat zone
6. Deccan Plateau zone
7. Gangetic Plain zone
8. Coastal zone
9. North-east zone
10. Indian Islands of Andaman, Nicobar and Lakshadweep

1. Trans-Himalayan zone is the northern most area of the country (Ladakh, Jammu and Kashmir, Himachal Pradesh and Lahul) with around an area of 1,86,200 km^2, at an altitude range of 4,500–6,000 m around the Himalayas. This zone is not related to mountains, but is the area surrounding the mountains. Irregular vegetation is characteristic of this zone. Specialized mammals of this zone are the snow leopard (*Panthera uncia* or *Uncia uncial* Schreber, 1775), Tibetan ass *(Equus kiang* Moorcroft, 1841), wild yak (*Bos grunniens* Linnaeus, 1766) and the Palas cat (*Otocolobus manul* Pallas, 1776). This zone has the wild sheep, producing best quality wool, and superior quality goat in the world. Migratory birds such as black-necked crane (*Grus nigricollis* Przhevalsky, 1876) and the endangered great Indian bustard (*Ardeotis nigriceps* Vigors, 1831) are seen in this zone.

2. Himalayan zone (north-eastern, western, central and eastern regions of the Himalayas) is comprised of a hilly region with its own flora and fauna, exhibiting high biodiversity, in an area around 2,36,300 km^2. There are 56 protected areas in this zone with a number of endemic species.

3. Desert zone (Kutch – Gujarat and Thar – Rajasthan) stretches around 2,50,000 km^2 comprising the region from where the great 'Thar desert' starts, including the areas such as Jaisalmer surrounded by sand dunes. This represents a very sensitive ecosystem where an area of 89 km^2 is under protective coverage. Wild ass *(Equus hemionus* Pallas, 1775) and Desert cat (*Felis* species) are present in this zone.

4. Semi-arid zone represents comparatively dry areas in Rajasthan and Gujarat, and extends upto some areas in Punjab and Haryana comprising an extent of 5,08,000 km^2, of which an extent of 11,675 km^2 is under protective coverage, including Gir forests, Ranthambore and Sariska. In this area, both the ground and surface water is less, and hence dry xerophytic vegetation is

predominant. The faunal composition is also comparatively less, but the Gir forest is the only natural habitat of *P. leo* in India.

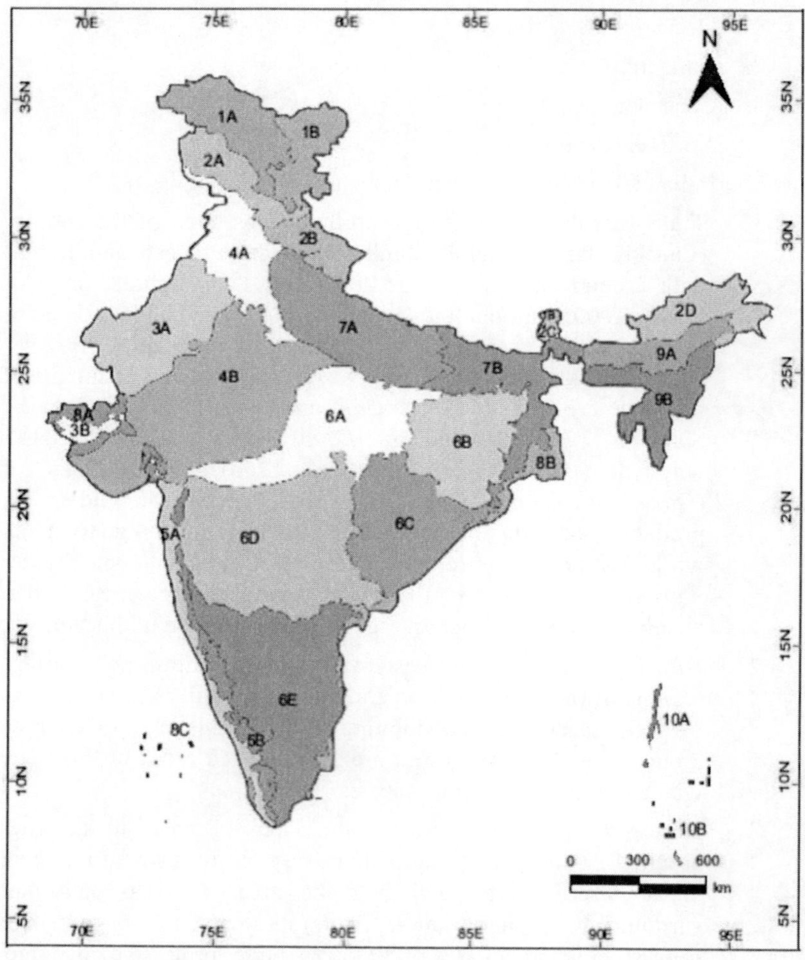

1A-Trans-Himalaya-Ladakh Mins
1B-Trans-Himalaya-Tibetan Plateau
2A- Himalaya-North-West Himalaya
2B-Himalya-West Himalaya
2C-Himalaya-Central Himalaya
2D-Himalaya-East HIMALAYA
3A-Desert-Thar
3B-Desert-Katchchh
4A-Semi-Arid-Punjab plains
4B-Semi-Arid Gujarat-Rajaputana
5A-Western Ghats-Malabar Plains
5B-Western Ghats-Western Ghats Mns
6A-Deccan Peninsula-Central Highlands
6B-Deccan Peninsula-Chotta Nagpur
6C-Deccan Peninsula-Eastern Highlands
6D-Deccan Peninsula-Central plateau
6E-Deccan Peninsula-Deccan South
7A-Gangetic Plain-Upper Gangetic Plain
7B-Gangetic Plain-Lower Gangetic Plain
8A-Coasts-West Coast
8B-Coasts-East Coast
8C-Coasts-Lakshadweep
9A-North East-Brahamaputra Valley
9B-North East-North East Hills
10A-Islands-Andamans
10B-Islands-Nicobars

Fig 11. Zoogeographic zones of India
(Modified from: Rodgers and Panwar 1988)

5. Western Ghats represent the mountainous western zone of south peninsular India having rich flora and fauna with tropical rainforests extending from the Konkan region of the State of Maharashtra through the western parts of Karnataka and Kerala, and some of the areas of Tamil Nadu representing 15,90,002 km^2 area, of which an area of 15,955 km^2 is protected as Wildlife Sanctuaries, National Parks and under the Nilgiri Biosphere Reserve. Wild relatives of cultivated plants such as banana, mango, citrus and black pepper are found abundantly in this zone. The region of Western Ghats is one of the biological hotspots of the world (Myers et al 2000). This zone is the home of several endemic plants and animals, including the lion-tailed macaque (*Macaca silenus* Linnaeus, 1758) and the Nilgiri tahr (*Nilgiritragus hylocrius* Ogilby, 1838).
6. The Deccan Plateau is extended through the south-central, north-eastern (Chota Nagpur) and the central highlands of India, through the States of Madhya Pradesh, Maharashtra, Odisha, Andhra Pradesh, Tamil Nadu, Karnataka and Kerala, representing an area of around 14,21,000 km^2, out of which an area of 4,610 km^2 is under protective coverage. This is the largest biogeographical zone of India representing around 43% of the total land of the country. Some areas in this zone represent semi-arid type of vegetation. Rainfall occurs mainly by the south-west monsoon. Dense forest area is less in this zone.
7. Gangetic plain is spread over 3,59,400 km^2 area in Uttar Pradesh, Bihar, West Bengal and Odisha, where rainfall is moderate, and the soil is fertile. The vegetation communities in this zone are of moist and dry-deciduous types. Mangrove vegetation is common in the Gangetic delta (South Bengal). This zone is well known for agriculture in India.
8. Coastal zone represents the eastern and western coasts of India. The Sunderbans Tiger Reserve is located in the eastern coastal zone. The western coast is also known as the Malabar Coast, extending from Gujarat up to Cape Comorin in Tamil Nadu. Most of the areas in this region get heavy annual rainfall, and is characterized by the presence of tropical evergreen, semi-evergreen, montane forests and mangroves.
9. North-east zone of India is located in the Brahmaputra valley and Assam. It is also characterized by heavy rainfall with dense evergreen and subtropical forests. The extent is around 1,880 km^2, where 17 protected areas are situated. Kazhiranga of this

zone is famous for Indian Rhinoceros (*Rhinoceros unicornis* Linnaeus, 1758).

10. Indian Islands of Andaman, Nicobar and Lakshadweep consist of 384 islands where 101 protected areas are established in about 710 km^2. There are 225 animal species reported from these islands; of which 50% is endemic to the island/s where they are present.

In the transition zones of the high altitudes above the tree-line, the strip of grassy savannas between the mountains and arid plateau of Tibet is characterized by Oriental and Palaearctic regions. The Tibetan region constitutes the Tibetan plateau together with eastern Ladakh, which is a unique geographical area of wide range of altitudes and weather conditions with distinct vegetation types of both tropical and alpine, with corresponding faunal and floral combinations. It covers the barren plateau and uplands that lie beyond the Himalayas, but within the northern frontiers of India from Kashmir to Bhutan. This is the only region of the world where both desert and arctic conditions prevail. The Gangetic plain is with moderate rainfall with low winter temperatures. In the north-western regions of this plain, the vegetation gradually merges with that of the adjoining desert zone, and hence, a combination of desert animals are common in this region. The plains of East and West Bengal are more humid and the vegetation comprises of mangoes, figs and palms. In the east, the great plain is transformed into swampy forests of Sunderbans, with grassy savannas, mangroves and dense evergreen forests. Assam and the north-eastern States of India constitute a geographical region, which is more or less similar with the forest regions of the Himalayas. There is tropical and temperate vegetation in this region, with a variety of wild animals including those of the Malayan region. This is also one of the biodiversity hotspots of the world (Myers et al 2000).

These zones are further subdivided into 26 biogeographic provinces using GIS techniques for planning and developing wildlife protected areas in India (Rodgers et al 2002). This classification is based on factors such as altitude, topography, rainfall, moisture and other factors responsible for the existence of the great variety of biodiversity of India. Each of these zones is characterized by its own vegetation communities that support specific animal species, in addition to those that are more or less common in distribution. Species that are common to few or more zones are generalists with better adaptations to survive in different environmental conditions. They can depend upon a variety of food sources. Species that are specific to one or few zones are specialists, and they are more specialized to the respective environmental factors of their specific geographic zone. They are less adapted to survive in a variety of environmental conditions. They have more specialized feeding patterns.

1.5. ECOLOGY OF NATURAL COMMUNITIES

The living parts of an ecosystem, at any of the given times and places, is known as the biotic community or community (Bolen and Robinson 2003). Ecosystems consist of several distinctive communities, each having specific groups of plants and animals. For example, a forest ecosystem consists of stands of seedlings and mature trees of various species of different age groups. Shrublands and grasslands have specific composition of shrubs and grasses, respectively. There may also be other types of vegetation in such habitats, but we name the habitat based on major vegetation composition; e.g., wet evergreen, semi-evergreen, deciduous, dry-deciduous, shrubs, herbs or grasses (Figs 12 to 16). The herbaceous cover, invading a recently burned section of the forest, constitutes a developing pioneer community in the forest ecosystem. The vegetation bordering the banks of a stream (riverine vegetation) and the plants in the stream are different communities. Various kinds of animals are associated with each of these habitats, and their presence and interactions complete each of these communities.

When we analyze the meaning of the word 'ecology', we have to elaborate the concepts of ecology in terms of the interactions of organisms and related metabolic processes that they undergo. Life cannot exist all alone; no species can sustain without interactions with other species. In an ecosystem, interactions take place not only with living organisms, but also with non-living elements in the environment, which are also vital for the existence and maintenance of life. For example, carbon dioxide is an important natural element essential for photosynthetic activities of plants to synthesize food, whereas in the absence of oxygen, most of the animal species cannot sustain even for seconds. Oxygen is an important element in the metabolic processes of organisms. Similarly, minerals (e.g., sodium, iron, calcium etc.) are also essential to maintain metabolic processes of animals including human beings, even though the required amount in the blood stream is only marginal. All these living and non-living elements participate in the life processes, undergo changes and again enter the physical environment through the processes of respiration, excretion and finally decomposition of the organism itself, and thus, they are cycled between physical and biological environments. This is essential for sustenance of life of any kind on the planet Earth. Wildlife ecology deals with such exchanges between wild organisms and their natural habitats.

Ecosystems are modified by internal (natural) and external factors (man-made). Both of these operate concurrently. Natural factors are ageing (e.g., seedlings gradually become mature forests through time), long-term genetic responses of organisms to evolutionary adjustments in the community, floods, droughts and lightning. Man-made factors are external factors affecting the community and ecosystem. Nowadays, floods and droughts are more frequent in some parts of the world (e.g., African continent). Human-related issues are found to be responsible for such increased frequencies of these events. For example, deforestation of large extent of hilly forest terrain (Fig 17) would lead to soil erosion and result in both drought and floods in subsequent years.

Ecosystems can range in size from an area as small as a few square metres (e.g., a desert spring) to much or all of the Earth (e.g., tidal zones). Ponds, pastures and woodlots are small ecosystems. The Amazon rainforest is an example of a large ecosystem. The biosphere, the part of the Earth that extends from a few hundred metres beneath the surface to several kilometres into the atmosphere, is a vast area where life sustains. These days, each of the ecosystems is influenced by human activities at various levels. Ecosystems are generally self-sustaining, but require external source of energy that almost comes from the Sun.

Fig 12. A view of the peak from Syranthri in Silent Valley; part of the evergreen habitats on the slopes and grasslands on the peak are seen (Photo: S Anilkumar)

Fig 13. An overview of the top canopy of a typical evergreen vegetation composition during the dry season (Photo: Sharon M Balakrishnan)

Fig 14. A riverine habitat on the banks of Awash River (Photo: M Balakrishnan)

Fig 15. A mangrove forest in Kerala, a wetland habitat, rich in biodiversity (Photo: M Balakrishnan)

Fig 16. Dead and decaying trees also form special microhabitats for a number of minute organisms as seen in this picture (Photo: M Balakrishnan)

Fig 17. Agricultural expansion through shifting cultivation in a hilly forest area (Photo: M Balakrishnan)

1.6. NICHE IN WILDLIFE ECOLOGY

Each of the organisms in a wild habitat has specific roles to play to sustain in the habitat and to retain the habitat qualities. Niche or simply the ecological role of organisms in a community is closely linked to food habits and feeding behaviour of organisms. Each of the organisms, whether small or large, competes and interacts with other species in the ecosystem for essential resources. Based on the genetic differences among them, their abilities ultimately lead to be successful in acquiring resources in the competitive world, which leads to specializations through evolutionary successes. During this process of natural selection, those individuals with more competitive genetic features would be successful to pass on such genetic characteristics or genetic traits to the next generation, while those with inferior features would fail to be selected for the generations to come, and thus, less favourable combinations of genes would be gradually eliminated from future generations. As a consequence of such processes, certain species may also become restricted to specific habitat types as 'specialists', whereas others would develop as more 'generalist' species. A specialist species has restricted range of tolerance, whereas a generalist species has wider ranges of tolerance, and are distributed in a variety of habitat types.

Diverse species occupy the same area as they have difference in their niche and also in their behavioural patterns such as diurnal and nocturnal adaptations. Also, there are differences in the roles of each of the organisms in relation to the seasonality patterns in the habitat. In relation to such seasonality patterns, there may also be changes in the role of the species that occupy the area. Some of the species of animals, including larger mammals, migrate from one area or region to another based on the phenological changes in the habitat. Migration is also related to the climatic conditions in the area during specific time periods. Closely-related species such as artiodactyls occupy the same habitat, but competition between them is limited due to the differences in their feeding habits such as feeding on different species and feeding on different parts of a species or a combination of species in the area. For example, *C. taurinus* eats short green sprouting leaves of some of the plants as those eat the topis (*Damaliscua lunatus* Ogilby, 1837), but topis depends mainly on mature plants. Thus, these antelopes eat more or less the same plants, but of different stages of growth, and reduce the competition between them. The ability of habitats to provide essential resources for a species to be active in its niche is important in wildlife management. Habitats with intensive human activities, and damaged and altered habitats often lack essential resources for the variety of natural organisms of the area, especially for those species with narrow tolerance limits such as most of the endemics. Hence, these essential ecological principles should form guidelines while establishing wildlife conservation and management programmes.

1.7. ECOLOGICAL SUCCESSION

Natural communities show patterns in time and space. Many features of communities clearly reflect the regional climate. For example, shrubs that are separated by bare soil and are arranged in a widely spaced, regular pattern are typical of desert conditions. Spines, spongy cells, thickened cuticles and other specialized features for water conservation characterize desert plants. The species differ among desert communities, but their structural and physiological adaptations are alike (so are ecological equivalents).

In terms of time, the composition of natural communities is realized in a sequential process known as ecological succession or simply succession. A pioneer community is the first step in succession, and a climax community is the final community in the sequence. Some kind of disturbance, whether natural (e.g., a lightning fire, glacier, volcanic eruption) or caused by humans (e.g., ploughing, lumbering, grazing), initiates the development of a pioneer community on the exposed soil.

Primary succession occurs where no community previously existed, as on a newly-formed volcanic island. Secondary succession takes place where there are remnants of a previous community, as in the case of a burned-over field or a clear-cut forest. In either primary or secondary succession, a series of different communities gradually and sequentially occupy a site until a climax community is established. At climax, the community is self-perpetuating in a state of dynamic equilibrium; i.e., although there is a constant turnover of individuals within the community, the species remain the same, and the overall composition of the climax community remains essentially unchanged. Thus, climax communities persist in both time and space until they are disturbed. At present, wildlife species of early to mid-succession stages are most abundant in most of the ecosystems because of the many ways to keep succession away from reaching the climax. This is clearly due to the involvement of human activities. A degraded climax community cannot be re-established by human beings.

Natural ecosystems are self nourishing, self perpetuating and self sustaining, using energy from the Sun, water from soils, and moisture, oxygen and carbon dioxide from the surrounding atmosphere. Other essential food and mineral requirements are also met with from the surroundings; i.e., from air, water and soils. Each of the organisms in the ecosystem is self perpetuating, which forms the basis of sustainability of ecosystems. Some ecosystems require more time to recover from damages caused by external factors such as human-induced damages, but some are able to resolve such problems more quickly. For example, a climax ecosystem (e.g., tropical evergreen forest) if damaged cannot be replenished easily even if left without further human interference for decades, but an ecosystem in the early to mid-succession stage of growth and development can replenish if left unharmed further for a couple of years. Ecosystems in the early stages of development require more energy for growth than an ecosystem that has attained a climax stage of development.

Various types of modifications take place in the ecosystems by natural and external factors. Human-induced factors are examples of external factors affecting ecosystems. Probably, this has been one of the most dangerous of all adverse factors affecting ecosystems during the past few centuries all over the world. Ageing is an internal natural factor. Long-term genetic responses of organisms help them to adjust themselves to the changes in the environment through time and through generations, leading to the process of survival of the fittest. In the absence of such adjustments and evolutionary development processes, many of the organisms and individuals may not achieve the objectives of life (i.e., successful reproduction to support generations of own species for long). Weathering is a process by which soil is formed as a result of both internal (e.g. parent material, vegetation) and external factors such as weather conditions.

1.8. ASSOCIATION BETWEEN SPECIES

There may be close associations between two or more unrelated species in the community. Mutualism is a common type of symbiotic association in which two or more species benefit from each other (Box 1.4 – Figs 18 to 22). Even though animals depend upon food resources, the plants also get benefitted by animals by way of pollination when it is in the flowering state. Some animals eat fruits and deposit seeds unharmed far away from the mother plant, and thus, act as seed-dispersal agents. Both these are vital for the continued existence of ecosystems. Some mutualistic associations are required for the survival of either member of the pair (obligative mutualism), whereas the bond may be broken in other cases without fatal results (facultative mutualism). In areas with large number of endemic animals and plants, such mutualistic interactions have major and specific roles to play. Many such organisms are species-specific in their food requirements as well as agents for pollination and seed dispersal. There are two types of mutualisms — symbiotic and non-symbiotic. Those species involved in the former live in close physical contacts, whereas those involved in the later do not live in close physical contacts. There are also divisions of mutualisms such as seed-dispersal mutualism, pollination mutualism, digestive mutualism and protective mutualism (Janzen 1985). Each of these is closely related to the meaning of the respective terms. Several species of tropical plants are pollinated by insects, birds and bats. Flowers develop special characters such as shape, colour and scent to attract the most suitable animal pollinator. For example, those flowers pollinated by moths and bats (nocturnal animals) are mostly white or drab to attract their suitable pollinators. On the other hand, those flowers pollinated by diurnal animals are more or less red, white or blue. The respective pollinators also develop morphological, physiological and sensory features to be more effective pollinators.

Box 1.4

Mutualism – A Case Study in the Western Ghats of Kerala, India

S Anilkumar

Jawaharlal Nehru Tropical Botanic Garden and Research Institute, Palode, Thiruvananthapuram, 695 562, Kerala, India

(E-mail: anilks73@yahoo.com)

In an article addressed to the Royal Academy of Belgium on December 16, 1873, van Beneden, the renowned Belgian Zoologist, used the term 'mutualism' for the first time in the biological context. Earlier, it was used only in human social contexts. Mutualism is the net result of positive interactions between species at individual and population levels. Mutualistic interactions play crucial role in determining the abundance and distribution of organisms and in maintaining the balance of nature (Boucher 1985, Bascompte et al 2006). Mutualistic interactions are the first to be extinguished by man's onslaught and the last to be lamented. Mutualism can be classified into symbiotic and non-symbiotic mutualisms. In symbiotic mutualism, the participants live in close physical contacts, and in non-symbiotic mutualism, they do not live in close physical contact. Janzen (1985) has classified mutualisms into four types. These are seed-dispersal mutualism, pollination mutualism, digestive mutualism and protective mutualism. Non-symbiotic mutualism between plants and animals occurs in the contexts of pollination, seed dispersal and protection. Among them, pollination mutualism and seed dispersal mutualism are the most studied. They are obligate and facultative. If mutualistic interactions between two species are interdependent, then it is obligate, and if individuals can survive in the absence of the other, then it is facultative. One of the challenges of modern ecology is to acquire and understand the means by which plants and animals exploit each other for food and shelter. In this process, pollination and seed dispersal are affected, which are so vital for the survival of the existing ecosystems.

It was during the late 19th and early 20th centuries that several researchers have concentrated on studies on mutualism. Classical examples of mutualistic interactions studied were between Ocellaris clownfish *(Amphiprion ocellaris*) and Ritteri sea anemone (*Heteractis magnifica*), and between Sapotaceous tree (*Sideroxylum grandiflorum*) and the extinct dodo (*Raphus cucullatus*). Recent studies include mutualism between a leguminous tree (*Daniellia pynaertii*) and large African monkeys (*Cercopithecus ascanius*, *C. wolfi*, *C. nigroviridis* and *Cercocebus aterrimus*) from Zaire Basin (Gautier-Hion and Maisels 1994). Obligate mutualistic relationships of figs with their species-specific pollinating wasps have been well studied. Travset and Riera (2005) reported the disruption of mutualistic interactions between two perennial shrubs and the frugivorous lizard (*Podarcis lilfordi*) in the Balearic Islands. Whitmer and Cheke (1991) have revealed facultative mutualism between the Sapotaceous tree (*Sideroxylum grandiflorum*) and the dodo (*Raphus cucullatus*).

In the biogeographical region such as the Western Ghats, these types of interactions are highly relevant as the area is one among the 34 biodiversity hotspots along with Sri Lanka, and harbours several endemic plants and animals. One among the endemic plants, *Cullenia exarillata* (Bombacaceae), the sole representative of the Genus *Cullenia* in India, is distributed in the southern part of the Western Ghats and stands as a keystone species by providing food resource to a number of animal species during the food scarcity period (August–November) in the region. The study on mutualism between *C. exarillata* and the vertebrate community in the Silent Valley forest ecosystem of Kerala revealed the existence of facultative pollination mutualism in the area (Anilkumar 2009). In Silent Valley, general flowering and fruiting is during January–March. In *C. exarillata*, immediately when fruiting ceases in late August, cauliflorous flowering starts. *Cullenia exarillata* profusely flowers during September–November. Only *Cullenia* will be in flowering and all the frugivorous vertebrates are attracted towards it during this time. Flowers are ramiflorous and densely fasciculated on protuberances of stem and old branches. They are brownish white and 6–7 cm long. Calyx is 3–4 cm long, tubular and five-lobed at apex. Petals are absent. Staminal tube is about 5 cm long, exerted, five-lobed at the apex with 7–9 stamens along the margins. From these lobes arise small connectives and from the head region arises individual stalk-bearing theca in red or yellow spherical clusters. There are small sticky villai-like structures on the surface of the thecae. Pollen grains, which are white in colour, are located inside the theca. The sticky villai-like structure and the sticky pollens are the adaptations for attaching to the fur or feathers of the vertebrates. An average-size tree produces thousands of flowers and the number of flowers produced per cluster ranges from 2–90. Even though, flower buds appear in late August, flowers open only by late October and early November, and it will take 35–45 days for maturity. There is no specific time for flower opening and is indicated by protruded stigma head, which is lemon yellow in colour, and will be receptive for the next 48 hours. After about 24 hours only, the stamina tube with fully-dehisced anthers will expose and the base of the calyx will be swollen with accumulated nectar. This adaptation of stigma protruding first and later anther dehiscence is an adaptation in plants to avoid self pollination and to favour cross pollination.

Fig 18. *Cullenia exarillata* – a keystone species in Silent Valley forest of the Western Ghats, which provides food to frugivorous vertebrates during the food scarce period

Among the 28 species of mammals recorded from Silent Valley forests, ten species were seen interacting with *C. exarillata* for flowers and fruits. They are Fulvous fruit bat (*Rousettus leschenaultia),* Bonnet macaque (*Macaca radiata*), Lion-tailed macaque (*M. silenus*), Common langur (*Semnopithecus entellus*), Nilgiri langur (*Trachypithecus johnii*), Sloth bear (*Melursus ursinus*), Nilgiri marten (*Martes gwatkinsi*), Common palm civet (*Paradoxurus hermaphrodites*), Malabar giant squirrel (*Ratufa indica*) and Indian giant flying squirrel (*Petaurista philippensis*). Out of the 87 species of birds observed in Silent Valley, 15 were seen interacting with *C. exarillata* for nectar. They are Indian lorikeet (*Loriculus vernalis*), Malabar Crimson-throated barbet (*Megalaima rubricapilla*), Small green barbet (*M. viridis*), Oriole (*Oriolus oriolus*), southern grackle (*Gracula religiosa*), southern jungle myna (*Acridotheres fuscus*), black bulbul (*Hypsipetes madagascariensis)*, grey-headed bulbul (*Pycnonotus periocephalus*), southern redwhiskered bulbul (*P. jocosus*), yellowbrowed bulbul (*Hypsipetes indicus*), Nilgiri quaker babbler (*Alcippe poioicephala*), Bourdillon's blackbird (*Turdus merula*), small sunbird (*Nectarinia minima*), Indian purplerumped sunbird (*N. zeylonica*) and Nilgiri white-eye (*Zosterops palpebrosa*).

Fig 19. A branch of *Cullenia exarillata* showing cauliflorous nature of flowering

Fig 20. A cluster bearing different stages of flowers of *Cullenia exarillata*

Fig 21. Malabar giant squirrel (*Ratufa indica*) relishing the nectar resource of *Cullenia exrillata* and in turn aids in cross pollination

Vertebrates adopt various foraging techniques to exploit natural resources. Primates such as the lion-tailed macaque, Nilgiri langur, common langur and bonnet macaque pluck flowers at the stage when nectar is present at its optimum quantity. They pluck individual flowers, bite at the base portion, sip the nectar, and discard the rest. The fulvous fruit bat removes individual flowers with nectar and chews the succulent base of the flower leisurely, where nectar is stored. The juicy portion is taken and the rest is discarded. Semi-digested flowers of all stages in the scats of sloth bear and civet suggest that they too voraciously fed the flower regardless of the maturity stage. The Malabar giant squirrel was seen relishing the floral resources right from the bud stage. It adopted three different foraging strategies for feeding the floral resources during the three different stages of flowers. In the first two stages, the flowers were totally destroyed as they took individual flowers, holding them in the palm, and longitudinal slits made for luring the nectar. In the third stage flowers (older flowers) at times the mode of feeding was not total destruction. It just nibbles or licks the stage three flowers for nectar. Nectarivorous and frugivorous birds usually make slits/pores at the base of flowers and take nectar. Often sun birds insert the beak through the tip and then lure nectar.

While moving over the branches bearing flowers for feeding, the fur or feathers of the animals will be dusted with pollen. The villai-like structure of the anther and the sticky nature of the pollen will enable them to get attached to the fur or feathers of vertebrates. This pollen will get attached to the receptive stigma head resulting in pollination. Hence, animals are provided with flower resources during the food scarcity period and the flowers, in turn, are pollinated. Thus, there exists a facultative mutualism between *C. exarillata* and the interacting vertebrates. Flowers of *C. exarillata* open at any time during the day or night, and this is indicated by protrusion of the head of the stigma. The stigma is receptive for up to 48 h from the time of protrusion. Hence, both nocturnal and diurnal animals, those who visit during this stage of flowers, with pollen from other flowers might act as pollination agents. While plucking the flowers, at times only calyx was detached, which was sipped for nectar and then discarded. The remaining portion (style and stigma) of the flower remained on the tree itself. Hence, if the flower was already pollinated, there was a chance to set fruit. The results of pollination studies also revealed cross pollination and geitonogamous pollen transfer (= transfer of pollen from flowers of the same plant), and no autogamy in *C. exarillata.* Percentage of fruits set in natural condition and in experimental conditions of exclusion of Malabar giant squirrel and primates, were almost equal.

Floral syndromes of pollination by bats (chiropterophily), non-flying mammals and birds categorized by Faegri and van der Pijl (1979) and Procter et al (1996) are those flowers, which open both during the day and night, with copious amount of nectar, plenty of pollen, good display, prolonged flowering, small or large, and held strongly by short and thick stalk. These characters are also possessed by *C. exarillata*. After successful pollination, fruit initiation starts by December; and fruits (capsules) will mature enough for vertebrate feeding by June–August, next year. They are spiny and globose, 8–14 cm in diameter. Each seed contains 6–9 seeds that are shining, angular, dark brownish in colour and about 4–5 cm long and with pseudo arils. Due to the spiny nature of the fruits, absence of pulp, large seeds with pseudo aril being the only resource, only specialized vertebrates like Nilgiri langur, common langur, lion-tailed macaque, Malabar giant squirrel, three-striped palm squirrel and terrestrial rodents have been observed to feed on the seeds of *C. exarillata* in higher proportion, adopting various foraging strategies. Primates like Nilgiri langur, common langur and lion-tailed macaque bite and tear the fruit rind and then relish the endosperm. The Malabar giant squirrel gnaws the thorny rind and relishes the endosperm. Fruit feeding behaviour of the vertebrates have revealed that Nilgiri langur, lion-tailed macaque, common langur and Malabar giant squirrel are pre-dispersal seed predators of *C. exarillata* rather than primary seed dispersal agents. No bird species was seen eating the seeds of *C. exarillata.*

The dispersal mode in *Cullenia* is barochory; i.e., when the fruit gets matured, the thorny rind splits into 4–5 halves and the seeds fall beneath the canopy of the mother tree due to gravity. Dispersal studies also revealed that seeds were removed at a rate of 5–10 from each of the study quadrates by rodents, which acted as post-dispersal seed predators rather than secondary seed dispersers. This was more evident from the seedling demography studies, which revealed the absence of *C. exarillata* saplings beyond 20 m away from the mother plant. Seed dispersal by scatter hoarding is also ruled out. Even though the percentage of seed germination was high (>70%), after heavy pre-dispersal seed predation by arboreal mammals and post-dispersal seed predation by rodents, only a small percentage (around 5%) will get established.

However, there is no seed dispersal mutualism between vertebrates and *Cullenia* (Anilkumar 2009). Generalist pollinators are predicted to be sensitive to human-caused disruption, and their absence might trigger a "cascade effect" among the multiple plant species that they pollinate, which may ultimately result in biodiversity loss. If the mutualistic pollinators such as insects, birds and mammals fail to visit, many plant species, either fail to reproduce, or produce only fewer seeds or produce seeds of lower quantity.

Genetic variation within the population alaysed using random amplified polymorphic DNA analysis (RAPD) technique has revealed little genetic differentiation, showing high level of gene flow. This is due to the gene flow through pollination mutualism in *C. exarillata* and the vertebrate community in the Silent Valley forest ecosystem, even though there is no gene flow through seed dispersal.

Hence, pollinators can be crucial drivers of plant population and community dynamics. Any of the factors, which contribute to seed and seedling mortality, can have a profound impact on forest community composition, structure and diversity at later stages. Plants might be pushed to extinction in the absence of pollinators and seed dispersers leading to ecological degradation of the habitat especially in the tropics, where mutually interspecific interactions are more common.

Fig 22. Barochory mode dispersal in *Cullenia exarillata*

References and Additional Readings

Anilkumar S 2009 Mutualism between Vertebrates and *Cullenia exarillata* in Silent Valley Forest Ecosystems, Kerala. PhD Dissertation, Thiruvananthapuram, University of Kerala

Anstett MC, Hossaert-McKey M and Kjellberg F 1997 Figs and fig pollinators: evolutionary conflicts in coevolved mutualisms, Trends in Ecology and Evolution **12** 94–99

Bascompte J, Jordano P and Olesen JM 2006 Asymmetric co-evolutionary networks facilitate biodiversity maintenance, Science **312** 431–433

Boucher D H 1985 The Biology of Mutualism, New York: Oxford University Press

Faegri K and van der Pijl L 1979 The Principles of Pollination Ecology, Oxford: Pergamon Press

Gautier-Hion A and Maisels F 1994 Mutualism between a leguminous tree and large African monkeys as pollinators, Behavioural Ecology and Sociobiology **34** 203–210

Herrera CM and Pellmyr O (eds) 2002 Plant–Animal Interactions – An Evolutionary Approach, Oxford: Blackwell Publishing Company

Janzen D H 1977 Co-evolution of mutualism between ants and *Accacia* in Central America, Evolution **20** 249–275

Janzen D H 1985 On ecological fitting, Oikos **45** 308–310

Proctor M, Yeo P and Lack A 1996 The Natural History of Pollination, London: Harper & Collins

Traveset A and Riera N 2005 Disruption of a plant–lizard seed dispersal system and its ecological consequences for a threatened endemic plant in the Balearic Islands, Conservation Biology **19** 421–431

Whitmer MC and Cheke AS 1991 The dodo and the tambalacoque tree: an obligate mutualism reconsidered, Oikos **61** 133–137

Several species of wild mammals and birds act as seed-dispersal agents. The fruits develop attractive colours at the time of maturity to attract a suitable dispersal agent. Suitable animals are attracted to the flesh of the fruit and its taste, which ultimately results in the process of seed dispersal; i.e., a functional response of fruit-eating animals. They carry seeds unharmed as they pass through the alimentary canal and get deposited along the faeces far away from the mother plant, a process by which seeds are deposited far away from the 'seed shadow'. This process helps the seedlings to have better chances for growth and development (Paulsen et al 2012). It also helps in the maintenance of heterogeneity of the habitat, which is also essential for habitats to sustain.

1.9. DIVERSITY, STABILITY AND RESILIENCE

Tropical habitats are known for the rich biodiversity. Each habitat has its specialized forms of plants and animals that live and sustain depending upon each other. It is this dependence that forms the basis of ecosystem maintenance. When such links are broken, the food web is affected and the sustenance of the ecosystem comes under threat. The number of species in a community reflects abundance and diversity. Abundance denotes the number of individuals of a species in an area or in the habitat of observation. It is a numerical measure of a population size. As it varies considerably with the extent of habitat, it is better to denote as density (individuals per unit area). Biomass is another way of measuring abundance in ecosystems. It is expressed by the total weight of the organisms, or the total weight of a species in the study area, which reflects the biomass of that particular species in the area. Stability denotes the relative constancy in the abundance of populations. Diversity denotes species composition in the area. Populations of wildlife in any area or habitat would change from time to time, mostly based on the seasonal phenological changes. For example, plant biomass in an area may be very rich during the wet season, and hence, the dependence of animals in the area for better diet is also high during the wet season. When the weather changes, especially during the dry season, there is reduction in the biomass of vegetation communities. Hence, animals migrate to other suitable areas in search of better fodder and water sources. This results in the reduction of abundance of populations, and it also reflects on the density and animal biomass in areas from where they have migrated. However, the reduction in the density in such cases may not affect the stability of the population because such migratory activities, in fact, support the stability of the population for its survival.

Is it diversity that produces stability in natural communities or stability that produces diversity? Ecologists argue about the importance of diversity in maintaining stable communities. Simple communities are unstable as is the case of communities with short-term life in farmlands. Such communities as well as other pioneer communities are not stable as major changes take place during the initial development process in the ecosystem. Diversity will be less in such stages of development. Diversity increases along with the development, leading to the establishment of more stable ecosystems with characteristic species composition. Physical environments with relatively constant features (e.g., precipitation, temperature, day length) create conditions that ultimately enrich natural communities. The logic behind the view that diversity produces stability seems most acceptable. Communities simplified by human influences result in instability. Hence, it is important to eliminate human activities in natural habitats because such activities can cause the depletion of habitat qualities including richness of biodiversity, ultimately leading to major changes in wildlife populations.

Natural disturbances are part of normal functioning of all communities, but vary in their degree of stability and resilience. However, when natural ecosystems are affected by intensive human impacts, stability and resilience of ecosystems are disturbed. Stability is the tendency of an ecosystem to reach and maintain an equilibrium condition, either a steady state or a state of stable oscillation. If the ecosystem is highly stable, it resists departures from a steady state, and if disturbed, it recovers rapidly. Resilience is the ability of an ecosystem to maintain its structure and pattern of behaviour in the event of disturbance. It is a measure of the ability of an ecosystem to absorb changes and still persist (Scheffer et al 2001, Folke et al 2004).

Questions for Review Exercise

1. Have you understood definition and meaning of the following terms?

   a. Wildlife b. Wildlife ecology

   c. Biomes d. Biogeography

   e. Niche f. Ecological succession

   g. Ecological stability h. Resilience

   i. Endangered organisms j. Endemic organisms

   k. Mutualism

2. What do you think are the key differences between 'wildlife' and 'biodiversity'? Are they different at all? Can you substantiate your view with few examples drawn from your region?

2. Establish the association between any two or more species in your locality with which you are familiar or in contact, such as parasites and symbionts. Relate such association between populations of any species in natural habitats in your locality.

3. What are the major biomes of the world? What are your perceptions of biomes? What do you understand as the characteristics of each biome? Identify few wildlife elements in the biomes you are familiar with.

4. What have you learned from Boxed items 1.1–1.4?

   i. Are caecilians wildlife? Is this not contrary to your earlier information that only large animals are wildlife forms? Based on the definition of wildlife, are you sure that caecilians similar to other small animals are also wildlife forms? They are not domestic, laboratory or farm animals. Hence, they are definitely a part of wildlife. Just like earthworms, they mostly live under soil, but earthworms are invertebrates whereas caecilians are vertebrates (amphibians). They prefer moist soils. So, what are your thoughts about them?

ii. Barbets are beautifully-coloured small birds. Can you appreciate their diversity? How do their niches differ? Can you explain it to yourself?

iii. Mountain nyala is one of the large artiodactyls. They are endangered and endemic only to certain higher altitudinal areas in Ethiopia. Try to identify an endangered and an endemic wildlife (it can be an animal or a plant seen in natural habitats) from your region. Write a short account in 2–3 pages on each of them. Your write-up should contain the common name of the species, its binomial name (scientific name: remember to write the scientific name in *italics* as it is Latin in origin; the Genus name should start with capital letter whereas that of the species is always with small letter; write name of the person who first described, known as the Authority, and the year in which the first formal description was made; e.g., Asiatic lion, *Panthera leo* Linnaeus, 1758), brief description of the characteristics of the species, its general ecology, behaviour, food habits, distribution and conservation status (as per the latest IUCN threatened category of species).

iv. Mutualism is a natural phenomenon of wildlife. It is the associations and interactions between the organisms, which make habitats dynamic. Identify such interactions between any two species of organisms in your locality. It can be between an animal and a plant species. Mention advantages of these associations with respect to both the species. Write down your observations briefly and present to your group of students during a seminar.

Wildlife Populations and Habitats 2

Types of habitats, habitats and food resources for wildlife, habitat association of wildlife, disturbances in natural habitats, natural agents of disturbances, human-induced impacts, climate change and its effects on wildlife, ozone depletion

---

2.1. TYPES OF HABITATS

Habitat is the place where organisms live and sustain. Habitat denotes biogeography of the area with the most suitable conditions including availability of most of the resources from time to time. It consists of both biotic (plants and animals including all microorganisms) and abiotic (terrain, slope, weather, water, temperature and edaphic) factors (Smith 1974). Habitat can be a wider extent of an area with generalized characteristics of the biotic and abiotic factors. It can also be a smaller area with specific environmental factors. For example, a microhabitat is much different from the surrounding habitats of wider extent. Microhabitats are characterized by specific environmental factors, and hence, are characterized by specificity of organisms that occupy such habitats. Wetland habitats of various types are also biodiversity-rich habitats (Box 2.1 – Figs 23 to 28). Wetlands play a major ecological role in maintaining its own habitats, and also, they are responsible for the sustenance of the surrounding habitats.

Box 2.1

Wetland Ecosystems, Wildlife and Conservation

*B Gopal**

Centre for Inland Waters in South Asia, Jaipur, Rajasthan
(E-mail: brij44@gmail.com)
(* Formerly: Professor of School of Environmental Sciences, Jawaharlal Nehru University, New Delhi)

Wetland is the term used since late 1960s for a large diversity of habitats, which were earlier called as bog, fen, mire, marsh and swamp. These habitats are called 'Anoop' in Sanskrit language, and they are known in various local names such as tal, sar, sarovar, beel, bheel, chaur, maun, phat, kere and kol in different parts of India (Gopal and Sah 1995).

Wetlands occur in all climates, across a wide range of latitudes, and from sea level to more than 5,000 m altitude (as in the Himalaya). They are characterized by water-saturation or submergence of land under shallow water for a period ranging from several weeks to the whole year, and the presence of vegetation, which tolerates such prolonged waterlogging or flooding condition. Wetlands often constitute the transition between deepwater and terrestrial habitats. They may depend for their water either entirely upon the precipitation (bogs) or on the surface water from the rivers (riparian/floodplain wetlands), lakes (littoral zones) or the oceans (coastal wetlands such as mangroves and beds of marine algae). Wetlands also develop where the groundwater is discharged naturally onto the surface (e.g., fens). Wetlands are usually grouped into freshwater and marine wetlands or into inland and coastal wetlands. The inland wetlands, occurring above the mean sea level, also include those developing in saline waters (salt lakes) besides those experiencing estuarine/brackish water conditions (lagoons and backwaters).

Early human civilizations developed and grew only around wetlands. Agriculture had its beginning in the floodplain wetlands. Extensive references to lotus in Hindu and Buddhist scriptures and literature reflect the importance of wetlands in the Asian region. Wetland animals such as fish, turtle and swan are associated with Gods in Hindu religion. Floodplains are major grazing grounds for cattle. Reeds are extensively used in most parts of the world. The discovery of paper owes to papyrus (*Cyperus papyrus*), which was also used for making boats. Rice and fish constitute the staple food of more than half of the world's human population today. Yet, in Europe, wetlands were treated as wastelands (an attitude reflected in the English terms used for them), which were drained and reclaimed. This view was propagated throughout the world by the colonialists over the past few centuries.

Around the middle of the 20th century, naturalists drew attention to the importance of wetlands as habitats of wildlife, especially of waterfowl, which use them in spectacular numbers and migrate over very long distances (occasionally, some of them even cross the continents) between different wetlands. Important migratory waterfowl are flamingos, geese, open bill storks, pelicans and pintails. Besides waterfowl, cranes, alligators, turtles, beavers, otters, rhinoceros and swamp deers are the important examples of wildlife supported by wetlands. Some of such wetland-dependent species are currently threatened with extinction due to human manipulations and resulting impacts on their habitats.

Fig 23. A view from Keoladeo National Park, Bharatpur, India

Fig 24. Papyrus (*Cyperus papyrus*) marsh

Fig 25. Common reed (*Phragmites karka*)

During the second half of the last century, several studies have revealed that wetlands are highly productive ecosystems, often matching with the tropical rainforests. Numerous studies, mostly in Europe and North America, highlighted several other important functional attributes of wetlands; for example, they are the hotspots of biodiversity. Faunal diversity is exceptionally high considering the fact that wetland animals not only include residents only in wetlands, but also include migrants from deepwater habitats, terrestrial uplands, occasional visitors from other wetlands and those indirectly dependent on wetland biota. On a regional scale, the beta-diversity of wetlands is quite high as 15–20% of all biota of the region.

Studies during the past three decades have brought to light many important functions of wetland ecosystems such as high production and support to biodiversity, and regulation of hydrology, climate and water quality (MEA 2005). These functions are translated into ecosystem services such as flood and drought mitigation, groundwater recharge, climate moderation and treatment of wastewaters. Many wetlands, especially marshes, swamps and shallow water bodies have also high aesthetic and cultural values.

Fig 26. A view of Pantanal wetland, Brazil

Fig 27. The Sarus Crane (*Grus antigone*) found in parts of south-east Asia and Australia is the tallest, non-flying and wetland-dependent bird

Fig 28. *Rhinoceros unicornis* in Kaziranga National Park, Assam

Wetlands, especially inland freshwater wetlands are impacted by human activities, which alter the hydrological regimes directly or indirectly. Diversion and impoundment of surface water flows for irrigation, hydropower and other human uses affect the downstream wetlands. Water regimes are also affected by the abstraction of surface and ground water, drainage and landfills, and discharge of wastewater. Other major threats to wetlands and their biota are from the introduction of exotic invasive species, mostly plants (e.g., water hyacinth) and over-exploitation of biota.

During recent years, many efforts have been made at national and international levels to conserve wetlands. These efforts have been made not only to protect them against degradation, but also to restore the degraded wetlands. The Ramsar Convention on Wetlands of International Importance, which was agreed upon in 1971 and came into force in 1975, is the only global convention dealing with a specific type of ecosystem. India and all other countries of South Asia are now parties to this convention and have designated many important wetlands within their territories. Yet, wetlands require greater attention essential to conserve a vast variety of biodiversity.

References and Additional Readings

Gopal B and Sah M 1995 Inventory and classification of wetlands in India, *Vegetation* **118** 39–48

Gopal B (compiler) 1995 *Handbook of Wetland Management*, New Delhi: WWF – India www.ramsar.org/

Keddy PA 2010 *Wetland Ecology: Principles and Conservation*, New York: Cambridge University Press

MEA (Millennium Ecosystem Assessment) 2005 *Ecosystems and Human Well-being: Wetlands and Water Synthesis,* Washington DC: World Resources Institute

Misch WJ and Gosselink JG 2007 *Wetlands* 4th edition, New York: John Wiley

Odum EP 1971 *Fundamentals of Ecology* 3rd edition, Philadelphia: WB Saunders

Diverse species of plants and animals coexist in any type of habitat. This is based on the niche of each of these organisms having species-specific and diverse roles to play in the habitat, and as a result of which, both the species and habitat sustain naturally. Vegetation is the most important element in the habitat, by which one can recognize the type of habitat. The specific vegetation in an area is the cumulative result of growth and development in response to the factors such as the parent material, elevation, hydrologic regime, humidity, slop, soil characteristics, and the seasonal events of flood and fire. The habitat provides food and shelter for the wildlife. There are several species of wild animals that are habitat-specific. For example, *P. uncia* of the mountainous regions in the Himalayan areas, and *N. hylocrius* (Fig 29) of the Western Ghats of India have narrow ranges of distribution in relation to their adaptability to specific types of habitat. On the other hand, there are also animals that have wider distribution ranges such as the Sambar deer (*Rusa unicolor* Kerr 1792) (Fig 30), the largest and most widely distributed among artiodactyls in the Indian sub-continent.

Fig 29. The endemic Nilgir tahr in Eravikiulam National Park of South India, which is a habitat specialist (Photo: S Anilkumar)

Fig 30. Sambar deer (doe), the largest among artiodactyls in the Indian sub-continent, which has wide distribution as a habitat generalist species (Photo: S Anilkumar)

2.2. HABITAT AND FOOD RESOURCES FOR WILDLIFE

Wildlife habitat includes all resources and conditions present in an area of occupancy, survival and constant reproduction of the species concerned (Hall et al 1997). In this respect, habitat is species-specific. The suitable habitat condition of one species is far different from that of the other. However, a variety of species of different trophic levels may occupy any one type of habitat based on the differences in their niche. Thus, a variety of species can survive and reproduce in a given habitat without competing with each other; rather they develop cohesive mutualistic behaviour; i.e., one species depends on the other and *vice-versa*.

Nowadays, most of the wildlife habitats in the tropical nations are not in good status due to extensive human interactions. As a general understanding, existing wildlife habitats in the developing tropics are either fragmented and reduced into smaller extents, or are degraded and encroached by invasive species. Such situations may restrict the original species of wildlife to exist in such areas. However, these days, such problems are not strictly confined only to the tropical nations, but even rich and well-developed nations face various environmental problems affecting a variety of wildlife. A typical example of affected fish population in Norway is described in Box 2.2 (Figs 31 to 33), and Box 2.3 (Figs 34 to

37) describes the impact of human-related issues including habitat fragmentation leading to behavioural changes and new adaptations of the evergreen dwelling lion-tailed macaque in South India. Wildlife managers have to analyze such situations thoroughly and find out means by which habitat restoration can be made, so as to support wildlife populations to have enough food, shelter and other essential resources. Habitat restoration generally includes provision of water holes by digging ponds in areas of water shortage. Small ponds can also be created in streams, which may dry up seasonally, but otherwise can store some water for wild animals for consumption and for wallowing. Availability of water in the area helps the soil to retain moisture, and thus, provide opportunity for vegetation communities to sustain. There will be more animal movements through such areas because of the presence of both water and fodder. Suitable fodder plants that are naturally occurring in the area can also be planted to replenish the habitat. When the habitat is almost restored to the original status, most of the extirpated wildlife is re-established and the habitat is redeveloped into the original status, if left without further human exploitation. Habitat restoration also helps wild animals to have enough cover for nesting and protection, which are also essential components of any ecosystem for the survival of animals.

Box 2.2

Fish in High Mountain Lakes of Norway – Heavily Influenced by Human Activities

R Borgstrøm

Department of Ecology and Natural Resource Management,
Norwegian University of Life Sciences, Aas, Norway
(E-mail: reidar.borgstrom@umb.no)

During the last ice age, Scandinavia was covered by a thick ice cap, and not until this huge inland ice started to melt around 15,000 years ago, the first plants and animals could invade both freshwater and terrestrial areas from ice-free areas in the South and East. As a consequence, the freshwater fauna in this area is young, and with a low biodiversity. In Norway, only 32 freshwater fish species have been able to invade the country, mostly from watercourses in Sweden and Finland, but some anadromous species were able to migrate from the ocean (Huitfeldt-Kaas 1918). In the mountain areas in southern Norway, at elevations mainly between 1,000 and 1,400 m asl, there are thousands of lakes inhabited by one or a few fish species, mainly allopatric brown trout. All fish in these high mountain localities are probably established after introduction by humans because high waterfalls prevent a natural spread from lowland areas. Written documentation of such introductions has existed for a long time; i.e., by a runic inscription on a large stone from about 1100 AC, telling that "Eilif Elg carried fish to the lake Raudsjøen" (Huitfeldt-Kaas 1918, Eknaes 1979). This most probably was brown trout (*Salmo trutta*) as it was the only fish species in the lake Raudsjøen until 1860, when pike *Esox lucius* was stocked as revenge. Bones of brown trout, estimated to be around 6000 years old, have been found in dungs at Stone Age camp sites on the Hardangervidda mountain plateau. It has also been documented that fish has been in these mountain areas for a long time, and used by man as well. To date, brown trout is the only fish species over large parts of the mountain areas of Southern Norway, but the dominance is threatened due to an extensive spread of exotic species. In addition, populations are affected due to hydel projects, resulting in large annual fluctuations in water level. Further, all populations are exploited by fishing, even in national parks, and may have a large effect on fish production, as well as changing the age and length structure of the populations.

Fig 31. Brown trout captured in a small lake on the Hardangervidda mountain plateau. This species is still allopatric in thousands of high mountain localities in southern Norway (Photo: R Borgstrom)

The lake Øvre Heimdalsvatn at 1,088 m asl on the eastern slope of the Jotunheimen Mountains was the main locality in the International Biological Program. This freshwater ecosystem was thoroughly studied during the period 1968–1972 (Larsson et al 1978). The lake is ice-covered from the end of October to mid-June, with summer temperatures usually low, and seldom above 12–14°C.

The lake area is 0.78 km^2 with maximum depth of 13 m. This is a typically high mountain lake in Norway. According to the energy budget for this system, the brown trout production was estimated to 7 KJ m^{-2} $year^{-1}$, while the total input of particulate allochthonous organic matter was 920 KJ m^{-2} $year^{-1}$, and the total autochthonous plant production was in the range of 960–1,210 KJ m^{-2} $year^{-1}$. Fish production was mainly based on benthic invertebrates, and to some extent, on zooplankton and terrestrial insects. Energetically, most important benthic invertebrates were *Lepidurus arcticus* (tadpole shrimp), *Gammarus lacustris*, Tricoptera, Chironomidae and Coleoptera. Fish was not found as prey. When Jensen started his study of the brown trout population in this lake, brown trout was the only fish species present, and the biomass of fish older than three years was estimated to be 19.5 kg ha^{-1} (in 1958), but after a heavy fishing, the biomass was reduced to 8.2 kg ha^{-1} (in 1963). According to Jensen (1978), the predicted sustained yield of brown trout would be 5.7 kg ha^{-1}, with mean weight 269 g, given that the yearly fishing effort was 1,600 gillnet nights with mesh size 32 mm (knot to knot). Thus, the harvest of brown trout from this lake agrees with the obtained harvests from other high mountain lakes in southern Norway, commonly in the range 1–5 kg/ha. With an effort of 20 gillnet nights per ha, the annual fishing mortality of the population in the lake Øvre Heimdalsvatn was estimated to be 60% (Jensen 1985). By reducing the fishing effort to 10 gillnet nights per ha, the annual fishing mortality would have been reduced to 37%. The reason for the high gillnet effort needed for obtaining an annual mortality in the range 37–60% is probably related to the fish density and swimming activity of the fish (Borgstrøm and Plahte 1992). At low densities, fish has a very high swimming activity; and thus, a much higher catch probability by gillnetting, as observed in small lakes on the Hardangervidda mountain plateau. This is an important condition because large individuals of brown trout are most attractive for fishermen. However, large individuals of brown trout in allopatric populations are mainly obtained at low population densities; i.e., in populations, which are vulnerable to gillnetting. Thus, such populations may easily become over-exploited.

In high mountain areas of Norway, brown trout may obtain a high age, with 38 years as a recorded maximum (Svalastog 1991). Similar or even higher age of freshwater fish species such as Arctic charr (*Salvelinus alpinus*), Lake charr (*Salvelinus namaycush*) and Lake whitefish (*Coregonus clupeaformis*) are recorded from Arctic areas (Svenning 1993). In exploited populations, the probability of obtaining such high ages becomes very low, and fish with ages above 10–15 y become rare, even at a relatively low exploitation rate.

Fig 32. The subalpine lake Øvre Heimdalsvatn has been thoroughly studied during the last fifty years, and has become a reference ecosystem (Photo: R Borgstrøm)

The total number of freshwater fish species in Norway having a natural immigration history is about 32, but in addition, at least 10 species have been imported by man from other European countries and from North America. Some exotic species have recently been introduced to Norway, and they have established their populations such as the gudgeon (*Gobio gobio*) and the sunbleak (*Leucaspius delineatu*). In addition, the natural distribution area of native species has increased considerably during the last hundred years due to spread by man. In this way, the sovereignty of brown trout in the high mountain areas has been broken by human introductions of other species, mainly Artic charr, whitefish (*C. lavaretus*) and European minnow (*Phoxinus phoxinus*), although brown trout is still allopatric over large areas. Overall, these new introductions had negative consequences for brown trout production and harvest, mainly because of competition for common food resources. When Arctic charr are established, individual growth rate may be high in the first generation, giving large fish, but after a few generations, population density will increase due to high annual recruitment, and individual growth rate will decline rapidly. At the same time, growth rate of brown trout will decline due to the high overlap in feeding between the two species. Thus, a good brown trout fishery may be destroyed after introduction of Arctic charr (Svärdson, 1976). Similarly, brown trout recruitment and individual growth rate have declined after establishment of European minnow, partly because of a dietary overlap, both in the littoral zone and in the streams (Borgstrøm et al 2010, Museth et al 2010).

Fig 33. The reservoir Ringedalsmagasinet on the western border of the Hardangervidda mountain plateau is one of several reservoirs surrounding the Hardangervidda National Park. Water level in this reservoir may fluctuate 94 m, giving a large and steep draw down zone (Photo: R Borgstrøm)

In Norway, there are more than 800 reservoirs for production of electricity. Most of the reservoirs are made by damming natural lakes, and more than 1,100 km^2 of land areas surrounding the lakes has become part of the reservoirs. Reservoirs are filled mainly from spring to autumn, and the stored water is used for production of electricity during the winter when demand for energy is at the highest and precipitation comes as snow. This means that water level fluctuates annually in some reservoirs with more than hundred meter from lowest regulated level to the highest, creating large ecological changes, with erosion and freezing of the draw-down zone, giving loss of most benthic invertebrates. Many reservoirs are found in the high mountain areas where brown trout often dominates. The

Hardangervidda National Park is more or less surrounded by hydroelectric reservoirs. Due to damming of lakes, the outlet river becomes closed as spawning area for the lacustrine fish species, resulting in decreased natural recruitment. Usually, fish production is heavily reduced in reservoirs with high annual water level amplitudes (Garnås and Hesthagen 1982). However, the supposed loss in recruitment has commonly been compensated by stocking.

References and Additional Readings

Borgstrøm R 1992 Effect of population density on gillnet catchability in four allopatric populations of brown trout (*Salmo trutta*), *Canadian Journal of Fisheries and Aquatic Sciences* **49** 1539–1545

Borgstrøm R, Museth J and Brittain JE 2010 The brown trout (*Salmo trutta*) in the lake, Øvre Heimdalsvatn: long-term changes in population dynamics due to exploitation and the invasive species, European minnow (*Phoxinus phoxinus*), *Hydrobiologia* **642** 81–91

Borgstrøm R Plahte E 1992 Gillnet selectivity and a model for capture probabilities for a stunted brown trout (*Salmo trutta*) population, *Canadian Journal of Fisheries and Aquatic Sciences* **49** 1546–1554

Eken M and Borgstrøm R 1994 First report of *Gobio gobio* from Norway, *Fauna* **47** 120 123

Hesthagen T and Sandlund OT 2007 Non native fish species in Norway: history, consequences and perspectives, *Journal of Fish Biology* **71** (Supplement D) 173–183

Eknæs Å 1979 *Inland Fisheries*, Oslo: Det Norske Samlaget (in Norwegian)

Jensen KW 1978 The dynamics and exploitation of the brown trout population of Øvre Heimdalsvatn, *Holarctic Ecology* **1** 277–278

Jensen KW 1985 Fisheries management, in: *Jakt, Fiske, Friluftsliv* Volume 4, eds R Frislid and K. Rom, Oslo: Tiden Norsk Forlag (in Norwegian) 352–373 pages

Larsson P, Brittain J E, Lien L, Lillehammer A and Tangen K 1978 The lake ecosystem of Øvre Heimdalsvatn, *Holarctic Ecology* **1** 304–320

Museth J, Borgstrøm R and Brittain JE 2010 Diet overlap between introduced European minnow (*Phoxinus phoxinus*) and young brown trout (*Salmo tyrutta*) in the lake, Øvre Heimdalsvatn: a result of abundant resources or forced niche overlap?, *Hydrobiologia* **642** 93–100

Svalastog D 1991 A note on maximum age of brown trout, *Salmo trutta* L., *Journal of Fish Biology* **38** 967–968

Svenning MA 1993 Life history variation and polymorphism in Arctic charr, *Salvelinus alpinus* (L.) on Svalbard and in northern Norway Dr. Scient. Dissertation, Tromsø: Norwegian College of Fishery Science

Svärdson G 1976 Interspecific population dominance in fish communities of Scandinavian lakes, *Report of the Institute of Freshwater Research Drottningholm* **55** 144–171

Vik R 1978 The Lake Øvre Heimdalsvatn - a subalpine freshwater ecosystem: Introduction, *Holarctic Ecology* **1** 84–88

Natural habitats provide food, shelter and cover for wild organisms. Shelter is made available by providing nesting sites and nesting materials. In addition to nursing the young ones, the shelter is used by wild animals to rest and sleep. Cover denotes the canopy of the surroundings where the organism can get concealment from the adverse environmental conditions such as high temperature and extreme cold weather. Also, they can conceal themselves for predatory purpose (e.g., predators have to conceal themselves under the cover to be unnoticed by the prey in the vicinity) as well as to escape from the predators (e.g., prey species have to conceal themselves to hide from the foraging predators).

2.3. HABITAT ASSOCIATION OF WILDLIFE

Habitat association of wild animals depends mainly on the food preferences of each of the species. There are animals such as the Koala bear (*Phascolarctos cinereus* Goldfuss, 1817), which consumes about 1 kg per day of vegetation from few species of gum trees (*Eucalyptus* species). *Eucalyptus* occurs in large blocks of forest in which other species are uncommon. Koalas exploit a simple food base with a good deal of efficiency. Thus, habitat specificity is also related to the food specificity. Another example is the association of *M. silenus* (Fig 38) in South Indian forests with the tree *C. exarillata*. The Silent Valley forest of Kerala, which is part of the Nilgiri Biosphere Reserve, is well known for its lion-tailed macaque populations (Balakrishnan 1984, Anilkumar 2009). The Ethiopian wolf (*Canis simensis* Rüppell, 1840) is a habitat specialist, and is seen only in the high-altitude Afro-alpine habitats in Ethiopia (Fig 39), where they depend mostly on the giant mole rats for their food. Such organisms can survive only in areas where their specific food items are abundant. They also restrict their ranges within such habitats (limited range). Hence, it is essential to maintain habitats of such endemic species, even though the habitat is less diverse.

Box 2.3

The Lion-tailed Macaque: A Case Study on the Impacts of Forest Fragmentation

*A Kumar and G Umapathy**

Wildlife Conservation Society – India Program, 269 5th Main,
Canara Bank Layout, Bangalore, Karnataka, India
(E mail: ajith@ncbs.res.in)
*Laboratory for the Conservation of Endangered Species,
Centre for Cellular and Molecular Biology,
Hyderabad, Andhra Pradesh, India

The lion-tailed macaque, a species endemic to the tropical rainforest of the Western Ghats mountain ranges, is a typical example of how evolutionary history and ecological adaptations can make it vulnerable to the habitat loss and fragmentation. Tropical rainforests covered most of India when the first macaque, perhaps *Macaca paleoindica*, arrived in India from northern Africa through Europe, nearly five million years ago. It is believed that the lion-tailed macaque is the earliest descendent of this ancestor. This ancestor of all Asian macaques had to adapt to a diet of fruits, seeds and foliage invertebrates. Leaves of most tropical rainforest plants are inedible due to the presence of poisonous secondary metabolites, unless the animal has a specialized digestive system for dealing with them, as the colobines do. Tropical rainforests are the most stable forests with a year-round availability of fruits, although there are seasonal fluctuations. Although, once widely distributed through most of peninsular India, as the monsoonal climate set in, the tropical rainforest in the plains quickly gave way to more seasonal dry forests, and became restricted to the Western Ghats known for its high rainfall. Adaptations to a diet of fruits and invertebrates have had major implications for the social organization and life history of the lion-tailed macaque. As fruits are never abundant, they live in relatively small groups of about 20 animals, mostly with only one adult male and 5–6 adult females; the rest being immature animals. For the same reason, they have large home ranges of up to 5 km^2, and therefore, occur in low densities. Throughout the year, one group of lion-tailed macaques feeds on fruits of nearly 100 plant species, out of around 200 plant food species reported through its range (Krishnamani and Kumar 2000). Females give birth only from six years of age as compared to 3–4 y in other macaques. More importantly, they give birth only once in 2.5 y. Hence, the birth rate is only about 0.3 infants per year per female.

However, they have a remarkably high survival rate. Nearly, 80% of the newborns reach adulthood. Males become adults only at about eight years of age, and they have lower survival rate than females. Therefore, there are always fewer adult males in the population, explaining adult ratio more skewed in favour of females than most other macaques.

These adaptations have had major consequences to the survival of the lion-tailed when massive loss and fragmentation of habitats began in the Western Ghats, especially from mid 19th century. It is estimated that even between 1920 and 1990, there was a 40% decline in the forest area and four-fold increase in the number of forest fragments in the Western Ghats (Menon and Bawa 1997). This was initially due to roads, then for cardamom, coffee and tea plantations, and later for dams, forestry plantations and government-sponsored agriculture. These activities also brought in large human settlements in the proximity of remnant forests. Presently, the region of Western Ghats is among the most fragmented and densely-populated biodiversity hotspots in the world (Cincotta et al 2000).

The present wild population of the lion-tailed macaque is probably of 4,000 individuals, distributed as nearly 47 sub-populations occurring in isolated forest fragments. Only less than 10 of these populations have more than 100 animals; the largest being the Ashambu Hills, with about 600 animals (Kumar et al 2008). Nearly 30% of the remaining populations occur in privately-owned coffee and cardamom plantations. Thus, the lion-tailed macaque represents an ideal model to examine behavioural and other consequences of population and habitat fragmentation.

As has been found elsewhere, the fragmented habitats of the lion-tailed have become degraded due to logging, edge effect and poor regeneration. As a result, groups in forest fragments spend less time feeding, and feed on fewer plant species. More importantly, they also feed less on foliage invertebrates (Umapathy and Kumar 2003). To some extent, they seem to compensate this by feeding opportunistically on invertebrate and vertebrate fauna such as lizards on the forest floor, where they spent a lot more time compared to contiguous forests (Kumar et al 2000).

Their behaviour in one coffee plantation (Puthuthottam Estate, near Valaprai town in the Anamalai Hills) reveals how they have become increasingly adapted to human presence. Beginning with raiding vegetable gardens, which the estate labourers started nearly two decades back, the lion-tailed macaques now regularly go into labour sheds and tourist vehicles, and some even visit the local market. Although this patch of forest was isolated in early 1900s and people have always lived inside the estate, it is only in the last decade or so that such drastic change in the behaviour has happened. This offers an interesting opportunity to examine the process of adaptation among primates to human presence.

Fig 34. Lion-tailed macaques have recently started raiding houses and kitchen gardens in places they occur near human settlements. A scene from Puthuthottam Estate in Valparai Town, Anamalai Hills (Photo: Kalhan Varma)

Fig 35. Lion-tailed macaques raiding house (Photo: Kalhan Varma)

Increasing demographic stochasticity is a major predicted property of isolated populations. This is clearly demonstrated in the case of the lion-tailed macaque. Group size and adult sex ratio are highly variable as the fragment size becomes smaller and birth rate declines (see Fig 36). Long-term data on populations from fragment show drastic changes in population, unlike populations in contiguous forests. Long-term monitoring of forest patches of Anamalai hills shows that decreased birth and growth rates, and the percentage of immature animals in the small forest fragments may be due to shortage of resources, and higher mortality due to a variety of reasons (Umapathy et al 2011).

Habitat fragmentation brings in new parasites and pathogens that can wipe out the remnant populations. Carriers of these parasites and pathogens are often domesticated animals and human commensals such as rats and mice. Proximity of human settlements and the relatively long perimeter of the forest fragments facilitate the invasion by these carriers. Poor nutritional condition, higher host density in the confined space and more time on the ground make the lion-tailed macaque more susceptible to parasitic infestation and higher mortality rate. Hussain et al (2013) collected faecal samples from groups in nine forest fragments and examined the occurrence of intestinal parasites. Overall 75.8% of 91 samples had at least one parasite; the incidence in a group varying from 42% of individuals to 100%. Ten parasite taxa were detected: six nematodes, two cestodes, one each of trematode and protozoan. Strongyloides was the most prevalent parasite found in eight of the nine forest fragments, followed by *Trichuris* and *Ancylostoma* in seven forest fragments. Both the cestodes were found in only one forest fragment. Out of the nine taxa identified, six occurred only near human settlements, and three occurred in both. Of the three occurring in both, two had higher prevalence near human settlements and all the rare ones were found only in fragments near settlements. Presence of human settlements was found to be the best predictor of richness of parasitic species as well as the occurrence of some parasitic groups. Group size, which was higher in small fragments, was the next best predictor. This clearly demonstrates the impact of habitat fragmentation and associated human impacts on parasitic load in the lion-tailed macaque.

A comparison of the mitochondrial DNA of lion-tailed macaques in the fragmented population in the Anamalai hills with that in other areas of the Western Ghats shows a high loss of genetic diversity in the former. A total of 111 sequences were examined from nine forest patches of the Anamalai hills using DNA extracted from faecal samples. All sequences of cytochrome-b region were similar and formed a single haplotype. However, the hvr-I (hyper variable region-I) sequences had a single major haplotype (H1) shared by many individuals and two sub-haplotypes.

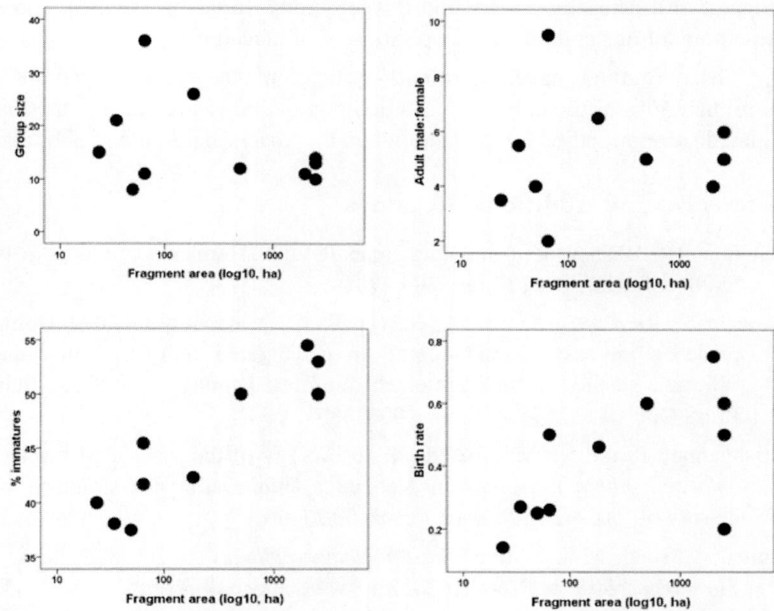

Fig 36. Some demographic parameters of the lion-tailed macaque in relation to that of the forest fragment

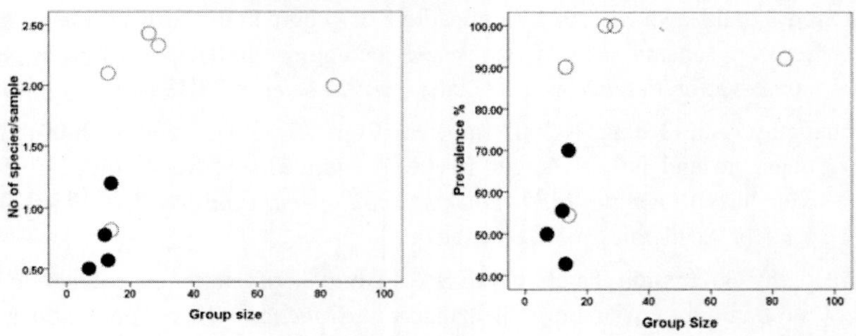

Fig 37. The relationship of number of parasitic taxa/fecal sample and prevalance of parasites in lion-tailed macaques with group size and humans near forest fragment
(present = hollow circle, absent = filled circle)

The mean genetic and haplotype diversity (Hd) in hvr-I region were 0.000219 and 0.087, respectively, indicating a highly homogenous population in terms of mitochondrial DNA variation (Umapathy et al in preparation). The reasons for the low diversity could be the isolation of these populations and the female philopatry of macaques.

These findings have important implications for the management of more than 50% of the current lion-tailed population, which occurs as small isolated sub-populations, especially when they occur near human habitats.

References and Additional Readings

Cincotta RP, Wisnewski J and Engelman R 2000 Human population in the biodiversity hotspots, *Nature* **404** 990–992

Hussain S, Ram MS, Kumar A, Shivaji S and Umapathy G 2013 Human presence increases parasitic load in endangered lion-tailed macaques (*Macaca silenus*) in its fragmented rainforest habitats in southern India, PLoS One 10.1371/journal.pone.0063685.

Krishnamani R and Kumar A 2000 Phyto-ecology of the lion-tailed macaque (*Macaca silenus*) habitats in Karnataka, India: Floristic structure and density of food-trees, *Primate Report* **58** 27–56

Kumar A, Singh M and Molur S 2008 *Macaca silenus*, in *IUCN Red List of Threatened Species*, Version 2012.1 <www.iucnredlist.org>. Downloaded on 25 June 2012

Kumara HN, Singh ME, Sharma AK, Singh MR and Kumar AM 2000 Faunal components in the diet of lion-tailed macaque, *Primate Report* **58** 57–65

Menon S and Bawa KS 1997 Applications of geographic information systems, remote sensing and a landscape ecology approach to biodiversity conservation in the Western Ghats, *Current Science* **73** 134–145

Umapathy A and Kumar A 2003 Impacts of forest fragmentation on lion-tailed macaque and Nilgiri langur in the Western Ghats, South India, in LK Marsh (ed) *Primates in Fragments: Ecology and Conservation, New York:* Kluwer Academic, 163–189 pages

Umapathy G, Hussain S and Shivaji S 2011 Impact of habitat fragmentation on the demography of lion-tailed macaque (*Macaca silenus*) populations in the rainforests of Anamalai Hills, Western Ghats, India, *International Journal of Primatology* **32** 879–900

Also, there are animals that depend upon a wide variety of food items. For example, the Sambar deer of the Asian sub-continent, which is also the largest and most widely distributed among artiodactyls, has a wider food choice (Balakrishnan 2002). They are not restricted to a few variety of vegetation; rather they sample several species of vegetation present in their habitats. This phenomenon of less specificity with food helps them to survive in wider habitats and also in harsh conditions as they can find out some vegetation in the habitats where they range from time to time. Thus, to some extent, the success of life depends on the ability of the organisms to sustain on a variety of food items without much specificity.

Fig 38. A lion-tailed macaque in close contact with a Malabar giant squirrel; a rare observation from a fragmented and degraded evergreen habitat at Puthuthottam, South India (Photo: Davika Ramachandran)

Range tolerance of organisms is based on the restrictions of essential ecological requirements of the concerned species in the habitat. Geographical locations of higher altitudes above around 4,000 m elevation are known to contain no tall vegetation, based on the phenomenon known as tree limit. Hence, higher altitude can act as a physical barrier for forest habitats and for the presence of a variety of forest-depending animals. Their ranges are restricted in areas of higher elevation in the absence of essential food sources and shelter requirements. The tolerance for biological factors as well as for the non-biological factors such as physical factors set limits in the distribution of organisms in different habitats. Temperature is one of the physical factors having major role in limiting distribution of organisms. Both lower and upper limits of tolerance of such factors are important in the distribution of organisms in their habitats. Even though few organisms (e.g., *S. salar*) have the ability to depend upon both sea and freshwater habitats at least for some time of their life, majority of the species do not have the ability to survive in such different salt-tolerance limits. Majority of the natural organisms are restricted in their ranging behaviour based on both physical and biological factors present in the habitats. These are fundamental concepts in wildlife ecology.

Fig 39. An Ethiopian wolf foraging in the Sanetti Plateau in the Bale Mountains National Park, Ethiopia (Photo: M Balakrishnan)

Each species of wildlife is seen in a specific type of habitat based on its preference for essential resources such as food and shelter, and for easy means to escape from the predators. This also features the process of animal migration, as animals migrate from one area to another during different seasons in search of better food resources. Even if there are several species of edible vegetation in any area, all species of animals may not sample all species of vegetation. Only few species will form the food of each species. This is more prominent in the case of endemic species of animals. They are limited in distribution, mainly due to the fact that their resource requirements are very specific and narrow in range. Only few species of animals in an area have wider preferences for food sources. Such differences in the pattern of selection of food and suitable locations for shelter provide a variety of species of the same ecological niche to inhabit in the same area without much competition for food and space. Also, such behavioural adaptations of organisms make them to spread all over with specific identity of each species of animal with a specific type of habitat. For example, *R. unicolor*, the most widespread species of the deer in the Indian sub-continent, has a wider preference for food. They are seen in most of the types of forests. On the other hand, the chital or spotted deer (*Axix axis* Erxleben, 1777) prefers mostly the drier zones and they do not disperse into dense forests. The lion-tailed macaque is habitat specific, adapted to rainforests, whereas the bonnet macaque (*Macaca radiate* É. Geoffroy, 1812) is widespread in most-deciduous and dry-deciduous forests, and in the boarders of forests and villages. They have developed as a noxius agricultural pest as they often intrude into farms. The Nilgiri langur (*Trachypithecus johnii* J. Fischer, 1829) is distributed in the semi-evergreen and moist-deciduous forests, whereas the Hanuman langur (*Simia entellus* Dufresne, 1797) is more of a dry-deciduous species. *Nilgiritragus hylocrius* is very specific in their pattern of distribution. It is found only in certain high altitudinal areas of the southern part of the Western Ghats. It prefers to rest in rocky peaks through which larger predators can not move fast. This is a special behavioural adaptation to escape from the predators. Such association of animals with a specific type of habitat or few habitats is known as habitat association or in general, habitat preferences, by which a species (e.g., *M. silenus*) adapts to be in its own selected habitat type. On the other hand, a species with its wider food choices (e.g., *M. radiata*) can have wider distribution patterns.

The patterns of habitat selection of wildlife give us ideas on the resource requirements of the species concerned. Wildlife managers, responsible for conservation and management of wildlife, should have a basic idea on such issues in the areas of their jurisdiction. In case a management involvement is required in the area, the manager should effectively implement programmes through which such essential resources are made available. A typical example of such habitat manipulation under management programmes is the provision of water holes for wild animals in areas where scarcity for water is common during the summer. If such provisions are made, animals need not have to wander longer distances in search of water. Wandering long distances in search of essential resources might make wild animals to encounter accidental predation pressure. This can be avoided in case their movements are more through known habitats around own home ranges.

2.4. DISTURBANCES IN NATURAL HABITATS

Various types of disturbances disrupt ecosystems, communities and populations, resulting in changes in substrates and availability of resources from time to time for the organisms to sustain. It may also create opportunities for new individuals or colonies to establish. Disturbances have both spatial and temporal characteristics such as the size of the area disturbed, frequency of disturbance, number of events per unit time, turnover, intensity and its impacts on populations or on the community as a whole.

The scale of disturbances may range from small and frequent events such as the death of a tree in a forest to large-scale rare events in which extensive areas are swept by fire, buried under volcanic ash, torn by landslides or denuded as a result of land-clearing processes. Among all these, the direct impact of human-related activities is the major type of disturbance to wildlife and their habitats. Imagine a situation when one of the trees dies in a natural forest. It would open up space in the area, which makes changes in the canopy leading to a new set of vegetation communities to develop in the presence of more sunlight on the ground. If a patch of trees dies in an area, a large area on the ground is exposed to direct sunlight, resulting in large-scale changes in the vegetation composition of the area in the near future. Such gaps result in patches of different stages of succession. Natural disturbances of low intensity make the ecosystem more dynamic and diversity-rich. Timber logging, poaching, forest-fire and land clearings induce large-scale disturbances. Such events invite colonization by opportunistic species. The type of species that colonize the disturbed area is influenced by seeds present around the site, seedlings and saplings present nearby, soil conditions, intensity of competition and other ecological conditions.

2.4.1. *Natural Agents of Disturbances*

Wind, moving water (flood), drought, fire, volcanic eruption and high population density of animals act as natural disturbances in ecosystems. Strong, high-speed and continuous wind affects canopies of trees, growth of vegetation, and uproots trees. Floodwater is a powerful agent of disturbance in natural ecosystems. Strong and intensive floods affect low-lying areas as a whole and its fauna and flora. They cut away river banks, change the course of streams and rivers, move and deposit sediments, and bury or carry aquatic organisms to far away locations. It also results in large-scale soil erosion both in the flooded areas and on the slopes of the surroundings. Prolonged drought can have serious effects on vegetation composition and structure in the affected ecosystem with consequent impacts on animal life. On the other hand, drought dries up wetlands, and causes famine and crisis conditions among wildlife. Drought in African savannas results in widespread deaths of herbivores due to dehydration and starvation. Forest-fire is both a destroyer and a regenerator of life. Three conditions are necessary for fire to assume ecological importance. These are as follows:

1. Accumulation of organic matter sufficient to burn
2. Dry weather conditions

3. Source of ignition

The only two sources of ignition are lightning and humans. Forest ecosystems appear to be burnt and developed under certain fire frequencies influenced by changes in climate. Fire affects not only the vegetation, but also the soil. Fire increases the soil temperature, but the moisture content of the soil provides insulation. Severe fires alter the habitat and eliminate several fire-prone species in the affected area. Forest-fires affect populations of ground-nesting birds, amphibians and reptiles. Fire also creates new habitat for a different set of species, especially of those that are favourable for open and shrubland habitats.

Natural populations of large mammals such as the African elephants and other large herbivores influence savannah ecosystems. When their numbers were in balance with the vegetation and their movements were not restricted, they were playing an important role in creating and maintaining forests. But when the number of elephants exceeded the capacity of the habitat to support them, either as a result of increased levels of reproduction or as a result of reduction in the extent of habitat (both result in increased population density), their feeding habits combined with summer fires devastate flora, fauna and soils. Elephant depredation on trees acts as a catalyst to fires, leading to the process of conversion of forest to grassland.

2.4.2. *Human-induced Impacts*

In the recent past, it is revealed by several studies that human beings act as a major threat to the existence of ecosystems. This has major reference to the explosion of human population during the past few centuries. Man as a forest-dweller, during millions of years of his evolution, remained as any other natural organism for a long time, taking whatever is required for his own subsistence from the surroundings, but later turned out to be one with powers to conquer nature with vested interests. For example, logging, cultivation, urbanization and mining – all these processes have major environmental impacts affecting a variety of organisms and their habitats.

2.4.2.1. *Logging*

Disturbances by timber logging depend upon the method employed such as selection-felling or some form of even-aged management (clear-felling). A single tree or groups of trees are removed in selection-felling. Selection-felling produces gaps in the forest canopy and favours reproduction of tolerant trees, but composition of the forest essentially remains unchanged. Clear-felling results in removal of forest in a large area and reversion to an early stage of succession. Large-scale clear-felling inevitably changes the composition and structure of the regenerating forest. Such areas eliminate habitat of canopy-dwelling species, both vertebrates and invertebrates. Such species may altogether vanish from affected areas.

2.4.2.2. *Poaching*

Human beings have been poaching wildlife since time immemorial. However, the rate of poaching has increased several folds in the past few decades. As wild meat is the only source of protein for large number of forest-dwellers and rural communities, poaching is widespread among the indigenous people living inside and nearby the forest areas. Villagers around the forest also depend upon wildlife for their protein requirements. Large mammals, birds and reptiles are widely used for food by such people. However, commercial poaching has a major effect on wildlife. It is responsible for depletion of populations of several species of large mammals such as elephants, rhinoceros, Indian cheetah (already extinct in the wild) and tiger. Several species of birds and reptiles have also been extracted due to commercial poaching for flesh, feathers and skin. The commercial value of the products of such animals is very high. Such species are costly in one way or the other, and hence human beings have vested interest for their flesh, skin, tusk, horn and other products.

2.4.2.3. *Cultivation*

The advancement of human beings from nomadic and hunting gatherer to settled and farming patterns of life has had immense effects on ecosystems. In the beginning, the requirements were only for their own subsistence, but later, as man advanced with large-scale, high-input and commercial agriculture, human impacts on ecosystems started becoming severe. All over the world, large extents of forests were clear-felled for agricultural expansion, particularly in the developing tropical regions as a result of high density of human population in the tropics. During this process, man was responsible for the destruction of extensive natural habitats with its rich fauna and flora, and for intensive soil erosion and spread of weedy plants, which finally affected the water cycle. Natural wetlands have been drained and converted for different land-use purposes. Tropical rainforests of South America have become grazing lands and settlements. Most of the rainforests of south-east Asia has been converted to plantations of rubber and palm nut. Even areas of forests, left as conservation areas, have become isolated and fragmented island-like habitats where wildlife can not sustain for long.

2.4.2.4. *Livestock Grazing*

Intensive livestock grazing affects wildlife habitats as the livestock and wildlife would compete for limited resources in the habitat. The presence of livestock in forest habitats can contribute for increased animal density. In such areas, limitations of the carrying capacity of the habitat affect the normal regeneration potential of the vegetation communities, and the availability of resources (fodder) for wild animals decreases. Hence, wild animals are affected due to the absence of sufficient vegetation and quality fodder in their habitats. Grazing also affects soil qualities, which will have its own impacts on growth and maintenance of natural vegetation in the area. Compaction of soil is the first adverse effect that one can visualize in intensively-grazed areas. This affects infiltration of water during the rainy season, and much of the rain water is lost as run-off. In the presence of ground vegetation, water run-off is limited, which helps in the process of infiltration resulting in retaining moisture in the soils. Soil moisture helps vegetation to sustain. The ultimate effects of intensive livestock grazing in wildlife habitats are deterioration of soil structure and quality, loss of vegetative cover, loss of litter and soil organic matter, leading to soil erosion.

2.4.2.5. *Urbanization*

The human hunter-gatherers with a loose social structure became agrarian and established villages. As villages became centres of major human activities including trade, they developed into cities. Over 10,000 years as agrarians, humans have become urban. Humans are now engaged in an ecological revolution, which they had experienced only once in the past when they abandoned the hunting way of life and shifted to an agrarian one. Along with the growth of human populations, urbanization has increased rapidly. At present, more than half of the population of the world lives in cities of all sizes. Urban revolution has totally displaced natural environment, wildlife and wilderness. Cities have spread out in all directions from the core. Sub-urban development including development of roads, industries, housing development, shopping malls, and diversion of rivers and water canals for the use in cities, air-polluting automobiles, effluents from industries and human wastes – all these have severe adverse effects on a variety of wildlife forms, ultimately leading to depletion of natural habitats and the rich wildlife. Ecologically, the urban environment is a different world than the world it displaced. The inner core is impoverished in biodiversity.

2.4.2.6. *Pollution*

Accompanying cultivation and urbanization is pollution of air, water and soils. Use of artificial fertilizers and chemical pesticides in croplands and home gardens, often in excessive amounts, contaminates soil and water. Pesticides, accumulated in soils and leached out into ponds, rivers and lakes, are taken up by plants and other lower organisms. These pesticides are incorporated into their tissues, and are transferred to larger animals including human beings through the ecological web. By the time these chemicals get accumulated in the tissues of tertiary carnivores including human beings, their concentrations increase up to lethal levels through the process of biomagnification.

2.4.2.7. *Mining*

Demands for energy to support urbanized societies have intensified disturbances to natural ecosystems all over the world, including developed nations. For example, mining of coal, iron, gold, etc. for which large areas of natural terrain are converted with intensive activities. The impacts of such activities are much pronounced in mountainous regions. The magnitude of damage varies with the region and with the degree and success of reclamation efforts.

2.4.2.8. *Dam Construction*

Dams, constructed for hydroelectricity, irrigation and drinking purposes, are located in valleys of rich biodiversity. In case of large dams, large areas of natural forests are submerged. Although percentage-wise the extent may be less, such submerged areas represent the most important wildlife habitats as they are located in the biodiversity-rich areas. Once the area is submerged under a dam, biodiversity of the habitat is lost forever. Several endemic species that are limited in their distribution patterns are directly affected by dams as their habitats are lost under submergence.

2.5. CLIMATE CHANGE AND ITS EFFECTS ON WILDLIFE

We know that the atmospheric temperature is increasing all over the world from time to time. The more constant increase has been recorded during the past few decades, which has lead to the diverse physical and biological changes in all types of habitats, affecting a variety of organisms including human beings. Anthropogenic activities such as deforestation, extensive extraction and use of fossil fuel, expansion of agriculture and industrial processes, and greenhouse gas emissions have been suggested as major causes of global warming (Meinshausen et al 2009). As human population has been on the increase since long, human-induced global warming has the potential to continue and make additional stress on ecosystem characteristics. The projections of climate-change model indicate that the global temperature can increase up to 6.4°C during the present century (IPCC 2007). If such major difference in the atmospheric temperature occurs, its direct and indirect effects are to be reflected on growth, development and activities of all organisms, all over the world. As one can easily imagine, the direct and immediate consequence would be the increase of sea level. Even though the climate of the Earth is dynamic and has been fluctuating, climate change models suggest that in the coming decades, the frequency and intensity of the changing weather patterns will accelerate the occurrence of repeated floods, droughts and storms, leading to intensive natural calamities and famine.

Increased emission level of greenhouse gases is pointed out as one of the major reasons of global climate change. Although there are over 30 greenhouse gases, Carbon dioxide (CO_2), Methane (CH_4), Nitrous oxide (N_2O), industrial gases such as Hydroflurocarbons (HFCs), Perfluorocarbons (PFCs) and Sulphur hexafluoride (SF_6), Ozone (O_3) and water vapour (H_2O) have direct effect on global warming. In the absence of greenhouse gases (gases that trap heat in the atmosphere), the planet Earth would have been far colder and largely covered by ice. However, increase in the amount of the greenhouse gases beyond a certain level as a result of burning large quantities of fossil fuel, deforestation, volcanic eruption, earthquake-induced fire, etc. would lead to higher atmospheric temperature and global warming. Average temperature around the world has risen by 0.5°C since the beginning of the 20th century. If the emissions of greenhouse gases continue at the current rate, a temperature rise of 1.5–4.5°C is expected by the year 2030.

Both natural and man-induced causes are suggested as responsible for the recent climate change and the processes leading to global warming. In addition to CO_2 and N_2O released in the process of metabolic activities of organisms, the reported natural factors are continental drift, volcanoes, ocean currents and meteorites. Separation of land masses changes the physical features of water bodies and the flow of ocean currents and winds, which affects the climate. Volcanoes throw out large amounts of SO_2, water vapour, dust and ash responsible for changes in climatic patterns. All over the world, human-induced factors are related to industrialization, agriculture, pollution and land-use pattern changes, and are responsible for the emission of large amounts of greenhouse gases. In fact, the effects of such factors are more intense in the developed part of the world, and hence, Kyoto Protocol insists developed nations to reduce their emissions. However, the increase in human population has direct and indirect effects, mostly in the developing part of the world, and more specifically in the tropical nations. The most important effect in this context is the increase in the rate of deforestation in consequence to the need for more food production to feed the increasing human population year by year.

The climate change, particularly global warming, has revealed its effects on distribution patterns, physiology and life cycles of a variety of wild organisms (Botkin 2007). It is revealed that the distribution ranges of some species of insects, birds and mammals have moved towards the polar region or into higher altitudes in the past in response to the shifting climatic zones (Thuiller 2008). Further, there are also evidences for changes in the time of flowering and fruiting of plants as well as breeding of animals in response to changes in atmospheric temperature. Global warming will have a direct effect on the polar ice caps. Polar ice caps will melt due to global warming, and hence, there will be increase of the sea level. As a result, coastal habitats will be lost or expanded into the boundaries of nearby habitats. This will shrink or expand certain types of habitats affecting diverse organisms. It is easy to guess the probable impacts of melting of polar ice caps to a large extent on Polar bear (*Ursus maritimus* Phipps, 1774), which is strictly habitat-specific. They will not find another suitable habitat to move away. Variations in atmospheric temperature also would directly affect vegetation communities and their respective species composition in the wild. Certain species may overcome such effects, but in majority of the cases, any such adjustment may not be possible if changes occur fast. Reduced availability of water in the habitat negatively affects availability of forage for wild animals. Vegetation structure is a major parameter of habitat status. Habitats with complex structures provide optimum conditions for a variety of wild animals to sustain. In the absence of resources, the structural quality of habitats will deplete, affecting the populations of associated animals. This is common in drought-prone areas, and the severity increases as a result of further climatic changes. In such situations, populations of wild animals tend to move from dispersed to clumped distribution pattern, restricting themselves within the areas of limited resources. Prevalence of diseases may also increase in relation to changes in the climatic patterns. Increased atmospheric temperature also induces forest-fire, with enhanced frequency and intensity affecting larger areas. Intensive forest-fires affect a variety of organisms such as ground-nesting birds, terrestrial organisms such as reptiles and amphibians that are unable to move faster than other organisms such as fast-running mammals and flying birds. Climate change may have both direct and indirect effects at individual, population and community levels affecting the growth, reproduction, dispersal and survival of a variety of species.

As already mentioned in this chapter, the impact of global warming is easy to explain in the context of polar bear. Seasonal longevity of oceanic ice caps in the polar region has been reducing as a result of the increase in the temperature. Stirling et al (1999) have studied polar bears in western Hudson Bay and revealed that they have comparatively less time for hunting their preferred prey, the pups of ringled seal (*Phoca hispida*), in the context of climate change. It is revealed that in the recent decades, global warming has resulted in the advancement of spring, which has adversely affected polar bears due to less availability of food resulting in poor health and reproductive conditions. Similar to the polar bear, the Emperor penguin (*Aptenodytes forsteri* Gray, 1844) will also face food shortage in relation to reduced oceanic ice caps, which will ultimately end up with higher mortality rates in their habitats in Antarctica. The supply of major food sources such as the krill (*Euphausia superba*) are affected by climate change, which results in the poor health and higher levels of mortality of penguins (Barbraud and Weimerskirch 2001).

During the late Pleistocene, all but one species of the gelada baboons went extinct, leaving only the extant gelada baboon *(Theropithecus gelada* Rüppell, 1835) as a relict population in a retreat habitat in the high montane grasslands of the Ethiopian Plateau (Dunbar 1998, Fashing et al 2010). Extinction of all populations throughout their ranges in the eastern and southern Africa outside Ethiopia's central highlands is related to the impacts of global warming. It is expected that one further short period of warming may be enough to vanish the last surviving species of this baboon. Using a systems model of gelada socioecology, Dunbar (1998) suggested that the lower altitudinal limit of the species would rise by approximately 500 m for every 2°C increase in global mean temperature. Thus, the land area of suitable habitats available for the species would effectively reduce and become more fragmented. With this assumption, a 6°C rise of atmospheric temperature by the turn of the present century would reduce the current total population of around 2,50,000 individuals of the gelada baboons to about 5,000, and would confine them to a few isolated mountain peaks where their long-term persistence would be doubtful (Chapman et al 2006).

In Canada, the red fox *(Vulpes vulpes* Linnaeus, 1758) has expanded northward over the past 75 years, while the arctic fox (*Vulpes logopus* Linnaeus, 1758) has contracted towards the Arctic Ocean. The timings of the boundary changes have tracked warming phases. Occasional accidental transplants of the arctic fox southward from its range limit had succeeded, provided that the red fox, which is competitively dominant, was locally absent. However, prior to the recent global warming, multiple accidental transplants of the red fox towards north of its range limit had failed. The red fox has physical attributes that make it less adapted to cold conditions than the arctic fox (e.g., longer ears and limbs). From this, it is inferred that the expansion of the range of the red fox is due to the recent warming trends, causing the competitively inferior arctic fox to retreat northward (Parmesan and Matthews 2005).

Some of the species of birds and insects have shown range shifts in their distribution patterns in the recent past. For example, the mountain pine beetle (*Dendroctonus ponderosae* Hopkins, 1902) feeds primarily on lodgepole pine (*Pinus contorta* Douglas) and occasionally on other species including whitebark pine (*Pinus albicaulis* Engelm). Until recent decades, the pine beetles have been restricted to the higher elevation, due to their inability to complete their full life cycle during the short growing season. It is revealed that with the effect of global warming, they are able to complete a full reproductive cycle in whitebark pine habitats, and as a result, they can time their breeding cycle to produce more eggs whose emerging larvae can overwhelm the defences not only of damaged trees, but also of healthy ones. Expansion of the habitats of the pine beetle correlates closely with local warming data. This beetle is reported to be moving up mountains, and also northward into Canada, where it is occupying some forest areas for the first time in the recorded history (Sheikh et al 2007). Butterflies have also provided information of effects of temperature on distribution and abundance (Parmesan et al 1999, Thomas et al 2001). A northward range expansion of *Papilio memnon* was reported in Japan (Yoshio and Ishii 2001).

McClean et al (2005) studied the impact of projected climate change on the distribution of 5,197 species of African plants. For 81–97% of these species, areas of suitable climate are projected to decrease in size and/or shift in location, many to higher altitudes. For 25–41% of the species, it is expected to lose all their area by the year 2085. The models suggest dramatic changes in the Guineo-Congolian region, which today encompasses some of the most extensive blocks of the remaining forest in Africa. A 32-year record of plant phenological patterns in Kibale National

Park, Uganda suggests that climate change has led to periods of food scarcity causing nutritional stress among primates (Chapman et al 2005).

2.6. OZONE DEPLETION

Ozone is a form of oxygen and is a pollutant in the troposphere (the layer of the atmosphere closest to the Earth's surface). However, it is a beneficial component of the stratosphere, which extends 10–50 km above the surface of the Earth. The stratospheric ozone layer has been protecting living organisms on the Earth from harmful effects of ultraviolet (UV-B) radiations from the Sun. However, in the recent past, the level of UV-B radiation reaching the surface of the Earth has increased as a result of depletion of ozone layer, which is due to the increased emissions of ozone depleting substances (ODSs) such as chloroflurocarbons. This has resulted in the harmful UV-B radiation falling on the surface of organisms at higher levels, much above the levels that they can contain within their threshold limits. UV-B radiations (280–315 nm) are essential for organisms to sustain, but if their level is beyond the threshold limit, it makes adverse effects including prevalence of cancer. Wildlife such as lower plants and animals including amphibians, which are more sensitive to such effects, face major threats from the effects of increased UV-B radiations (Blaustein et al 2011).

Questions for Review Exercise

1. Have you understood the definitions and meaning of the following terms?

   a. Wildlife habitats

   b. Food specificity of wild animals

   c. Habitat association of wild animals

   d. Man-made disturbances in wildlife habitats

   e. Natural disturbances in wildlife habitats

   f. Effects of climate change on wildlife

   g. Ozone depletion

2. What do you think are the key differences between human and wildlife habitats? Are they really different in their basic characteristics? Can you substantiate your thinking about different types of wildlife habitats? Can you draw different habitat types from your region?

3. Are you living near a forest area? Have you visited any forest area? What are the characteristics of natural habitats you have seen? Identify some of the wild animals you have seen in different habitats.
4. What is your understanding about the presence of animals in different habitats? Are all animals present in all types of habitats? What did you understand from the topic 'Habitat Association of Wildlife'?
5. Some of the wild animals (e.g., Sambar deer) have wider distribution in a variety of habitats, whereas some other animals (e.g., Lion-tailed macaque) have restricted areas of distribution. What are the major reasons for this difference? What are your thoughts on the distribution patterns of animals in the context of their food habits?
6. What are your perceptions of sustainability of habitats and wildlife? Are all habitats and wildlife expected to remain as such for hundreds or even thousands of years? What do you think is the role of evolution on the dynamics of ecosystems and wildlife? Can evolution affect ecosystems and wildlife on a long-term basis?
7. What is your opinion about the fast-changing natural habitats and declining wildlife populations in the developing tropics? Are these the results of evolutionary changes? Can you expect to visualize evolutionary changes so fast? If not, can you visualize human-related activities as a major reason for the depletion of wildlife in the tropics?
8. Are there natural agents of disturbances in forests? If so, can you substantiate your thinking with few examples?
9. What do you think are human-induced factors affecting natural habitats, and what are the impacts of such changes on wildlife populations?
10. What are your perceptions of climate change? How does it affect wildlife populations? How can it affect human beings in the long run?
11. What have you learned from Boxed items 2.1–2.3?
12. What is the importance of wetland as an ecosystem? Is there any wetland near your locality? Describe the characteristics of any of the wetland ecosystems you have seen. What are the major species of animals associated with that ecosystem? Have you

seen migratory birds in wetland habitats? Have you seen any of the endangered and/or endemic wildlife in wetlands? Have you witnessed any of the human activities directly affecting wetland habitats?

13. Have you heard of Keoladeo National Park, Bharatpur, India? What is your perception on its role in supporting migratory birds? Are there migratory birds visiting your region? If so, when do they visit? Can you identify few such species of birds? Is there any special season for their visit?
14. There is a general understanding that environmental problems are more serious in the developing tropical parts of the world. A typical example of human impacts on high mountain lakes in one of the most environmentally-conscious nations, Norway is given in Boxed item 2.2. What are your perceptions on similar problems in your country?
15. What did you learn about the impacts of introduction of exotic fish species on native fish species in the Norwegian lakes? Can you substantiate your thinking with a few specific examples drawn from your region or State?
16. What is your understanding about the roles played by evolutionary history and ecological adaptations of wildlife? What are your perceptions on habitat loss and fragmentation in the context of lion-tailed macaques in the Western Ghats of South India?

Manuscript 2.1

The Population Density and Structure of Asian Elephants in Parambikulam Wildlife Sanctuary, Kerala, India

PS Easa[1] and M Balakrishnan[2]

[1]Division of Wildlife Biology, Kerala Forest Research Institute, Peechi, Kerala 680 653, India
[2]Department of Zoology, University of Kerala, Kariyavattom, Trivandrum, Kerala 695 581, India

Abstract: The population of Asian elephants in Parambikulam Wildlife Sanctuary, Kerala, India was studied during 1981–1983. The sanctuary with an area of about 270 sq km has both natural forest and plantations. A total count indicated about 114 elephants with an ecological density of about 0.5 animals/sq km. The biomass was about 905 kg/sq km. The variation in herd size showed a polymodal distribution. The herd size of eleven was more frequent. The basic family unit in the population was around five. There was no significant seasonal or monthly difference in the herd size. Forty three per cent of the herds were without males. About 66% of the adult males observed were solitary. The sex ratio of 1:6.8 (male:female) indicates a slight increase in male mortality. The proportion of juveniles and calves (21%) indicates a high percentage of breeding females in the population.

Keywords: Asian elephant, *Elephas maximus*, Parambikulam Wildlife Sanctuary, population, biomass, density, group composition, solitary elephant

Reference: Journal of Bombay Natural History Society **92** 225–229, 1995

1. Read the abstract presented above and come up with your understanding about the Asiatic elephant population in the study area. Are the status of elephant population and age and sex structure satisfactory in this study area? What is your view on the sustainability of this population?
2. What are the conclusions you can draw from this study?
3. Are you familiar with the keywords presented at the bottom of the abstract?
4. Have you visited any of the forest areas where you have noticed the presence of elephants? Are you confident of carrying out a similar population ecology study of elephants?
5. What are the difficulties you may anticipate in making such investigations?

Manuscript 2.2

Habitat Preferences of the Larger Mammals in the Parambikulam Wildlife Sanctuary, Kerala, India

*M Balakrishnan and PS Easa**

Department of Zoology, University of Kerala,
Kariyavattom 695 581, Trivandrum, Kerala, India
*Division of Wildlife Biology, Kerala Forest Research Institute,
Peechi 680 653, Trichur, Kerala, India

Abstract: The habitat preferences of large and medium-sized mammals in the Parambikulam Wildlife Sanctuary were studied. The information about 24 species was recorded. Most of the species prefer to live in specific forest types. In the case of primates, *Macaca radiata* was more abundant in forest plantations and in deciduous forests. *Macaca silenus* was confined to the evergreen forests alone. *Presbytis entellus* was more numerous in deciduous forests. *Presbytis johni* was more or less evenly distributed between the west-coast tropical evergreen and moist-deciduous forests. Herebivores and carnivores were most abundant in grasslands and in the *Tectona grandis* plantations.

Method of study

The observer walked through the forest along a path or animal track and recorded the following details on a check sheet: (1) Species of animal seen, (2) Time of observation, (3) Activity of the animal/troop/herd/pack at the time of observation, (4) Number of animals in the herd/troop/pack, (5) Indirect evidences such as spoor, scats, pellets, pugmarks, calls and (6) The habitat type where the animals were located. Pugmarks of tiger and leopard were drawn using a pugmark tracer whenever sighted. In case of elephant, gaur and wild boar, direct evidences alone were taken into consideration. The field observations were made during 0730 h–1800 h. The whole area within walking distance from a given camp was surveyed before the next camp was set up. The camp sites were restricted to forest rest houses, observation towers and the camps of Forest Department field staff in the sanctuary. Individuals and troops of various wild mammals could be identified by the various identification marks on their bodies, size of pugmarks/footprints, sex and the area where they were sighted. Repeated observations were made in some of the areas so as to make a complete census, and the animals counted once were excluded from subsequent counts.

The relative frequency of each species occurring in a vegetation type was calculated and the density of a species in that particular vegetation type was compared to the total density of the species in the sanctuary (Lavieren and Esser 1979). A goodness-of-fit test (chi-square) was used to evaluate the habitat preferences of the species of larger mammals in this sanctuary. As the area of different habitats varied, the expected frequencies were adjusted proportionately. The lower limit of expected frequencies permitted in any case was unity. The exact probability of the computed chi-square value was calculated by actual enumeration whenever the total number was low.

Results: None of the carnivores showed a preference for the natural forests compared with plantations. *Elephas maximus*, *Axis axis* and *Muntiacus muntjak* were found more in grasslands, and *Bos gaurus* and *Cervus unicolor* more in plantations. However, *Axis Axis*, *Muntiacus munjak* and *Tragulus meminna* did not show any significant preference. The rodents, *Ratufa indica* and *Hystrix indica* were observed more in deciduous forests and evergreen forests, respectively. *Sus scrofa* had a wide distribution in all habitat types.

Reference: Biological Conservation **37** 191–200, 1986

(Tables, figures and references are not included)

Some part of one of the published papers dealing with habitat preferences of wild mammals is presented for the benefit of students to read and understand the first-hand information provided. Students have to critically analyze the presented text and come up with new thoughts on the subject.

1. What are your perceptions on habitat preferences of wild animals? Are they all seen only in such special habitat types or seen in other habitats also; but mostly concentrate in special habitat types? Evaluate the terms 'habitat of occurrence'/ 'habitat preferences' of wild animals.
2. What did you learn about habitat specialists (restricted only to specific habitat types)? Can you cite few examples of such animals from your country?
3. What is your understanding about habitat specialists? What about their distribution patterns? How can we come to an understanding that *Sus scrofa* has wider distribution?

4. What are the methodologies used in the study reported in this publication? Can you identify individuals of a species of wild animals such as elephants in a herd or different herds of elephants in a study area? What are the means by which we can identify individuals of herds?
5. Check the statistical methods used to analyze the data presented in the above publication and the relevance of each of it in the context of the parameters studied.
6. Some of the names of species presented in the text above are changed since its publication. Can you locate these species and give the present binomial (species names)?

Predator – Prey Interactions 3

Behavioural ecology, food and feeding, foraging as a major activity of animals, predation and predator–prey interactions, Optimal Foraging Theory (OFT), aggregation in relation to food availability, food as a factor responsible for distribution of animals, mammalian foraging and seed dispersal, predation as a factor responsible for maintaining healthy habitats

---

3.1. BEHAVIOURAL ECOLOGY

Behaviour of an animal at a given time is the external reflection of its internal physiological status in relation to the environmental factors acting on it at that time. Internal physiological status of animals varies from time to time in relation to each of its functional systems. For example, the level of motivation in relation to feeding at a time is different from that of social or reproductive activities at the same time. Hence, motivational status concerned with a specific factor induces the animal to involve in a specific act to satisfy the concerned requirement, and the animal initiates that specific act. However, such acts vary in relation to the environmental circumstances. For example, an animal, which is hungry, has to search for food sources in the surroundings. If food sources of the types on which that species depend upon (for example, differences in the food of herbivores and carnivores) are available in plenty around, the animal can directly start the process of feeding. On the other hand, if food resources are not available around, it has to wander for distances until it can find out a good food resource. Thus, behavioural acts are different in these two occasions. In this example, we can find the sequence of behaviour as the stimulation from the internal cues concerned with the deficiency of nutrients in the blood stream, which the animal feels as hunger. This initiates the process of searching for food in the surroundings, and then the actual process of feeding until it is satiated. The process of searching for food varies based on its availability or non-availability in the immediate surroundings. Hence, environmental cues and related stimuli are also important parameters of animal behaviour. In case of wildlife, there can also be a difference in the sequence of behavioural acts as wild animals have to respond from time to time in relation to the events in their surroundings. For example, a prey animal has to escape from a site immediately when it senses the presence of a predator. Here, sensing the presence of a predator (external stimulus) initiates the related behaviour. Hence, the sequence is as follows:

1. The stimulus from the surrounding environment is sensed by the animal when it is transmitted to its central nervous system through sense organs and nerve impulses.
2. The stimulus is processed in the concerned motivational centre of the hypothalamus of the brain and the motor system is signalled for an appropriate act.
3. The animal appropriately acts (motor functions – behaviour) in response to the instructions from the central nervous system.

In either of these cases, behaviour is initiated as a cumulative act in response to the internal and environmental stimuli that are processed in the central nervous system, and action is done as per the instructions from the hypothalamic centres (stimulatory and inhibitory).

Behavioural ecology deals with each and every act of an animal in its natural habitat in relation to the changing abiotic and biotic factors. This includes behaviour of wild animals in relation to all environmental factors with specific functional relevance. One can consider any behavioural act related to feeding, social or reproduction in response to the environmental factors around and the changes that may be reflected in the behavioural patterns in relation to changes in the environmental conditions. For example, even though there is a species-specific pattern of feeding in animals, in response to the phenological changes in the habitat and the corresponding changes in the availability of food resources, the pattern of feeding would be different from season to season. During the wet season, when food is available in plenty, artiodactyls congregate in larger herds, but during the dry season when food resources are scarce; they form smaller herds of family units and disperse widely to make use of all available food resources.

Thus, there is a change in the strategy for the optimum use of resources available in the surroundings. Hence, behaviour is also a combination of acts, which is the outcome of both innate and learned processes through the experience of the individual. Innate relates to the information, which is not learned through experience, but are instinctive and inherited through generations; and hence, innate processes are of genetic characteristics. Such behavioural acts are species-specific. For example, the feeding patterns of a species are different from that of another species. Generally, rodents have the adaptations to gnaw, whereas larger herbivores are grazers and browsers. Social and reproductive strategies also vary from species to species. All such processes are less developed in younger individuals of any species, whereas highly developed and adapted among older individuals of the same species. Such advancements in behaviour are results of learned processes by own experience. Fawns of artiodactyls are less adapted with the escape strategy from predation. On the other hand, a grown up individual is well adapted with the strategies to escape from its common predators in the surroundings. Such strategies are developed through experience, and hence, are learned or acquired. In general, the ability of a species or an individual to learn also varies as it involves with the neural processes of transformation and retention of the facts one has experienced. As neural functioning depends on the structural development of the nervous system, it also relates to the evolutionary processes. Organisms at higher levels of evolution are more advanced with the abilities to learn and to retain memory than those at lower levels of evolutionary development. Primates, in general, have more advanced nervous system with efficient learning and memory processes. It is well known that human nervous system has advanced more than that of any other organism, and human beings have developed various skills, which none of the other organisms possess.

3.2. FOOD AND FEEDING

Food is the source of nutrients and energy for organisms of all kinds. Each species of animals has its own preferred food. In most of the natural habitats, availability of food for wildlife may vary from season to season. Phenology governs the availability of food resources based on seasonality. Wild animals have to depend on the available food resources in their habitat, and hence, seasonal changes in the food items are common among the wildlife. They have to gather what is available in the surroundings based on their feeding habits. Herbivores can find out some food resources everywhere if climatic conditions are favourable, whereas carnivores have to seek their prey for which they have to wander long distances based on the population density of prey in their habitats.

Feeding is the most important activity of any of the organisms in nature, for which they spend most of their active time. Feeding involves activities concerned with locating food source, choosing the best from the available source, manipulation of the food and ingestion. The behaviour of eating is in relation to the feeding apparatus or the mouth organs of the species concerned. In higher groups of organisms such as mammals, modifications of teeth are involved with the feeding adaptations of the concerned species. For example, herbivores and carnivores are adapted to herbivorous and carnivorous modes of life, respectively, for which their teeth are adapted.

In case of unicellular organisms such as amoeba, pseudopodial movements help to engulf food particles. Among ciliates, the ciliary movements are responsible for bringing food closer and ultimately engulfing it. The ciliary feeding involves filtering minute organisms from a current of water drawn through or towards the animal by cilia. Among coelenterates, movements of tentacles perform this function. In case of higher organisms, special body organs perform each function, starting from searching for prey, entangling, killing, chewing, engulfing and so on. Animals are clever in trapping their food organisms. For example, spider webs are species-specific, in which the prey organisms such as small insects are trapped accidentally. The animal world is so fascinating, particularly their feeding-related behaviours. There are also animals such as rodents that are capable of hoarding food (like human behaviour) when available in plenty. They use this food in conditions of food scarcity.

3.3. FORAGING AS A MAJOR ACTIVITY OF ANIMALS

In natural habitats, all animals spend most of their active time in foraging to locate food sources, and in feeding activities. Feeding-related activities constitute majority of the active time of both diurnal and nocturnal animals. Animal migration from one part of the world to the other during adverse environmental conditions is related to non-availability of food items in the original habitat, and the shortage of day time for foraging. Hence, during the winter season, when availability of food and feeding time are limited in the northern hemisphere, birds migrate to environmentally-safe areas in the southern hemisphere, where food is available in plenty and more day-time is available for foraging. Food and water are the most important requirements for the survival of organisms. Food and water are not readily available in all cases and in all habitat types, and hence, animals are forced to wander considerable distances over time in search of these most essential life requirements. This makes animals busy in their activity schedule. Case studies have revealed that animals spend up to 80% of their active time in searching for food and in feeding activities (Ancona and Loughry 2009).

3.4. PREDATION AND PREDATOR–PREY INTERACTIONS

Predation can be defined as the phenomenon where individuals of one species eat individuals of another species. This does not include the behaviour of detritivores and scavengers eating carcasses. Predation need not always be the one involved in carnivores killing and eating herbivores. It can also be the involvement of herbivores eating vegetation. Broadly, predation can be classified into following four types:

3.4.1. Herbivory

Herbivory deals with the eating of plant parts, fruits or seeds by plant-eating animals. In this type of feeding, plants are not killed by herbivores. Even though seed predators (granivores) and fruit eaters (frugivores) often kill some of the seeds, some seeds require passage through the mammalian digestive system for early germination (Sreekumar and Balakrishnan 2002). Herbivory is common in all types of habitat. For example, in aquatic habitats, small animals including fish depend on vegetation for their food, whereas in case of forests, several species of herbivores including small and large animals ranging from insects through rodents up to the largest of all terrestrial forms (the elephant) depend upon various types of vegetation for their food.

3.4.2. Parasitism

Parasitism relates to the behaviour of parasitizing one species by another. Parasites do not always kill the host, but it happens in rare cases. Parasitism is more common among animals, but also seen in plants.

3.4.3. Carnivory

This is the classical type of predation, where the predator kills and eats the prey animal. There are several examples from wildlife to cite for carnivory. A typical example is a tiger or a lion predating on a deer in natural habitats. However, they kill a prey animal only when it is required. Otherwise, even if there are plenty of prey animals around, a carnivore will not attempt to kill its prey species.

3.4.4. Cannibalism

Cannibalism is a special case of predation in which the predator and prey are of the same species. Cannibalism may be seen in special circumstances, when there is considerable shortage of natural food, such as those in captivity.

The feeding behaviour of animals is regulated on the basis of the food requirements of the concerned species. Each species of animal and each individual of the species has its own special requirements based on the metabolic status of the species and of the concerned individual. Even in the same individual, age is a major factor, which regulates the requirements of standard living activities. For example, younger and growing populations need more energy than the aging and older groups. Bolen and Robinson (1983) have described the differences in the levels of use of energy by animals of different body sizes, citing the following formula of Brody (1945): Daily energy requirement (kcal) = 140 x (Body weight, kg $^{-3/4}$).

According to this, a small animal such as a shrew weighing 5 g (0.005 kg) requires 2.63 kcal/day, an active man weighing 68 kg requires 3,300 kcal and a deer weighing 544 kg requires 15,800 kcal. This shows that small mammals such as shrews require about 526 kcal/kg of body weight/day, whereas a large mammal such as a deer requires only 28 kcal/kg of body weight/day. The reason for this major difference in the requirements of energy by different groups of animals is due to the differences in their metabolic activities. The level of activities of a small mammal is much higher than the level of activities of a large mammal. This is also being reflected by way of their behaviour; a shrew is more active during most of its nocturnal phase of activity, whereas a diurnal deer is more active during the day time and rests more time during the night. When we compare the energy requirements of mammals with that of birds, the later would require more energy as they are more active and have to maintain higher metabolic rates. For example, two species of smaller birds weighing about 4.0 and 3.5 g require about 1,400 and 1,600 kcal/kg/day, respectively, compared to 526 kcal/kg requirement of shrews (for details, Bolen and Robinson 1983).

The behavioural activities of predators and their prey are important in studying predator–prey interactions. This includes:

1. behaviour of predators in response to changes in the density of prey
2. behaviour of predators in response to changes in the density of the predator species itself
3. behaviour of predators in response to the differences in the degree of clumping of prey
4. behaviour of the prey species

In areas where predators are absent or experimentally removed, the prey population is expected to increase considerably for certain period until they are depleted as a result of shortage of food. The populations of prey also regulate the population of predator. In areas where prey population is less, predator populations cannot increase beyond certain limits. When food supply is limited, predator populations cannot increase constantly. The carrying capacity of any organism in its natural habitat is regulated by ecological factors in the habitat. Food is a major limiting factor in controlling populations of any species. It is also one of the major factors limiting distribution patterns of organisms. Thus, predator populations are responsible for controlling prey populations and *vice-versa.*

The response of predators to different prey densities, which depends on the feeding behaviour of individual predator, is called the functional response; and the response of the predator population through reproduction, immigration and emigration is called the numerical response. Suppose there are two prey species, A and B in the habitat. If A is rare and B is common, predators will concentrate on B and ignore A. Predators may change their search from B to A (switching) when B becomes rare as a result of predation. Thus, they switch from one prey to the other based on more abundant prey species at a characteristic density of the former. The success with which predators catch prey depends on the density of the predator population. Predators usually react to the presence of other individuals of their own species by dispersing. When the density of the predator species increases in the habitat, the efficiency to search and find out food reduces. Mammal and bird predators are usually territorial and they aggressively chase away other individuals of the species from the area once the space is fully occupied.

Prey usually live in small patches of high density with larger areas of low density in between (clumped distribution) as in the case of Impala

(*Aepyceros melampus* Lichtenstein, 1812). Predators concentrate their search (appetitive behaviour) in patches of high prey density. Thus, they have a regulating effect on prey. As prey density increases, more predators survive and reproduce. These two effects (survival and fecundity) result in an increase in the predator population, which in turn, affects the prey population as the predator population eats more prey. Territoriality results in dispersal so as to stabilize resident numbers (e.g., wolves at high density have high dispersal rate, around 20% of adults and 50% of juveniles). If prey species can migrate beyond the range of their main predators, their populations can escape from the predator regulation. With the slow-growing, non-precocial young, predators are forced to stay within a small area to breed and guard their young. In contrast, the ungulate prey with its precocial young does not require to stay in one place because the young can follow the mother within an hour or so of birth. Thus, the prey can follow a changing food supply while the predators cannot. For example, wildebeests in the Serengeti (Tanzania) and in Kruger National Park (South Africa) can follow seasonal changes in food, and are regulated by abundance of food. Meanwhile, their predators such as lion and hyena with their non-precocial young are restricted in their movements (Jarman 1979, Estes 1991).

Some prey species synchronize their reproduction to reduce the predation rate on their young. This behaviour is called predator swamping. For example, several artiodactyls including wildebeest in Africa have seasonal fawning (birth of young). Their breeding behaviour is regulated in such a way that fawning takes place in the beginning of the rainy season. In most of the grassland ecosystems, fire is common at the end of the summer season, and hence, the first few showers help in the process of regeneration of the vegetation in burned areas. As a result, in addition to the effect of predatory swamping, the fawns will have plenty of fresh nutritious offshoots to feed upon to support their energy requirements.

3.5. LOTKA-VOLTERRA MODEL OF PREDATOR–PREY INTERACTIONS

The Lotka-Volterra model shows that the prey population grows exponentially and the reproduction in the predator population is a function of the number of prey consumed. This shows that when the predator population increases, the prey population decreases to a point, at which the trend is reversed. Decrease in the prey populations would affect the predator population adversely, and hence, their population will come down in the absence of sufficient food resources. This would help the prey populations to increase in response to the decreased predation pressure. As a result, the prey population increases, followed by an increase in the predator population. Thus, the two populations rise and fall, oscillating. As the predator population increases, it consumes a progressively larger number of prey until the prey population begins to decline. In turn, the abundance of prey influences the reproductive rate of the predator population as a result of which the number of predators overshoots the availability of prey, and the predator population declines to a point where the reproduction of prey more than balances its losses through predation. The population of the prey then increases, which is followed by an increase in the population of predators. The cycle may continue indefinitely provided all other ecological factors remain suitable for both the prey and the predator populations.

This model is applicable to the feeding interactions between all trophic levels of organisms. For example, the dependence on primary producers by small herbivores such as insects, or small fish in a pond, dependence on grasses, herbs and shrubs by artiodactyls, dependence on small animal species such as arthropods and small fish by larger fish in lake and river ecosystems, and dependence on deer populations by larger carnivores or dependence on rodents and reptiles including snakes by birds of prey (Fig 40).

Models of interactions between predator and prey suggest two distinct responses of the predator to changes in prey density. One is for the individual predator to eat more prey as the prey population increases (functional response). The other is for predators to become more numerous through increased reproduction or immigration (numerical response). Regulation of populations of organisms of all trophic levels can be clearly revealed through these models. It can also help us to evaluate the roles of organisms of each of the trophic levels in controlling the growth and maintenance of organisms at lower levels.

3.6. OPTIMAL FORAGING THEORY (OFT)

The optimal foraging theory states that when an animal is feeding, it makes its decisions in such a way that its net energy intake is maximized. This can be explained by the behaviour of birds moving from one branch of a tree to another branch or from one tree to another during its foraging activity. When it eats up the fruits available in a given patch of vegetation, it will be less worth for the bird to search for food again in the same patch. The bird can move on to another patch, which has not been depleted as far as the food resource is concerned. So, the decision to move on to the next act in the sequence of behaviour of feeding depends upon the balance of energy considerations between staying back and flying away. If the next patch is far away, and requires a long flight, a foraging bird may actually do more for its energy reserves by staying back and feeding in the depleted patch than flying to a distant patch. As the bird continues to eat, the food becomes more and more scarce, and ultimately becomes severely depleted, when it will be worth for the bird to move to another patch located at a far away place, which requires a long flight. As per the predictions of OFT, an animal should stay back in one area until the net rate of energy intake drops to the average for the habitat.

Why a tiger or a lion does not hunt a rat or a mouse? What about the energy balance, if a tiger hunts a rat? Is it going to be benefited from such an act? If tigers and lions hunt rats, they will be the losers. Animals have to take appropriate decisions from time to time so as to achieve positive benefits from each of their acts.

3.7. AGGREGATION IN RESPONSE TO THE AVAILABILITY OF FOOD

Animals aggregate in areas where food is available in plenty. Rodent population increases in agricultural fields during the cropping season or when the crop is almost ready for harvesting (Workneh et al 2004, 2006). Rodents migrate from nearby areas to farms where food is available in plenty. Also, in natural habitats, population of wild animals would be more in areas with large quantity of food than in areas where food is scarce. Among artiodactyls, seasonal fawning is a common phenomenon. As already explained, synchronization of fawning occurs in the beginning of the rainy season, when fresh offshoots are available as food for the newborn fawns.

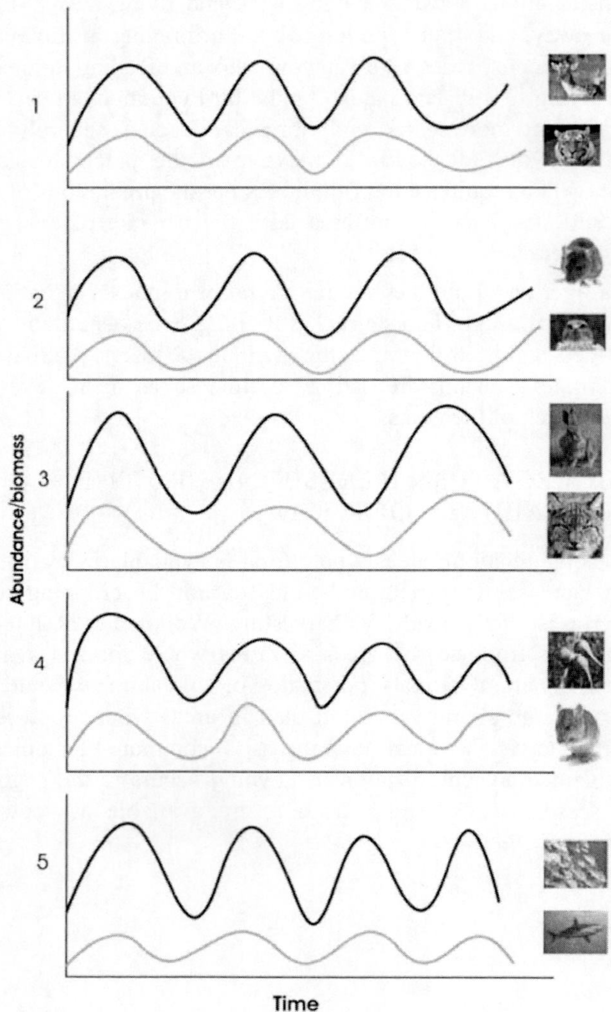

Fig 40. A schematic representation of predator–prey interactions showing various food webs in natural ecosystems; 1. Tiger as predator and sambar deer as prey; 2. Birds predating on rats; 3. Jungle cat as predator and rabbit as prey; 4. Rat predating on vegetative matters; 5. Large fish depends on smaller fish

3.8. FOOD AS A FACTOR RESPONSIBLE FOR DISTRIBUTION OF ANIMALS

Distribution of organisms in natural habitats is governed by the availability of food items. Animals restrict their movements through areas where preferred food items are available in plenty. This is also revealed by the habitat selection of animals. Carnivores are more abundant in areas where herbivorous prey animals are available in plenty. For example, the Afro-alpine habitat of the Bale Mountains is abundant with the giant mole rat (*Spalax giganteus* Nehring, 1898), the preferred prey of the Ethiopian wolf. Endemic species usually depend on less variety of food items. Thus, they become endemic to a limited area where their preferred food items are available. This is disadvantageous for the concerned species because the chances of their survival on a long-term basis are at risk unless their habitat is specially conserved with protective measures. This is the fundamental reason for selecting areas of endemic organisms for conservation measures.

Animals with wider distribution pattern show a better adaptive nature with a variety of food items on which they depend upon. This behaviour helps them to have wider areas of distribution, which is also advantageous for their survival. As suitable food resources are distributed in wider areas, animals naturally tend to forage through all such areas where their preferred food sources are distributed. This behaviour of animals also helps habitats to sustain for long with the resources as those animals that depend upon the resources will not stay for long; instead they move through larger areas as their food sources are widely distributed. For example, as stated earlier, *R. unicolor* has a wide choice of food items. On the other hand, if a foraging herd of elephants restricts its movements in a smaller area and depends upon fewer resources, the resources would deplete faster and the habitat recovery would become difficult.

3.9. MAMMALIAN FORAGING AND SEED DISPERSAL

Frugivorous mammals (e.g., sloth bear, *Melursus ursinus* Shaw, 1791) act as seed dispersing agents (Sreekumar and Balakrishnan 2002). They feed upon fruits, and in certain cases, the seeds are not damaged while passing through the intestine. Rather, this process helps in the germination of seeds after its deposition with faeces. Some frugivores, such as the spider monkey (*Simia paniscus* Linnaeus, 1758) in Bolivia, are found to be fast movers. The digestive processes would take about 4 h and by that time the animal must have moved several kilometres away from the mother plant, and seeds are deposited far away from the seed shadow. Such feeding and related activities of frugivores are responsible for maintaining heterogeneity of habitats (Pacheco and Simonetti 2000).

3.10. FEEDING BEHAVIOUR OF ANIMALS AS A FACTOR RESPONSIBLE FOR MAINTAINING HEALTHY HABITATS

The food web of the habitat is based on the food habits of the interacting organisms. In a natural habitat, herbivores are responsible for controlling overgrowth of several species of vegetation communities. Herbivores forage in large extent of areas and go on moving from one place to another. This helps the vegetation communities to recoup from the effects of herbivore foraging. Populations of herbivores are regulated by carnivores that depend upon them for food. Predatory behaviour of carnivores also helps to eliminate the old and unhealthy individuals among the herbivores, thus contributing considerably for the maintenance of a healthy habitat. The example of spider monkey also explains the specific role of a primate in maintaining a healthy heterogeneous habitat (Pacheco and Simonetti 2000). Thus, feeding behaviour and related activities of animals not only satisfy the nutritional requirements of the concerned animal, but also help to sustainably maintain the habitat and the organisms on which they depend upon.

Questions for Review/Exercise

1. Have you understood definitions and meanings of the following terms?

   a. Eco-behaviour b. Optimal foraging theory

   c. Seed dispersal d. Animal distribution patterns

2. What do you think about foraging and feeding activities of wild animals? How much of the active time of a wild animal is used in feeding-related activities? Do you know that some of the wild animals spend even >80% of their active time for feeding-related activities? What is your perception on feeding-related activities of human beings? Why is it so different from that of wild animals?

3. Identify any of the prey and predators (remember that prey can be both plants and animals — herbivores prey upon vegetation, and predators include both herbivores and carnivores — herbivores predate upon vegetation) in natural habitats, and understand their interactions (e.g., Have you seen butterflies

visiting flowering plants and birds feeding on fruiting trees?). Make brief accounts on such interactions and try to present during class seminars and discussions. Try to take photographs of few such interactions that you feel unique.

4. What are the different types of predation? How does feeding behaviour (predation) of a tiger differ from that of a deer?
5. What is your perception on Lotka-Volterra model of predator–prey interactions? See the examples given in Figure 40 and try to establish this concept using suitable prey and predators from your locality.
6. Observe the foraging patterns of any of the species of birds in your locality when there are plenty of food resources. Record the behavioural sequences including movements from one branch of a fruiting tree to another or from one tree to another tree. Is it possible for you to establish the concept of Optimal Foraging Theory from your observations?
7. In case you observe any animal aggregation anywhere, try to make close observations and find out the reasons. When (temporal) and where (spatial) essential resources are more, animals aggregate around. Food and water are such major resources. Resources-oriented aggregations can be seen seasonally in some of the habitats, or even at specific times of the day (e.g., animals aggregate near water resources during noon hours). What are your thoughts on such aggregation patterns of wild animals?
8. Record different species of animals foraging on a single species of tree, carrying away and dispersing seeds in different localities. What are your perceptions on the role of animals in seed dispersal?
9. What are your perceptions on predation as a factor responsible for maintaining healthy habitats? Use an herbivore (e.g., elephant) and a carnivore (e.g., tiger) as predators in the habitat, and find out their roles in maintaining a healthy vegetation structure and a healthy herbivore population, respectively, in the habitat.
10. A short publication from BIOTROPICA is presented below (Tables and References are not included) for your reading. What is your perception on the role of animals in the dispersal of seeds? What is your inference on the positive effects of passage

of seeds through the digestive track of animals? Does it help to speed up the process of seed germination?

Short Publication 3.1

Seed Dispersal by the Sloth Bear *(Melursus ursinus)* in South India

PG Sreekumar and M Balakrishnan

Department of Zoology, University of Kerala, Kariavattom, Thiruvananthapuram 695 581, Kerala, India

Abstract: Previous studies have shown that sloth bears are mainly myrmecophagous, but in this study, 39% and 83% (wet and dry season, respectively) of scats surveyed in South India contained fruits. Seeds from three of six plant species collected from scats germinated faster than seeds not passing through bears; germination percentage was not higher in any of the six species. These results suggest that the sloth bear may play an important role in the population dynamics of fleshy-fruited species of the region.

Keywords: *Melursus ursinus*, scat analysis, seed dispersal, seed germination, sloth bear, tropical forest

Ursids are sometimes important seed dispersers in rain forest where fruits form a major part of their diets (Willson 1993). Seed dispersal is generally advantageous because conspecific recruitment is often poor under parent trees (Howe 1984, Howe and Miriti 2000). Further, seed passage through the gut of mammalian frugivores may enhance the probability of seed germination (Willson 1993). The sloth bear (*Melursus ursinus*), distributed in forested hills in India, Nepal, and Sri Lanka, is a medium-sized mammal weighing between 127 and 145 kg. It can be distinguished by a narrow snout, protrusible mobile lips, coarse coat, long tongue and a whitish V-shaped breast patch (Prater 1971). Despite specific myrmecophagous adaptations, sloth bears are omnivorous and also feed on a variety of fruits. Recently, Joshi et al (1997) studied habitat-related diets of sloth bears in Nepal and found that insects, particularly termites, formed the major part of their diet. However, in South India, fruit comprises a major part of their diet (Bhaskaran et al 1997). Little is known about the role of the sloth bear in seed dispersal. Being highly mobile, ursids such as the sloth bear can potentially disperse seeds to large distances from a parent plant. The fate of seeds following passage through the digestive tract of animals is one qualitative aspect of disperser effectiveness (Schupp 1993). However, few studies have revealed the fate of seeds after passing through the guts of mammals (Lieberman and Lieberman 1986, Lieberman et al 1987, Nogales et al 1995, Traveset and Wdlson 1997, McConkey and Galetti 1999). To better understand the role of the sloth bear as a seed dispersal agent in Indian forests, we surveyed the occurrence of fleshy fruits and seeds in its faeces and assessed the germination potential of these seeds.

Frugivory by the sloth bear was evaluated by collecting its scats in Neyyar Wildlife Sanctuary (8'17'–8'53'N, 76'40'–77'17'E), a 122 km^2 protected area in Kerala State of India. Dominant vegetation types of this sanctuary include southern tropical wet evergreen forests and southern tropical moist-deciduous forests with an undergrowth of grasses and grasslands. Scats were collected from the study area (a total of 33 km transect length) once in a month from April 1996 to May 1998. A total of 285 scats were collected. All scats, except those used for germination studies, were preserved in 10% formalin. These samples were washed in water and sieved (1 X 1 mm mesh size) to recover seeds. Seed identities in the faeces were verified through comparison with seeds obtained from known plants. The percent frequency of occurrence for a given seed species in scats was estimated as the fraction of scats containing seeds of that species out of the total number of scats. Fruit availability in the study area varies seasonally (Sreekumar 2000), seed species frequencies in scats were determined for May–December (wet season) and January–April (dry season).

To assess whether passage of seeds through sloth bears affected germination, seeds of *Artocarpus integrifolia, A. hirsuta, Zizyphus oenoplina, Mangifera indica, Cassia fistula,* and *Syzigium cumini,* retrieved from sloth bear scats were planted in 5 m^2 plots. Ten seeds of each species were planted 10 cm apart and replicated five times (a total of 50 seeds for each species). Unpassed seeds collected from intact fruits of each species were germinated in an identical fashion. For each species, germination was observed for four months. Adequate soil moisture was maintained throughout the germination trial. Appearance of cotyledons was regarded as the sign of germination. The percentage of seeds that germinated and the time taken for each seed to germinate were determined.

Overall, fleshy fruits predominated in the food of the sloth bear, although animal material occurred more frequently in the wet season (61.2% of scats); fruits comprised the major diet in the dry season (82.7%).

Ants and termites constituted most of the animal food, irrespective of the season. Seeds of all plant species found in the faeces were intact. However, the fleshy exocarps or mesocarps were sloughed off, loosened, or completely macerated. Seeds from 14 plant species were recovered from the scats, in addition to those of undetermined *Ficus* species. In the dry season, scats were dominated by seeds of *C. fituh, M. indica*, and *A. integrijilia*. In the wet season, fruits were less frequent in scats, but those of *C. fistula, S. cumini*, and *A. hirsuta* predominated (Table 1 from referenced Short Publication 3.1).

Maximum germination occurred in *A. integrijilia* (82%), followed by *S. cumini* (58%), *A. hirsuta* (58%), *C. fistula* (54%), *Z. oenoplina* (46%) and *M. indica* (44%). None of these germination percentages differed significantly from those of control seeds that had not passed through sloth bears (binomial proportion tests, all $P > 0.05$). However, germination time of the seeds, going through seed passage, decreased significantly for three of the species examined: *C. fistula, A. hirsuta* and *Z. oenoplina* (Table 2 from the referenced Short Publication 3.1).

Intact seeds from fleshy-fruited species occurred frequently in the scats of sloth bears. Our data showed that fruits eaten by sloth bears varied widely in colour and fruit type, an observation that corroborates Herrera (1989), who reported similar findings for European mammals. Bhaskaran et al (1997) reported the presence of seeds from 20 species in the scats of sloth bears in Mudumalai Wildlife Sanctuary, whereas in Nepal, Joshi et al (1997) reported 12 species. In our study, the highest occurrence of fruits in scats was observed in the dry season, which was the peak fruiting season in the study area (Sreekumar 2000).

The presence of insects in the diet may vary depending on location and season (Joshi et al 1997). Similar diet composition was also reported earlier for sloth bears elsewhere in South India (Davidar 1983, Bhaskaran et al 1997). Seed passage through the sloth bear appeared to enhance the seed germination rate of a few species, but had no detrimental effect on the germination behaviours of any species we examined. A differential effect of seed passage through sloth bears on time to germination may be attributable to such factors as fruit retention time, fruit morphology, fruit size, seed coat thickness, and chemical constituents of the fruit pulp (Barnea et al 1990; Murray et al 1994). Enhanced germination of seeds of some species in response to passage through the alimentary canal of ursids and other mammals has been reported elsewhere (Lieberman et al 1987, Traveset and Wdson 1997, McConkey and Galetti 1999). Thus, the sloth bear may play an important role in the dispersal of seeds and influence the regeneration of some plant species across its distribution in India.

Reference: Biotropica **34** 474–477, 2002

(Tables and references are not included)

1. Read the manuscript and prepare a new abstract of the manuscript in around 200 words.
2. What are your inferences on the conclusion of the findings?
3. What do you know about the distribution and body size of sloth bear?
4. Can you reveal the characteristics of sloth bear, which may make this species a potential seed-dispersal agent?
5. Where was this study conducted? What are the major vegetation types of this area?
6. Can you design a similar experiment to find out the role of any of the animals in seed dispersal?
7. Can you briefly analyze the major findings of the research work presented in this short publication?
8. What are the conclusions arrived at the end of the discussion of this publication?
9. What are your perceptions on the few keywords presented after the abstract? What are the purposes of keywords?

❑ ❑ ❑

Ecological Principles in Wildlife Management and Conservation 4

Goal of wildlife management, human interactions in natural habitats, habitat loss and fragmentation, habitat restoration, population structure and regulation, wildlife stocking, competition between wildlife and livestock, wildlife harvest, reintroduction, wildlife conservation and management, behavioural biology and wildlife management, problems and prospects in wildlife management, communication, scent-marking, urban wildlife

---

4.1. INTRODUCTION

As wildlife of all kinds is part and parcel of natural resources, they are responsible for the maintenance of balanced ecological processes and natural habitats. They are also able to sustain by themselves, if left alone in their habitats. However, as a result of human population explosion, most of the wildlife habitats are affected by human activities to various levels, leading to the reduction of the extent of habitats available for wildlife. This has created problems for safe movements and activities of wildlife in their habitats. As a result of changed circumstances in and around forest habitats, some of the individuals of wild animals have also developed as problem animals through their occasional visits to the villages around forests. The shortage of food in the habitats, and the competition between members of the same species as well as between different species make them to wander through larger areas, and on the way they also intrude into farmlands where they create nuisance. Hence, wildlife management and conservation activities have to take such issues into consideration and plan suitable programmes for each of the areas or regions. Every species of wildlife has a unique genetic identity, which should be conserved for future. Wildlife management is also oriented with the view that if managed properly, wildlife can also be harvested sustainably. At the same time, any damage that may occur as a result of activities of wildlife populations should be controlled, rather than killing the individuals or species involved in it (Taber and Saharia 1995).

4.2. GOAL OF WILDLIFE MANAGEMENT

Wildlife management is expected to achieve material benefits from wildlife, controlling wildlife damage, restoring wildlife ecosystems to their natural state, regulating and improving socio-economic practices of local people and integrating regional development planning with wildlife conservation and management. Unfortunately, wildlife conservation and management practices in the Indian subcontinent are not viewed and aimed for its utility point of view.

Management and conservation are costly affairs, and hence, direct benefits are expected from such activities. As wildlife is an excellent natural resource, which is also renewable, there is no doubt on indirect benefits out of wildlife conservation. It has a major role in maintaining ecological balance of nature. Wildlife resources are directly used as food and medicine in most parts of the world. It is a major resource for eco-tourism development. As tourism has developed as a major industry all over the world, eco-tourism has great potential to contribute for the development of the tropical nations where natural habitats and wildlife resources are abundant.

Wild animals are responsible for damaging crops in rural areas around forests. Crop damage by various animals such as primates, artiodactyls, bovids, rodents, elephants and birds are well known. Poor farmers cannot withstand such problems if their crop is raided by wild animals. Small carnivores raid poultry farms; and tigers and leopards lift cattle and other domestic animals from the villages located around forest areas. Foresters experience wildlife damage to seeds and seedlings of crop species, or bark damage to growing trees. Elephants and various species of artiodactyls have the behaviour of debarking trees to substitute their mineral requirements.

4.3. HUMAN INTERACTIONS IN NATURAL HABITATS

Human interactions in wildlife habitats are more when the human populations around forest areas are prone to malnutrition and related socio-economic problems. It is to find out solutions to such problems that human beings intrude into wildlife habitats. To control such human activities, the socio-economic status of the rural human populations is to be improved. They should have essential resources for their subsistence. Regional development plans of areas around wildlife habitats should be integrated with wildlife management and conservation. Wildlife management and conservation should be given priority because wildlife habitats are of direct and indirect benefits to human beings.

4.4. HABITAT LOSS AND FRAGMENTATION

Depletion of the extent and fragmentation of wildlife habitats are worldwide phenomena. However, it is more intensive in the developing parts of the world in the context of ever-increasing human population. The exponential increase in human population is in no way comparable with the growth of wildlife populations. When there is constant increase in the human population, there is also need for more and more land for agriculture and related activities. Thus, there is an ever-increasing pressure on the land resources. The easiest way to achieve this goal is encroachment into forest habitats, which leads to the reduction in the extent and fragmentation of natural habitats. Loss and fragmentation of forest habitats would create pressure on animals leading to their increased density and isolation. Thus, habitat loss and fragmentation would lead to the decline of wildlife populations.

Expansion of agriculture and other development activities have also resulted in fragmentation of habitats available for wildlife. When a habitat is fragmented, it would develop in the form of islands. The Island Biogeography Theory states that when the habitat is converted for development activities and only 10% of the total area is set aside for conservation, 50% of the original species composition would be lost

(Diamond 1975). Only larger contiguous habitats can retain large populations of wildlife. Inbreeding is a genetic problem of organisms in small islands. Hence, fragmentation of natural habitats would accelerate the process of inbreeding, leading to depletion of wildlife populations.

4.5. HABITAT RESTORATION

When natural ecosystems are disturbed by human activities, wildlife populations are in a distress condition. They are forced to move around when there is shortage of food, water and shelter in their natural habitats. During such occasions, they also tend to move into nearby farming areas where they cause damages. In such situations, it is essential to restore the ecosystem into its original status. Management activities are required to bring back a degraded ecosystem to its original state. Habitat restoration activities include development of water holes and re-establishment of fodder vegetation to the original status, including the original species in the area of management. To fulfil this objective, conservation areas are set aside to cover representative and viable samples of all significant biogeographic subdivisions with the respective natural biodiversity.

4.6. POPULATION STRUCTURE AND REGULATION

The natural population structure of wildlife, with a large proportion of young individuals means a growing population having better reproductive potentiality. On the other hand, a population with only few or less young individuals is an unsafe population structure showing stress on reproductive potentiality of the population. Further, the number of reproductively potential females is also a major factor contributing to the growth rate of the population.

When eco-requirements of the concerned species are satisfactorily met with, growth rate of the population is expected to be at a satisfactory level. On the other hand, when eco-requirements are not satisfactorily met with, reproductive potentiality of the population would be affected. In a healthy population, individuals of all age groups should be present. For example, in a forest ecosystem, if recruitment of any species of tree is affected for a long time, that would be reflected on the sustenance of the ecosystem in the future. On such instances, managerial interactions are essential to find out the reasons responsible for non-recruitment of that species, and the problem should be eliminated so as to facilitate and maintain normal reproductive potentiality of the species from time to time.

4.7. WILDLIFE STOCKING

Whenever there is a problem in the reproductive status of the wild animals, management interruptions are required to provide essential eco-requirements of the concerned species including food, water and shelter. Adverse environmental conditions and presence of stressful situations in the habitat may affect the normal life of wildlife forms. Such adverse factors should be eliminated from wildlife habitats for their normal behavioural activities including movement patterns. When favourable situations are established and if the nutritional requirements are fulfilled satisfactorily, wildlife populations would increase to satisfactory levels.

4.8. COMPETITION BETWEEN WILDLIFE AND LIVESTOCK

When wildlife habitat is occupied by livestock, which is a widespread phenomenon in the developing African nations, there is a chance to develop competition between wildlife and livestock. In fact, the livestock is encroaching into the wildlife habitats, and a competition would develop for essential resources such as space, food and water. Herders consider grazing as their right, and they even fight with ethnic groups for the right for grazing areas and waterholes. Livestock may also be responsible for spreading epidemics such as rinderpest, and foot and mouth disease among wildlife.

4.9. WILDLIFE HARVEST

Populations of wildlife in scientifically-managed habitats would show constant increase, and if so, the populations growing above the carrying capacity should be controlled. It is in this context that culling is advised to maintain populations of wildlife, particularly of large mammals in managed habitats. Culling is the extraction of limited number of individuals from the growing populations of wildlife in a managed area to maintain the habitat intact from adverse impacts of over utilization of the resources for better sustainability of the habitat and populations of wildlife therein. As per the culling practice, populations in the natural habitats are retained at half the carrying capacity and those growing above that level are earmarked for extraction (Fig 41). As a normal practice, females having reproductive potentiality, healthy males and young ones are not culled. Only the older ones, particularly males are marked for culling. Wildlife is a better resource for harvesting in the habitats of marginal farming ability. Meat, hide, feather, tusk, horn and bone of wild animals are used for various purposes. In African nations, hippos, impalas, warthogs, wildebeests and zebras are culled under extraction practices.

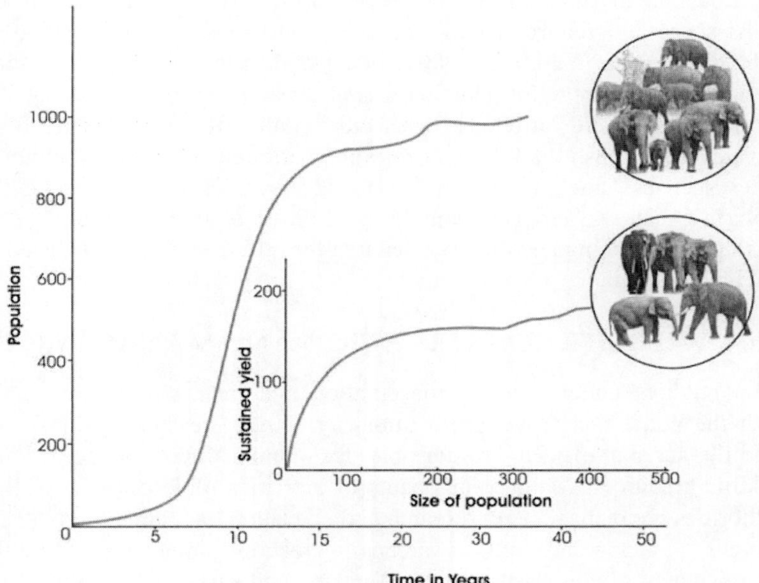

Fig 41. Carrying capacity of the extent of habitat can be calculated for each species and an optimum level of culling (sustained yield) can be made to retain the optimum population size at half the carrying capacity

4.10. REINTRODUCTION

Reintroduction is a practice for encouraging the population growth of endangered animals and plants. When their number is drastically reduced to a status of no hope for regeneration or redevelopment in the natural habitat (usually habitats where human interactions are severe), the remaining individuals of the species are to be trapped, maintained and managed in captivity. When they increase in number, reintroduction is possible in habitats suitable for them. Special management activities are essential for the welfare of such introduced individuals until they become familiarize with the new habitat, and achieve reproductive success and satisfactory population growth.

The Arabian Oryx or the white Oryx (*Oryx leucoryx* Pallas, 1777) is a medium-sized bovid with a distinct shoulder bump, long, straight horns and a tufted tail. It is the smallest member of the *Oryx* genus, and is native to desert and steppe areas of the Arabian Peninsula. It was extinct in the wild by the early 1972, but was saved in zoos and private preserves, and reintroduced into the wild in 1982. The 'IUCN Red List' categorized it as 'Endangered' in 1986, but in 2011 reverted to as 'Vulnerable', as a result of its successful reintroduction, and the populations grew to over 1,000 individuals. Also, there are 6,000–7,000 individuals of them in captivity, worldwide. The first reintroduction was in Oman from Phoenix Zoo. Later in 1986, they were introduced to Saudi Wildlife Research Centre from private collections. By 2009, there were reintroductions into additional states such as Israel, UAE and Jordan. However, the population of the Oryx in Oman was reduced from 450 in 1996 to around 65 in 2007. This was due to poaching and opening the area for oil exploration. It is listed in CITES Appendix I.

4.11. WILDLIFE CONSERVATION AND MANAGEMENT

Activities under wildlife conservation and management are vested with the concerned Government authority. There are special forest and wildlife service officials responsible for administration of forest and wildlife resources. Major components of activities of forest and wildlife authorities are management, research and education, including biodiversity surveys, assessment and awareness creation among the public, establishment of conservation areas, forestry, and sustainable use of forest and wildlife resources. Details of management activities are included in the chapter on 'Protected Areas and Wildlife Conservation'.

4.12. BEHAVIOURAL BIOLOGY AND WILDLIFE MANAGEMENT

An understanding of the behavioural patterns of wild animals is essential for proper planning of management activities. Habitat selection/habitat association, feeding ecology and territoriality patterns of wild animals are important parameters in the context of wildlife management. For example, bats are commonly known as agricultural pests due to frugivory, but they are useful animals as pollinators. Some of the bats are also excellent agents of insect pest control. Ecolocation is well known in bats (Box 4.1 – Figs 42 to 46). Social behaviour, migratory patterns, patterns of dispersal and rhythmic activity patterns are important parameters concerned with the specific lifestyle of animals, in relation with the environmental circumstances, and to make use of the resources at an optimum level. Such behavioural patterns and adaptability of organisms also help them to share the available habitats and resources, so as to be a part of the ecosystem links and to perform specific ecological functions.

Box 4.1

Biology and Behaviour of Bats

G Marimuthu

Department of Animal Behaviour and Physiology, School of Biological Sciences, Madurai Kamaraj University, Madurai – 625 021, India
(E-mail: emailboxgm@gmail.com)

Bats are the only flying mammals that have mastered flight much before the beginning of man's own lineage. Next to rodents, bats are the second largest group of mammals in the number of species, and the most abundant. About 1,100 species of bats are distributed around the world varying from the tropical 'flying fox' (*Pteropus* species) that weighs more than one kilogram to the 'tiny Kitty's hog-nosed' bat (*Craseonycteris thonglongyai*) of south-east Asia that weighs only about two grams.

Bats belong to the order Chiroptera, which means hand-winged. Their fingers are connected to one another by a web-like membrane that enables them to fly, and they crawl by their thumbs. The order Chiroptera is divided into two sub-orders: Microchiroptera and Megachiroptera. Bats roost in different kinds of habitats such as caves, caverns, crevices, cellars, branches of trees, hollows in tress, culverts, underside of bridges, temples, ruins and attics. Interestingly, a few species of bats modify large palm leaves by biting a series of holes across the centre of the leaf so that the leaf hangs like a tent (Kunz and Lumsden 2003), and they roost inside (e.g., *Cynopterus sphinx, Ectophylla alba, Uroderma bilobatum,* etc.).

Feeding habits of bats show remarkable dietary variations. Megachiropterans feed upon fruits such as grapes, mango, papaya, guava, custard apple, jack fruit, sapota, figs, etc. In addition, they visit flowers (e.g., banana, *Musa paradisiaca*, silk cotton, *Ceiba pentandra* and *Madhuca latifolia*) and lap nectar. Most of the microchiropterans prey mainly on insects such as moths, beetles, crickets, grasshoppers, mosquitoes, etc. However, there are bats belonging to the latter that feed upon additional food such as frogs, mice (e.g., *Megaderma lyra*) and fish (e.g., *Noctilio leporinus*). Vampire bats (e.g., *Desmodus rotundus*) that are distributed only in Central and South America feed upon blood of large mammals such as cow, horse and pig (Turner 1975).

It can be great fun to follow bats during nights in the field and gain an intimate picture of their foraging activity. By marking them with visual marker tags or chemiluminescent tags, it might be possible to follow bats in the open area. However, one has to visually locate the tagged individuals while using these light emitters. In this age of electronic gadgets, radio-tracking is a seductive technique to make a systematic study on the foraging behaviour of shy or elusive animals like bats. It needs a set of relatively-expensive equipments such as transmitters and receivers. By means of radio-tracking, it is possible to find out the direction and approximate distance covered by bats and to quantify the duration spent in flight and rest each day (e.g. Threlfall et al 2013).

Bats fascinate us because they 'see' with their ears (especially microchiropterans), hang themselves upside down to sleep during daytime, and can catch insects while flying even in the darkest of nights. Bats can thread their way dexterously through hill and dale and thickets and bushes in total darkness. They do it with a kind of sonar that challenges our highest technology. While flying, they emit trains of ultra-short pulses of patterned, high-frequency, high-intensity sounds through nose or mouth, and adroitly pinpoint the exact location of obstacles and of the smallest prey with the aid of the echoes coming back from them. This process is known as echolocation.

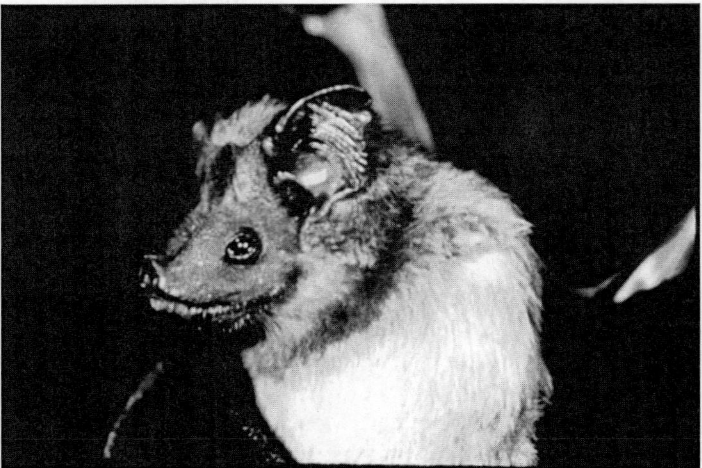

Fig 42. Mouse-tailed bat *Rhinopoma hardwickei* (Microchiroptera) (Photo: P Kumarasamy)

In 1920, a British Physiologist, Hamilton Hartridge, hypothesized that bats emit high-frequency sounds, above the range of human hearing, and use their echoes for orientation. In 1938, the American Biologist, Donald R. Griffin used instruments designed by George W Pierce and demonstrated that bats emit ultrasonic sounds with a frequency of more than 20,000 cycles per second (inaudible to human ears) and utilize the reflected echoes to detect obstacles and to navigate in their environment (Griffin 1958). Lazzaro Spallanzani, an Italian scientist demonstrated that bats can fly even when blinded. In Geneva, Ludwig Jurine added the significant discovery that bats could no longer perceive obstacles once their ears were plugged with wax.

Fig 43. Horse-shoe bat *Rhinolophus* species (Microchiroptera) (Photo: Bat Conservation International)

Echolocatory navigation is restricted only to microchiropteran bats. Megachiropteran bats have prominent eyes, and they depend on vision to reach the feeding areas and on the sense of smell to locate ripe fruits. Thus, it is clear that the microchiropteran bats are efficient gatherers of insects. A single bat may snatch two insects from air in less than five seconds, and as many as 500 insects in an hour. However, some of the insects have ear-like structures sensitive to ultrasounds, and are able to protect themselves from the marauding bats. Noctuid moths, which are serious agricultural pests, take evasive action by making a variety of flight patterns, loops, spirals and changes in both speed and direction or they simply fold their wings and drop-like pebbles when bats approach them. Arctiid moths emit a series of ultrasonic clicks to which are ascribed a variety of functions ranging from jamming the echolocatory system of bats, to startling them and to conveying a warning that they are distasteful (Jones and Rydell 2003).

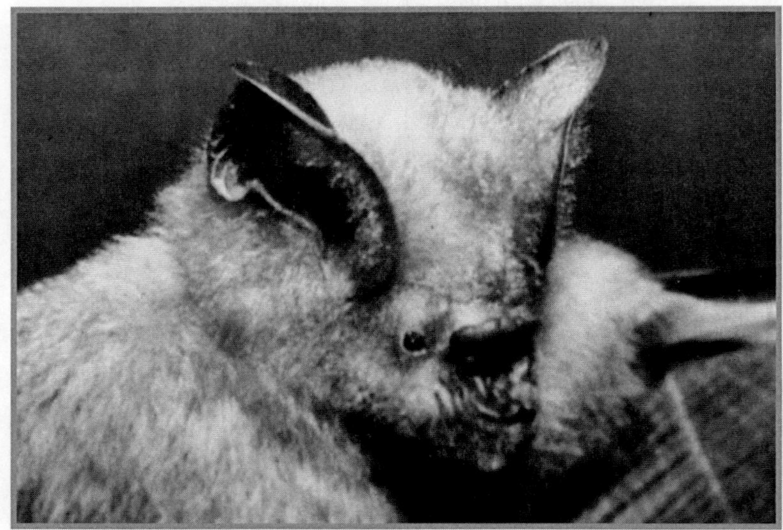

Fig 44. Nose-leaf bat *Hipposideros speoris* (Microchiroptera) (Photo: P Kumarasamy)

A few species of bats switch off their echolocatory machine to circumvent insect defences and rely on other cues. The fringe-lipped bats (*Trachops cirrhosus*) use the sounds produced by their prey, frogs (Ryan and Tuttle 1987). The Indian false vampire bats use the rustling noise created during the movement of their prey such as frogs, mice, geckoes and spiders. As they passively listen to the noise associated with the movement of the prey, it is known as passive method of prey detection (Marimuthu and Neuweiler 1987). However, a few individuals of *M. lyra* skim at the surface of the pond in darkness for a few minutes, apparently detect a stationary frog and in a dart captures the frog with its mouth (Marimuthu et al 1995). While skimming at the water surface, the bat emits ultrasounds, and thus, actively uses its echolocatory system. This method is known as active acoustic localization of prey in water. The returning echoes from the protruding targets (head of a stationary frog) in the water surface should be easily distinguished from the echoes returning from the (moving) ripples at the surface of the water. Thus, the water surface acts as an acoustic mirror. Interestingly, bats are not able to detect a freshly-killed frog if its head is submerged just below the water surface.

Fig 45. A captive Indian False Vampire bat *Megaderma lyra* (Microchiropteran) (Photo: G Marimuthu)

Being mammals, bats carry their young ones in the womb until they are ready to be born. The gestation period ranges from six weeks to eight months. Pregnancy does not begin soon after mating. A few hibernating bats living in temperate areas mate before the onset of hibernation, and the females store the sperm during the period of hibernation (about four months). Fertilization occurs after the termination of hibernation. None of the other mammals store sperm for such a long period.

Each female bat gives birth to a single young. The newborn babies are hairless, sightless and flightless, but with claws to grip for precious life. Mothers lavish meticulous care upon babies and carry the latter with them in the abdomen while flying night after night and out maneuvering hundreds of or even thousands of insects. Babies tenaciously cling to the body of their mothers on such wild rides. Sometimes, to lighten the load, mothers leave their babies at home. The infants wait for their mothers to return with teats full of warm milk. Remarkably, each mother is able to identify her own baby by the baby's sound and scent in the prevailing darkness. The adults of hipposiderid bats (*Hipposideros speoris*) emit constant frequency sounds at 135 kHz through a frequency-modulated tail with no harmonics. But, newborn babies emit multi-harmonic sounds with a combination of FM-CF-FM pattern. At prevolant and volant stages (about 30 days of age), most of the energy is concentrated in the second harmonic, which is elevated to the frequency range of adults' CF-FM pattern at the age about 6–7 weeks (Habersetzer and Marimuthu 1986). However, young bats wean after three months, begin to operate their own sonar system and become independent.

Fig 46. The Indian False Vampire bat feeding on a frog. Their prey-capture is by echolocation (Photo: P Kumarasamy)

Although bats are known to host a variety of microbial organisms, disease transmission to man is rare. The danger of rabies from bats has been vastly exaggerated. Bats are legally protected in Russia and in all European countries. In some areas of India, they are considered sacred.

However, unawareness and ignorance of the beneficial roles of fruit bats coupled with misconceptions and unjust persecution as crop pests (Singaravelan et al 2009) have resulted in their categorization as vermin in Schedule V of the Indian Wildlife (Protection) Act, 1972 (IWPA). However, recently, the IWPA shifted the Salim Ali's fruit bat (*Latidens salimalii*) and Wroughton's free-tailed bat (*Otomops wroughtoni*) to Schedule I by recognizing their rarity. Merlin D. Tuttle founded 'Bat Conservation International (BCI)' based at Austin, Texas, USA to disseminate information about bats to conservationists, health officials, educators and news media. The purpose of the BCI is to prevent extinction of bats, to ensure the survival of viable bat populations and to inform the public about their value. The Zoo Outreach Organization at Coimbatore (South India) founded a network called 'Chiroptera Conservation and Information Network of South Asia (CCINSA)'. The purpose of this network is to link together bat-field researchers and their knowledge throughout South Asia (Bangladesh, Bhutan, India, Maldives, Nepal, Pakistan and Sri Lanka) so that a clear picture can emerge leading to bat conservation actions. The network encourages and promotes the study of all bats, prioritizing species assessed as 'Data Deficient in India' for upcoming fieldwork.

References

Griffin DR 1958 *Listening in the Dark*, New Haven: Yale University Press

Habersetzer J and Marimuthu G 1986 Ontogeny of echolocation sounds in the echolocating bat *Hipposideros speoris*, *Journal of Comparative Physiology* **A158**: 247–257

Jones G and Rydell J 2003 Attack and defense: interactions between echolocating bats and their insect prey, in *Bat Ecology,* (eds) TH Kunz and MB Fenton, Chicago and London: The University of Chicago Press, 301–345 pages

Kunz TH and Lumsden LF 2003 Ecology of cavity and foliage roosting bats, in *Bat Ecology*), (eds) TH Kunz and MB Fenton, Chicago and London: The University of Chicago Press, 3–127 pages

Marimuthu G and Neuweiler G 1987 The use of acoustical cues for prey detection by the Indian false vampire bat, *Megaderma lyra*, *Journal of Comparative Physiology* **A160** 509–515

Marimuthu, G, Habersetzer J and Leippert D 1995 Active acoustic gleaning of water surface by the Indian false vampire bat, *Ethology* **99** 61–74

Ryan MJ and Tuttle MD 1987 The role of prey-generated sounds, vision and echolocation in prey localization by the African bat *Cardioderma cor* (Megadermatidae), *Journal of Comparative Physiology* **A161** 59–66

Singaravelan N, Marimuthu G and Racey PA 2009 Do fruit bats deserve to be listed as "vermin" in the Indian Wildlife (Protection) & Amended Act – A critical Review, *Oryx* **43** 608–613

Threlfall CG, Law B and Banks PB 2013 Roost selection in suburban bushland by the urban sensitive bat *Nyctophilus gouldi*, *Journal of Mammalogy* **94** 307–319

Turner DC 1975 *The Vampire Bat*, Baltimore and London: The Johns Hopkins University Press

4.13. COMMUNICATION

Communication is the means by which organisms use species-specific signals with specific messages from time to time, which will be perceived by individuals of the same species, and responded in a specific way. This is the backbone of social life of wildlife by which social responsibilities are exercised and social cohesion is maintained. There are both interspecific as well as intraspecific means of communications. However, most of the communication signals are to communicate between members of the same species. For example, to maintain social cohesion of the troop members, dominant individuals of the troop give signals to which all other members of the troop respond. It is by loud species-specific auditory signals in case of forest-dwelling primates, and in case of open-savanna dwellers, it is by visual signals. It is rarely that one communicates with other species; most of the times it is during predator–prey interactions.

Wild animals use a variety of means for communication. In general, these are grouped as visual, acoustic, tactile, olfactory and gustatory. Visual signals are signals concerned with sight, acoustic signals are with sound (hearing), tactile signals are with touch, gustatory signals are with taste and olfactory signals are with smell. Olfactory and gustatory signals together form chemical signals. It may be possible to deploy different means of signals at any time to convey the same message. This may help to convey the message more effectively during different social interactions. For example, when we speak, we also deploy facial expressions and signals by hands or other body parts, which make our presentation more effective. When someone is angry, in addition to how he speaks, we can also visualize the intensity of anger by observing his facial expressions.

Human beings became well equipped with the cultural advancements during the evolutionary processes when languages were developed. Thus, we have specific languages with written and spoken formats. Also, there are species of animals such as birds and mammals, which use species-specific auditory signals that are context-specific (e.g., calls of newborn babies, pups of dogs when hungry or when they huddle together during suckling, the response of nestlings when the mother bird returns from foraging with food to feed the nestlings). These are all species-specific and context-specific. Birds of each species have well-structured calls, by hearing which an experienced Ornithologist can identify the species and the context, even without sighting the individual that produces the sound. The hungry nestling produces specific signals when they sense the approach of the mother after foraging. However, among lower groups of organisms, such well-structured auditory signals may be absent; instead they depend upon other means of signals such as tactile. This shows that there is also an evolutionary development in the process of communication in the animal world.

4.14. SCENT-MARKING

Scent-marking is defined as a special pattern of behaviour whereby the animal transmits chemical signals of communication onto their immediate environmental objects (sign-posts) or on to other individuals of the same species. Scent-marking is a common behavioural pattern in many of the mammalian species (Ralls 1977). They mark environmental sign-posts with the scent-marking substances secreted from their specific body locations such as the flank glands of the Indian musk shrew (*Suncus murinus viridescens* Blyth) (Balakrishnan 1987), the chin glands of the European wild rabbit (*Oryctolagus cuniculus* Linnaeus, 1758) (Mykytowycz 1970), the ventral glands of the Mongolian gerbil (*Meriones unguiculatus* Milne-Edwards, 1867) (Thiessen et al 1971), the perineal glands of the small Indian civet (*Viverricula indica* E.Geoffroy Saint-Hilaire, 1803) (Balakrishnan and Sreedevi 2006) and the African civet (*Civettictis civetta* Schreber, 1776) (Bekele et al 2008, Wondmagegne et al 2011). During certain social interactions, more than one type of message is sent for more effective transfer of the signal. Urine marking, particularly on the boundaries of the territory, is common among many of the mammalian species. Some mammals (e.g., rabbits) mark the territories with their faeces and secretions of the specialized skin glands (e.g., secretions of the chin glands). Scent-marking is common among artiodactyls (Müller-Schwarze 1983). Most of them have well-developed pre-orbital glands, the secretions of which are used for marking environmental sign-posts such as twigs, branches of trees and shrubs. Pre-orbital marking, fore-head rubbing, tarsal marking, meta-tarsal marking and pedal marking are the most pronounced means of scent-marking patterns in artiodactyls. Self-marking and allo-marking are also common among them. They also sniff the marked sites of conspecifics during social interactions. Self-sniffing and allo-sniffing are common. Shrews scent-mark with their flank glands. Gerbils mark with the mid-ventral abdominal gland secretions. Felids scent-mark with their specific urine-spraying behaviour by which they mix the urine with the secretions of anal glands and spray over environmental sign-posts. Elephants have the temporal glands (musth gland) that occasionally enlarge and ooze out its secretion. During this period, the Asiatic elephant is known as in 'musth'. Hyenas scent-mark with the secretions of the bi-lobed anal glands. Civets have their perineal glands well developed in both sexes. They mark environmental objects with the glandular secretion known as 'civet' (Fig 47). The secretion of the perineal glands of the African civet is heavily used in the perfume industry. The secretion of the perineal glands of the small Indian civet is extensively used in traditional medicine (Ayurveda). Castoreum, the secretion of the large anal glands of Siberian and Canadian beavers and musk, the secretion of the abdominal glands of the male musk deer (*Moschus cupreus* Linnaeus, 1758) are other mammalian glandular secretions that are extensively used. The territorial marking in the European wild rabbit is well studied in Australia. They mark the boundaries of territory, repeatedly visited and reinforced by the dominant male and female individuals at regular intervals, by their dung piles. The pellets are coated with the secretions of the anal glands. They also chin at specific territorial boundaries regularly to reinforce them (Mykytowycz 1970).

Olfactory signalling is well studied in the European wild rabbit in connection with scent-markings. Rabbits live in small groups of 8–10 individuals with distinct territories. At the onset of breeding season, both males and females establish dominance hierarchy by fighting. The dominant male and female individuals rule over the group and their territory. The territory is defended by its occupants and recognized by rabbits of other groups. Chasing and submission constantly reinforce the rabbit hierarchy. A rabbit that crosses the border between its own territory and of a neighbouring one immediately becomes aware that it has done so. This can be revealed by its behaviour. In its own territory, it moves freely, it feeds and examines objects confidently, whereas in an alien territory, the intruder is always on alert.

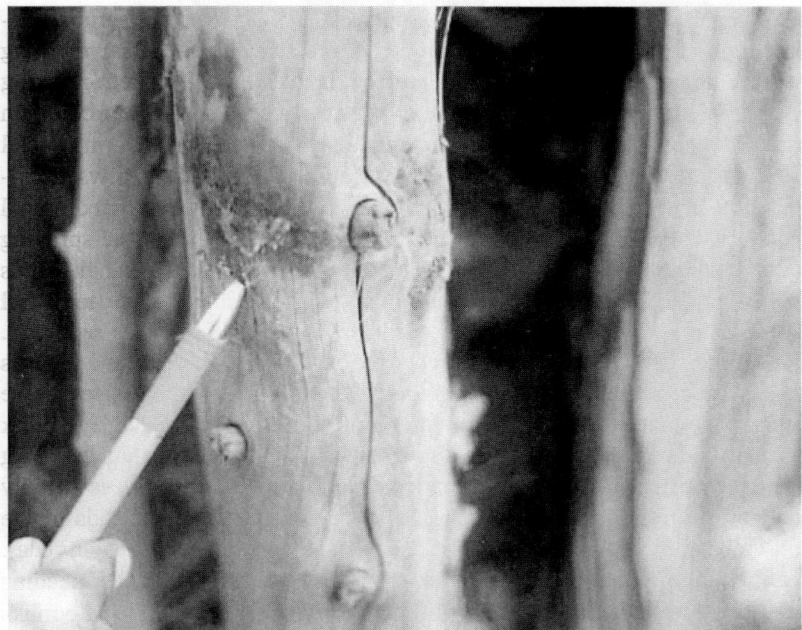

Fig 47. A scent-marking site of African civet on a smooth fencing stump in a human settlement area in Suba village, Ethiopia. The secretion adhered on the surface of the sign-post is shown with a pointer (Photo: M Balakrishnan)

A rabbit territory is saturated with its characteristic smell. Urine is one of the sources of smell in rabbit territories. They also urinate on each other during social interactions. Young ones are scent-marked by adults so as to recognize the group identity. The smell of urine of members of alien groups induces aggression. Faeces is another source of odour in the territory. Pheromones produced by specialized skin glands also possess species-specific odour. In rabbits, the anal gland consists of two clusters of brownish tissues forming a saddle-like mass around the end of the rectum. There is a significant relationship between the size of the anal gland and the hierarchical position of the rabbit in the warren; the larger the gland, the higher the social ranking. Dominant males possess larger glands. The secretary activity of the anal gland is also higher in the dominant ones. During the breeding season, the size and the secretary activity of the gland are the greatest. The secretion of the glands flows through few ducts into the rectum, where they coat the pellets of faeces passing out of anus. They deposit faeces repeatedly in a number of places in the territory. About 30 such small dung-hills are seen around a typical warren. During the breeding season, the territorial activity is most intense. When a stranger rabbit enters the territory, it encounters the dung-hills with the warning that the area is already occupied. Then, it starts the wary behaviour. Male rabbits also scent-mark environmental sign-posts, burrow entrances, females and young ones, faecal pellets of other rabbits and own weathered pellets using their chin gland secretion. In its own territory, a rabbit chins freely, whereas in an alien territory, it does not chin. When it confronts with faeces of individuals of other groups in its territory, it chins intensively. The most dominant individual chins more frequently. The size of the chin gland is also larger in dominant ones and it secrets more than that of the subordinate ones. The activity of the chin gland also fluctuates with season; secretes most during the breeding season. It is revealed that each of the marking patterns contain specific messages, which are species-specific and context-specific. The specialized integumentary glands and the frequency of scent-marking are under endocrine regulations. Scent glands develop during maturity and become more conspicuous in the dominant individuals. Sexual dimorphism of scent glands is common in most of the mammalian species. Such glandular activities are more during the breeding season. Territorial animals scent-mark more frequently to establish territories during the breeding season. But, all scent-markings are not territorial, and as many as 25 different types of messages are conveyed through mammalian scent-markings. Messages regarding individual, group, age, social status, sex, reproductive status, territoriality, warning, defence, alarm, submission, predator–prey, trail, gregariousness, aggressiveness, etc. are effectively conveyed by means of olfactory signals (Mykytowycz 1970).

Olfaction is affected through the odour produced by one individual in special social circumstances. Odours are highly volatile. Odour molecules are influenced by the movement of air and enter the nose of the recipient, as a result of which its olfactory system is triggered and the message is passed on to the brain centres. The brain centre elicits appropriate responses and the respective motor systems are activated for appropriate behavioural responses and actions.

4.15. PROBLEMS AND PROSPECTS IN WILDLIFE MANAGEMENT

In general, wildlife is facing several problems in most of the developing parts of the world. This situation is in one way or the other related to the increased human density in villages around forest habitats. Encroachment of human beings into natural habitats has reduced the extent and quality of wildlife habitats. Rural villagers have been depending upon forests for wildlife meat, firewood and thatching grass since time immemorial. Such practices cannot be halted suddenly. From past experiences in the developing tropical countries, it is clear that the cooperation of the local villagers is essential for wildlife conservation. To achieve this goal, the following parameters are to be considered under management programmes:

1. The conflict between villagers and wildlife management has originated from direct competition between man and wildlife, mainly for forage and water resources. There is need for a solution to this problem to convince the villagers and to get their support for the conservation of wildlife.
2. From the past experience, it is found that active cooperation of villagers is mandatory for effective protection of wildlife habitat; in the absence of which villagers may develop negative attitudes about wildlife conservation.
3. Villagers should be convinced regarding the need for wildlife conservation, and direct and indirect benefits out of it, so as to take up active roles in wildlife conservation.
4. Local people should have a role in wildlife conservation and management practices, so that they can become part and parcel of the management and conservation programmes.
5. All stakeholders of the wildlife area (villagers around, eco-tourism operators, hotel owners and managers, all government agencies in the nearby areas, non-governmental organizations among the village community, foresters and indigenous people, who depend upon forest for their livelihood) should be invited for discussions, and they should share various roles in conservation activities.

Wildlife conservation cannot be effectively achieved without the cooperation and support of local people. On the other hand, convincing the local people of their benefits from wildlife conservation and sustained yield, and when National Parks and Wildlife Service initiate programmes to promote such benefits to the local community, villagers are expected to be loyal to the Park administration, and report even anticipated poaching well in advance to take measures to control such illegal activities as evidenced from case studies in Africa (Lewis and Kaweche 1985, Balakrishnan and Ndhlovu 1992).

4.16. URBAN WILDLIFE

Primates are common and sometimes nuisance in some of the urban areas in India as well as in Africa. In India, macaques and langurs, and in Africa, baboons and vervets are common. They are well adapted to urbanization and also enter houses for food. The attractive peacock is often seen even in large cities in North India. Casual visits of larger carnivores such as leopards and accidental visits of elephants in townships have also been reported from urban areas in India and in Africa in the recent past.

Suitable urban habitats are commonly inhabited by wildlife, particularly rodent and birds, all over the world. Cemetery settings also allow wildlife to inhabit. In cities, in addition to diverse species of birds, small and medium-sized carnivores, musstelids and viverrids are also seen. The management of populations of them in urban areas requires special programmes, different from the management in wild habitats. Translocation of the problem animal is often required in urban areas.

Questions for Review/Exercise

1. Have you understood the definitions and meanings of the following terms?

   a. Habitat restoration b. Wildlife stocking

   c. Wildlife harvest (culling) d. Reintroduction of wildlife

   e. Scent-marking f. Urban wildlife

2. What do you think are the goals of wildlife management?
3. What are the different types of human activities you have witnessed in the wildlife habitats you have visited? Have you seen cattle grazing in areas where wild animals also forage? What are your perceptions of the adverse effects of such human activities in wildlife habitats?
4. How does forest fragmentation affect wildlife communities? Do you have any suggestion to overcome (mitigate) this problem?
5. Have you seen any of the mitigative measures undertaken by forest and wildlife officials of your region with respect to human impacts on wildlife habitats? Have you realized any positive effect of such mitigative measures in the region?
6. Do you think a degraded wildlife habitat can be restored to its original habitat type? How long does it take to restore an intensively-grazed (by domestic animals) forest habitat to its original vegetation characteristics? Is it that simple to restore a

degraded habitat? Write your views in brief and discuss with your colleagues, and find out their views.

7. How does competition between wildlife and livestock arise in a forest habitat? What is your opinion about the measures to solve this problem?
8. What are your perceptions on the measures undertaken to regulate the population of wildlife where recruitment is affected?
9. What did you understand from Fig 41? How can we arrive at an approximate rate for wildlife culling in an area? What is the role of habitat and its extent in estimating culling rate?
10. What did you understand by the terms cattle lifting and crop raiding by wildlife? Have you witnessed any of these problems in your region, locality or village? If so, try to assess this issue, and find out the probable causes and solutions. Discuss with the local people to get more information on the issue.
11. How can we integrate information from behavioural biology of wildlife into their management programme? Briefly describe the role of communication signals in maintaining social structure and organization of wildlife populations.
12. See Fig 47. Have you seen any such animal marking in any of the forest areas you have visited? Have you seen urine spraying on sign-posts by domestic dogs? What are the purposes for which wild animals scent-mark sign-posts in their habitats?
13. Can you try to elaborate the problems and prospects of wildlife management in your country?
14. What did you learn from Boxed item 4.1? You have learned that there are more than 1,100 species of bats in the world, and they are next to rodents in the diversity of species. Bats are the only mammals capable of flying. They are also well known for their ability to emit ultrasonic sounds during their flights. Bats are also one of the smallest among all the mammalian species; the smallest bat species weigh only around 2 g. Can you briefly elaborate the ecological role of bats from your experience in your locality or village?
15. What is known as echolocation? Can you cite some animals that depend upon echolocation during foraging?

Manuscript 4.1

Human Impact on the Plains Zebra (*Equus quagga*) Population in Nechisar Plains, Nechisar National Park, Ethiopia

Yisehak Doku, Afework Bekele and M Balakrishnan

Department of Biology, Addis Ababa University, PO Box 1176,
Addis Ababa, Ethiopia
*Corresponding author (E mail: balak212@yahoo.com)

Abstract: Socio-economic and physical surveys were conducted in eight Guji Oromo settlements in Nechisar plains, Nechisar National Park (NNP), southern Ethiopia during November 2001–April 2002 to assess human impact on the population of the plains zebra *Equus quagga*. Data collected since 1970 were also examined to indicate trends of human, livestock and plains zebra populations in the area. In 1980, the then military government forcefully evicted the local inhabitants. However, they returned to the area during the political change in 1991, which caused widespread destruction of the habitat and the wildlife in the area. The survey confirmed the presence of 250 households in eight settlements in the Nechisar plains, comprising 1,640 people and 7,080 heads of livestock. Cutting of grass and trees by local people has affected the population of plains zebra by decreasing the effective foraging area and cover. The population of plains zebra has declined since 1991, but there are signs of recovery in recent years. The current population of plains zebra in NNP was estimated to be around 4,500. Active measures have to be implemented to control human utilization of Nechisar plains to prevent further deterioration of the habitat and to safeguard the future of plains zebra.

Keywords: Plains zebra, Human impacts, Nechisar National Park, Ethiopia.

Introduction

Most wild mammals in developing countries are at risk due to a variety of factors in their natural habitats (Wilson et al 1996). It is important for conservationists to have detailed information on impacts of ever-expanding human populations in areas of rich wildlife populations. In Africa, most of the ranges of plains zebra, *Equus quagga* are threatened by poaching, settlement encroachment, agricultural expansion, civil war and livestock incursions (Graham et al 1996, Schloeder 1996, IUCN 1997). As a result, populations of plains zebra have to compete with human beings for habitat, pasture and water for survival. Even though, the problems involved in the conservation and management of wild equid populations are not well identified; predation and human impacts are considered as the most important factors (Duncan 1992).

The major ranges of plains zebra in Ethiopia are Mago, Omo and Nechisar National Parks and the Yabelo Wildlife Sanctuary (Bolton 1973, Kirubel Tesfaye 1985, Duncan 1992). Nechisar National Park (NNP) was primarily designated to conserve populations of endangered and endemic Swayne's hartebeest *Alcephalus bucelaphus swaynei* and Burchell's zebra *Equus burchelli*, which is now renamed as plains zebra. At present, the Nechisar plains support Ethiopia's largest concentration of populations of plains zebra. Plains zebras play an important role in maintaining the ecological balance in the Nechisar plains by using coarse tall grasses, which would not be grazed by smaller plain ungulates such as Grant's gazelle *Gazella granti* (IUCN 1997).

Initially, there were only few settlements of Guji Oromo pastoral tribe around the Nechisar plains (Bolton 1973). At present, the dependency of human populations on Nechisar plains for cultivation of crops and livestock grazing is increasing. Human impacts in the park include grazing of cattle by nomadic Guji and settled cultivation by the Kore and rampant poaching by Konso people (Kirubel Tesfaye 1985). In 1980, the previous military government evicted the local people who were living inside the park. However, they returned to the park area during the 1991 political change in the country. This has resulted in widespread destruction both to the habitat and wildlife in the area. Subsistence farming activities are presently concentrated along the Sermale River Valley alluvial soil. The river is used for irrigation. Rain-fed cultivation is also carried out in the area. Most Guji households collect wild honey from natural or manmade beehives in the park. Deliberate and frequent use of fire in the park has led to the destruction of trees (Duckworth et al 1992, Solomon Eshete 1996).

The main objective of the present investigation was to identify and compile human impacts in the Nechisar plains in NNP that are adversely affecting the largest population of plains zebra in Ethiopia.

The Study Area

The present study was carried out in Nechisar plains in NNP, Ethiopia. The NNP is located between 5° 51' to 6° 10' N latitudes and 37° 32' to 37° 48' E longitudes, 540 km south of Addis Ababa (Fig 1 from the referenced

Manuscript 4.1). The study area covers 270 km^2 of plains zebra range at altitudes ranging from 1,250 to 1,360 m. Annual rainfall is bimodal with a long rainy season from March to May, and a short rainy season from September to November. The mean annual rainfall is 895 mm. The mean annual maximum and minimum temperatures are 32.2°C and 17.4°C, respectively.

Many parts of the Nechisar plains are drainage basins. The vegetation of the study area is mainly savannah grassland with scattered *Acacia-Commiphora* trees and bushes (Bolton 1973). There are scattered trees and bushes that are invading the open grassy plains. In addition to plains zebra, the endangered Swayne's hartebeest *Alcelaphus buselaphus swayni*, Grant's gazelle *Gazella granti* and Greater kudu *Tragelaphus strepsiceros* are the diurnal larger wild herbivores in the area. Although the population status of the mammalian fauna in the area is not well known, buffalo has become locally extinct and Swayne's hartebeest is critically endangered (Kirubel Tesfaye 1985, Duckworth et al 1992).

Sample Design and Procedure

Socio-economic surveys in eight Guji Oromo settlements and transect sampling counts in two main habitats of Nechisar plains of NNP were conducted during November 2001–April 2002. From the lists of households of these settlements, a sample of 124 households was drawn randomly, representing half the number of households in each settlement. During ground surveys, data were collected about the livestock and number of houses in settlements. Ground survey revealed the number of wood and grass collectors, honey collectors, foot passengers, fishermen, herdsmen and the number of vehicles seen in the area per day. The extent of recent bush encroachments and degraded land were estimated in hectare. A questionnaire was used to conduct interviews from sampled households. The questionnaire had two types of questions; precise and closed questions, and broad and open-ended questions, giving the respondent an opportunity to express views freely (Balakrishnan and Ndhlovu 1992). It had two sections. The first section was designed to gather background of personal information. The second section sought information on the perception on population status of plains zebra, their population management and conservation.

Initial surveys were carried out in collaboration with the park staff. Prior to the survey, the selected eight settlements were visited and the intention of conducting interviews was explained to local leaders for their endorsement. A total of 124 informants of both sexes consisting of different age groups participated in the interview. The elderly people with good knowledge of wildlife, and traditional local informants (village headsman and local hunters) served as key informants. The attitude of the local people towards plains zebra and their conservation activities were evaluated through the questionnaire analysis.

The responses given in each completed questionnaire were numerically coded and analyzed. Data on the sample household characteristics were treated according to the settlements. Chi-square test was used to determine differences in the sample characteristics among the settlements. Other data on human activities that caused habitat destruction was pooled together for analysis. Responses to questions with options on a continuum scale were analyzed using chi-square test to assess the opinion of the local people. Questions with alternative opinions were analyzed using the same test. Population census data were analyzed using Distance Sampling Method (Buckland et al 1993) using 'DISTANCE' software programme.

Results

Settlement and Livestock

The abundance of Guji Oromo pasturalists and their livestock in eight settlements of Nechisar plains are given in Table 1 (from the referenced Manuscript 4.1). The Guji Oromo population in the 250 households comprised 1640. Among these, 770 (47%) were males and 870 (53%) were females. The average family size was 6.6 per household. The population of cattle and goats was 7,587. Cattle accounted for 93.3% and goats only 6.7%. The average number of cattle and goats per household was 28.30 and 2.03, respectively. The average size of cattle and goats per household was not different among the eight Guji Oromo settlements ($P>0.05$).

Resource Utilization and Human Disturbances

Human activities such as livestock grazing, grass cutting, wood cutting and wild honey collection have adverse effects on the ecology of plains zebra. On an average, 260 individuals were observed while cutting trees and collecting firewood per day. In the Nechisar plains, 236 herdsmen and thatchers were observed per day. Most of the Guji Oromo households were found to collect wild honey from natural or manmade beehives. An average of 15 wild honey collectors was observed daily. Honeybee hives were observed on branches of tall *Acacia* trees in the Kordae settlement. The honey collectors used to cut and remove the bark of big trees to prevent honey badger from climbing up. As a result, several big *Acacia* trees were found dried in the Nechisar plains. On an average, 65 foot-passengers and 5 vehicles were counted every day on the main tracks passing through the centre of the Nechisar plains from Arba Minch town to Amaro District. Herdsmen were also observed taking around 500 heads of livestock every day to Bole soil (salt lick), thereby disrupting zebra movements in their natural habitats.

Habitat Destruction

The major factors responsible for habitat destruction in the study area were recent bush encroachments, forest fire and trampling. During the last 30 years, around 9,000 ha (33%) of the suitable grassland of plains zebra has been disrupted in the Nechisar plains out of the total 27,000 ha (Figure 2 from the referenced Manuscript 4.1). Table 2 (from the referenced Manuscript 4.1) gives the extent of area degraded in each of the eight settlements in Nechisar plains as a result of human activities. Bush encroachment was the major cause of destruction in all settlements. This was intensive in Hare settlement. Fire-related destruction was recorded only in Watchole settlement.

Attitude of the Guji Oromo Pastoralists

Among the 124 informants, 87.9% expressed the view of increasing population of plains zebra in the area. Only 4.0% of the respondents opined that plains zebra population is decreasing. The suggested reasons for the decline are habitat destruction, direct shooting, poaching, effects of predators and parasites. Regarding the views of informants on conservation and management of plains zebra populations, there was no significant change among the eight Guji Oromo settlements in Nechisar plains ($P>0.05$). Among these respondents, 46.8% confirmed that the species has been neglected and 37.9% considered as a good species to be conserved. Only 15.3% considered it as a nuisance species. They responded that zebra is of no use whatsoever, and did not think worth conserving as they chase and kill newly-born livestock and compete with the cattle for the grazing land.

The age of the informants has been a factor responsible for grouping plains zebra as a 'good' or 'nuisance' species ($P<0.05$). Most of the elderly people (>60 years of age) considered this species as good while some respondents of age group 15–20 years considered as a nuisance species. Educated people and those who have lived there for several years showed positive attitude towards plains zebra ($P<0.05$). They were of the opinion that the plains zebra would be potential for eco-tourism. Some pointed out the reasons as cultural value and prey base for lion, and positively responded for conservation of the zebra.

Trends of Populations of Plains Zebra

The population trends of plains zebra in NNP during the past 30 years as recorded are given in Table 3 (from the referenced Manuscript 4.1). The maximum population size was 6,500 heads in 1985. The present population status of 4,500 shows a recovery from the depleted levels during mid-1990s.

Discussion

Duckworth et al (1992) and Schloeder (1996) have noted that the expansion of settlements and human activities in Nechisar plains have caused great disturbance to wild animals. This has been witnessed during the present study. The Guji Oromo people harvest various types of natural resources from Nechisar plains. Grass is the most important resource used by plains zebra and livestock. Local people also collect the same resource for thatching houses. Wood is collected for fuel and for construction. Such resource consumption by humans and their livestock directly affects the population of plains zebra resulting in depletion of their essential food and cover.

The local communities culturally consider plains zebra as pack animals called by the local name Hare-Dida (Wild donkey). Due to this belief, most local people show benign neglect to the conservation of this species. They are poached for meat in Kenya, Uganda, Rwanda, Sudan and Zambia (Duncan 1992, IUCN 1997). Most of the informants confirmed that Konso people from the South of the Nechisar plains poach them for meat during the dry season, when they disperse out of the park. The Gumayde and Menz people use their skin to decorate and to make furniture.

The main road from Arba Minch to Amaro District passes through the Nechisar central plain. Heavy use of this road by people and vehicles interrupts the movement patterns of plains zebra from Nechisar plains to River Sermele during the dry season. The daily livestock movement to the salt lick sites also negatively influences the population of plains zebra in the study area.

The three main factors responsible for depletion of grasslands in Nechisar plains are recent bush encroachment, forest fire and trampling.

The rapidly increasing livestock population of Guji Oromo pastoralists is a major cause of recent bush encroachment and habitat destruction. Recent bush encroachment has been expanded deliberately by planting trees around settlements for shade, flood catchments and as a livestock fence. Cultivation of crops takes up the dry season grazing area of livestock. The extent of grasslands has been reduced to about 33% during the past 30 years, resulting in grazing pressure. Frequent burning by pastoralists to remove tall dry grasses to secure new forage for livestock is also a major threat to the wildlife in the area.

The negative attitude of the Guji Oromo people towards wildlife conservation may have emanated from the fear of displacement from their settlements for the sake of wildlife conservation. The community that lives in Nechisar plains has been using diverse natural resources, other than wildlife meat. The local people claim that they have the right to utilize natural resources in their localities as long as they are pastoralists. The previous military government forcefully evicted them from the Nechisar plains; as a result of which, they have developed mistrust and revenge. Even though the local people agree to leave Nechisar plains, if the government improves their socio-economic conditions, they strongly oppose further displacement and exclusion from using the traditional resources they have been enjoying from time immemorial. However, with the present increase in human population, this cannot continue.

The Guji Oromo people are traditionally nomadic pastoral tribes. They are distributed in eight different areas of the Nechisar grassland to exploit the plains for grazing. In Nechisar plains, Kirubel Tesfaye (1985) counted 1,689 heads of cattle and 188 herdsmen. Solomon Eshete (1996) observed 190 households with 1,330 people and 3,534 cattle. The number of cattle more than doubled within ten years from 1985 to 1995. The present study showed that there is an increase in human and livestock populations in the area. The cattle population in the Nechisar plains has almost tripled within five years (1996–2001). However, as there was drastic decline in the population of plains zebra in NNP during the period 1991–1995, their population is not stable. Even though there has been an incipient population recovery in the area since 1995 as compared to the 1985 estimate of 6,500, their present population level is not satisfactory. The present study recommends that human activities in the NNP are to be controlled for the conservation of plains zebra population, the population trend is to be monitored and the negative human impacts are to be eliminated by resettling local people in areas outside the park. Alternative means for survival of the local people should be sought by the regional government to have a lasting solution to the problem.

Reference: International Journal of Ecology and Environmental Sciences **32** 137–142, 2006

(Figures, tables and references are not included)

One of the published papers dealing with human impacts on wildlife is presented for the benefit of students to read and understand the first-hand information provided. Students have to critically analyze the content and come up with new thoughts on human involvements in natural habitats and its impacts on wildlife habitats and the inhabiting wildlife.

1. What are your feelings about the enforcement of government decisions and their impacts in NNP as revealed in the abstract of the referenced manuscript? Have you experienced any such negative impacts on forest and wildlife in any of the wildlife areas with which you are familiar?
2. Make an assessment on various human activities on the plains zebra population in NNP. What have you learned about similar human activities affecting the populations of plains zebra in other parts of Africa?
3. What was the purpose for which NNP was established?
4. Briefly assess the ecological characteristics of the study area presented in the above manuscript.
5. What are the methods used to evaluate human impacts on plains zebra presented in the referenced manuscript? Can you design an investigation to assess human-related issues affecting wildlife populations in any of the areas with which you are familiar?
6. Briefly assess the intensity of human activities in wildlife areas presented in the result section of the above manuscript. Do you think that such activities can sustain for long in the context of increasing human population in the study area?
7. What are the views of local people (informants) on forest and wildlife in NNP revealed through the study presented? What are your perceptions on this subject?
8. What are the major factors responsible for depletion of grassland (zebra habitat) in NNP as revealed from the above investigation?
9. What are your views on participatory forest management and wildlife management programmes to amicably improve wildlife populations in wildlife areas and human socio-economic status?

10. Briefly evaluate the conclusion of the manuscript presented above. What are your views on alternative means of survival of local people suggested?

Manuscript 4.2

A Study on Olfactory Communication Signals in Sambar Deer, *Cervus unicolor*

Eugene Mary[1] *and M Balakrishnan*[2]

[1]Department of Zoology, Mar Ivanios College, Trivandrum 695 015, India
[2]Department of Zoology, University of Kerala, Kariyavattom, Trivandrum 695 581, India

Abstract: The olfactory communication signals in eleven individuals of sambar deer were studied. They marked the objects in the enclosure with the secretions of diverse specialized integumentary glands. The tendency to sniff the marked sites was more consistent among females than in males. Both sexes of this species of deer depend considerably on olfactory cues during various social interactions.

Keywords: *Cervus unicolor*, olfactory signals, scent-marking, pre-orbital marking, forehead rubbing, pedal marking

Behavioural Observations

Pre-orbital Marking

Both the sexes of the species have a pair of pre-orbital glands (Figure 1 from the referenced Manuscript 4.2). This gland is a pouch-like integumentary pocket extended forward from the anterior corner of each eye.

During pre-orbital rubbing, the animals were observed to examine the spot to be marked by sniffing twice or thrice, tilting its head forward and rubbing the pre-orbital glandular region on the object very lightly in a single forward movement. One such forward movement of the head was considered as one pre-orbital marking. Fence, trunk of the tamarind tree in the enclosure and dead twigs were marked. This behaviour was noticed at a higher frequency when they were moving around.

Forehead Rubbing

They rubbed their forehead against posts, bricks, fence and trunk of the trees (Figure 2 from the referenced Manuscript 4.2). The bucks rubbed their foreheads on overhead branches as they stood upright on their hind legs (preaching) to reach them. The foreheads were also rubbed against the inner regions of its own hind legs. This type of marking can be differentiated from similar behaviour in sambar deer to ease irritation wherein they use their antlers, lips or tongue. Rutting stags after rubbing the forehead in the soft mud, rubbed again on the basements and bricks seen in the enclosure. Forehead rubbing was more pronounced during various social interactions.

Rubbing with Hind Hoof

Bucks and hinds use their hind hooves to rub certain parts of their body such as pre-orbital region, ear pinnae, area around the ear pinnae and snout. During this behaviour, the interdigital area is pressed against various body regions. This can be differentiated from scratching wherein the tips of hooves are rubbed against various body sites. The ear pinnae and head area were rubbed more frequently. Usually this type of rubbing occurred twice at a time, and was noticed after eating bouts, and before and after reclining.

A strong association between the rubbing of forehead with hind hoof and the subsequent rubbing of forehead against objects were observed in males (Table 2 from the referenced Manuscript 4.2).

Pedal Marking (Stamping or Thumping)

The pedal glands were visible as brown thick projections in the posterior part of the interdigital area in the foreleg and in the anterior part of the hind leg. Thumping was observed when the animals were approached by the keeper for changing their enclosure. While running they stopped, stood still in an alert position and thumped the foreleg on the ground.

Discussion

Rubbing pre-orbital pouch, forehead and around antlers and ear pinnae with hind hooves enables sambar deer to transfer secretions from these specialized glandular areas on to the ground along with the pedal scent. At the same time, the pedal scent may also be transferred to these areas and marked along with the different chemical exudates of the gland concerned. This pattern of behaviour is described as 'self marking' (Ewer 1968) as there is a possibility of marking various body sites with glandular secretions from other body regions.

In sambar deer, the pedal scent leaves trails on the ground. Frequent sniffing of the ground by members of the group revealed the presence of these trails having olfactory cues.

During thumping behaviour, the interdigital glands may leave their secretions on to the ground. Probably, this scent may be used as an alarm signal. Stamping has also been recorded in pampas deer, *Blastoceros bezoarcticus* (Lungguth and Jackson 1980), spotted deer, *Axis axis* (Iyer 1979) and in males of pronghorn antelope, *Antelocapra americana* (Müller-Schwarze 1974).

The present study has revealed the various occasions during which sambar deer mark with their specific cutaneous glands. Thus, pre-orbital rubbing, forehead rubbing, rubbing with hind hooves and pedal markings have pronounced roles in maintaining environmental sign-posts, social signalling, self advertisement and in alarm signalling. However, these findings are to be confirmed using fractions of various chemical compounds seen in these glandular exudates.

Reference: Proceedings of the Indian Academy of Sciences (Animal Science) **93** 71–76, 1984

(Tables, figures and references are not included)

The abstract, part of observations and part of discussion of a scientific publication is presented in the above referenced proceeding for the students to read and come up with ideas on behavioural studies.

1. What is your understanding on olfactory communication signals in animals?
2. What are the different types of olfactory signals presented in the above manuscript?
3. Have you heard of the terms such as scent-marking, pre-orbital rubbing, forehead rubbing, rubbing with hind hooves and pedal marking earlier? If so, in what context? Do you know any other mammalian species, which also performs scent-marking?
4. Can you plan a similar study on any of the wild animals, either in natural habitats or in a seminatural habitat (e.g., a good zoo where animals are free living in a fenced open area) and study communication signals during various social interactions?
5. What are the functional roles attributed to each of the scent-marking types in sambar deer?

6. What are your perceptions on the conclusion of this investigation?
7. Are you familiar with the keywords presented at the bottom of the abstract?

❑❑❑

Conservation of Threatened Wildlife in India and Neighbouring Countries 5

India, project tiger, Gir lion project, snow leopard project, Asiatic elephant, Indian rhinoceros or the greater one-horned rhinoceros, threatened wildlife of India, threatened mammals of India, threatened birds of India, threatened reptiles of India, Nepal, lowland projects, mountain research, Sri Lanka, Pakistan, future prospects

---

5.1. INTRODUCTION

Even though investigations on wildlife populations in the Indian subcontinent have been progressing fast during the past few decades, there are several areas in the region where wildlife is yet to be studied in detail. There is a recent record of a new species of primate in Sri Lanka, which shows the region has even new mammal species to be located and studied. The climax ecosystems in the region include the snow-capped Himalayas, the Great Indian Desert and the north-eastern and southern regions of tropical wet evergreen and moist-deciduous forests, each with a wide range of climatic patterns that govern the distribution of diverse wildlife forms. The mammalian fauna in this region is one among the richest in the world. However, the rich forest and wildlife of the entire region have been under threat of intensive human pressure.

Following the well-known wildlife surveys and analyses of Gee (1964), Schaller (1967) and Krishnan (1975) in India; Caughley (1969), Mountfort (1969), Beg (1975) and Blower (1975) in Nepal; Eisenberg and Lockhart (1972) in Sri Lanka; and Schaller (1976 a, b) and Roberts (1977) in Pakistan, several in-depth investigations on several species of special interests with emphasis on their conservation have been initiated during the past four decades. Concurrently, more sophisticated conservation programmes have been launched by the governments in the region; i.e., Governments of India, Nepal, Pakistan and Sri Lanka.

5.2. INDIA

India has a long history of conservation activities. In the 3rd century BC, Emperor Ashoka forbade the injunction of pain on animals, and set up sanctuaries to protect wildlife. However, this tradition was later lost and replaced by a golden era of hunters. There are records of shooting up to 300 tigers by a single hunter, mainly the then regional rulers, to exhibit own vigour and the ability in shooting. Though sport-hunting continued during the British rule, the wild habitat as a whole was not much reduced. After the World War II, this situation was changed, and there was much concern for increasing agriculture and industries throughout the world. During the post-war period, large extents of natural forests all over India were slash-burned without any consideration for the wildlife and the rich genetic diversity that the region was supporting. However, later, the former hunters themselves realized the depletion of wildlife in a short span of time, and some of them were also responsible for initiating conservation measures in the region.

Volumes of 'Fauna of British India' have contributed as the pioneer records of wildlife surveys including animal systematics in India during the colonial period. In continuation to such studies and contribution by expatriates on wildlife studies in India, the Bombay Natural History Society and the Wildlife Preservation Society of India have been the two non-governmental national organizations most active in wildlife research and conservation activities. The well-known ornithologist of India, late Dr. Salim Ali was the person behind the recent conservation achievements in India. His contributions on birds of India and specifically on birds in different regions of India (e.g., Birds of Kerala) have motivated several youngsters to turn out as wildlife ecologists. Bombay Natural History Society has developed as an autonomous institute dealing with a variety of topics in the field of wildlife and conservation. Government agencies such as the Zoological Survey of India and the Botanical Survey of India are also contributing for wildlife-related developments throughout India. Currently, there are also several Universities in India where wildlife is the subject of research and teaching. However, the Wildlife Institute of India is the major centre in India responsible for several research projects and wildlife-related developments. State institutions such as the Kerala Forest Research Institute are also dealing with forest and wildlife research and conservation activities.

There are three major types of research activities in the field of wildlife. These are species-oriented, habitat-oriented, and conservation- and management-oriented research. All these types of research are being carried out in India with a thrust on conservation of endangered animals and their habitats. Even though international collaboration was pursued to begin with, India has been advancing with its own resources and capabilities in the recent years. At present, there are national parks and wildlife sanctuaries in India, constituted under Wildlife (Protection) Act, 1972. The total area under such protective measures is over 90,000 km^2 or roughly over 3% of the total land area. There are over 350 species of mammals and 1,200 species of birds reported from India.

One of the earliest detailed wildlife studies in India was the investigation on Indian primates under a joint Indo-Japan Project. This project was mainly concerned with the socio-ecological aspects of the Hanuman langur or the common langur in natural habitats in Dharwad, South India (Sugiyama 1965, Sugiyama et al 1965). The outstanding contributions of Rahman and Parthasarathy (1967, 1968, 1969a, b) on ecology and behaviour of the bonnet macaque followed the trend gained under Japanese collaboration. They observed the movement pattern, territorial behaviour and dominance hierarchy among bonnet macaques. They have reported that they have a core area within their home range where the troupe spreads out most of the time and gets protection against intruders from other troupes. Sugiyama (1976) also studied the ecology and social behaviour of the Himalayan langur (*Semnopithecus entellus* Desmarest, 1822), a subspecies of the common langur, distributed at higher altitudes and cold conditions at altitude ranges of 2,000–3,000 m asl. These investigations stimulated further investigations by Indian researchers on eco-behavioural aspects of Indian primates in other parts of India (Roonwal 1977, Singh et al 1980, 1987). One of such long-standing research is on a variety of topics on Hanuman langur in Rajasthan (Mohnot et al 1996, Rajpurohit et al 2001, 2003). The distribution of the lion-tailed macaque in South Indian forests was analyzed by Kumar (1985). It is alarming to note that the characteristic features of these forests have been lost in many areas as a result of human encroachment. Conservation of their entire habitat is required to protect this species from extinction.

Prasad et al (1978) and Johnsingh (1983) have reported on populations of various wild animals including elephants in the Bandipur National Park in Karnataka State. Even though the pressure of wild mammal population is high in this area, especially in the case of predators such as wild dog or dhole (*Cuon alpines* Pallas, 1811), leopard (*Panthera pardus* Linnaeus, 1758) and tiger (*Panthera tigris Linnaeus*, 1758), the habitat remains in good state. The main rutting season of the most common prey, the chital in Bandipur is from May to December and February. They form large herds in the monsoon months, but break up into smaller herds in the summer season. Mortality rate was about 92% in the first year and 25% in succeeding years (Sharatchandra and Gadgil 1975). The higher mortality rate was due to the presence of large population of predators such as tiger, leopard and wild dogs.

Wildlife populations in the Silent Valley and contiguous forests of South India were studied by Vijayan and Balakrishnan (1976). There are over 25 species of large and medium-sized mammals including two endangered species (the lion-tailed macaque and Nilgiri langur) in this forest habitat, which is considered as a virgin habitat in South India. The probable impacts of the then proposed Silent Valley Hydroelectric Project were also discussed and it was urged that the forest in this area should be conserved (Balakrishnan 1984). The Government of India later abandoned this project proposal and declared the area as the Silent Valley National Park.

Distribution, ecological requirements and resource availability of selected mammals in Periyar Tiger Reserve were studied by Nair et al (1985). Even though the intensity of human disturbances is quite high in this reserve, it remains an excellent habitat for the Nilgiri langur, elephant, sambar deer and the gaur (*Bos gaurus* Smith, 1827). Government of Kerala has developed this Reserve as a tourist centre, and hence, this is the best wildlife area in Kerala for animal observation. The Periyar Lake developed as a result of the Periyar dam constructed more than 100 years back attracts populations of wild animals including aquatic birds. Large herds of elephants of up to 60 individuals were observed in areas of inter-herd contacts in this reserve. This reserve is better known for observations of wild elephants and gaur. Also, sambar deer and wild boar are common.

The habitat preference of 24 large and medium-sized mammals in the Parambikulam Wildlife Sanctuary in South India was studied by Balakrishnan and Easa (1986). There is a good population of Asiatic elephants in Parambikulam Wildlife Sanctuary. Most of them have shown preferences to species-specific habitat types. In case of primates, bonnet macaque is more abundant in forest plantations and in deciduous forests. The lion-tailed macaque is confined to evergreen forests, and common langur to the deciduous forests. The Nilgiri langur is more or less evenly distributed between evergreen and moist-deciduous forests. Herbivores and carnivores are most abundant in grasslands and in the teak (*Tectona grandis*) plantations. Hence, it is important to conserve all types of natural habitats in any area as an integrated unit to conserve a variety of mammalian fauna.

The population of the southern subspecies of barasingha (*Rucervus duvaucelii* G. Cuviet, 1823) has declined to critically low numbers in its last stronghold in Kanha National Park in Central India. In this area, it is sympatric with chital, which has increased in numbers to above 10,000 individuals within the area where barasingha occurs. Barasingha depends almost exclusively on grasses, whereas the chital is more flexible and resort to browse when grass becomes scarce in the dry season. Over-use of grasslands by people around is probably the main threat to the barasingha (Schaller 1967).

The Kashmir stag (*Cervus elaphus hanglu*) is reported to be threatened with extinction, as its habitat has been drastically reduced through deforestation (Oza 1986). The State Government of Jammu and Kashmir has initiated an active programme to halt the decline of this species.

Population of the endangered Nilgiri tahr has only about 600 heads in the Eravikulam National Park in South India (Rice 1988a). Approximately an equal number of them are seen in the Nilgiris in Tamil Nadu. The governments of both these South Indian States have initiated measures for conservation of this species by conserving their beautiful, undulating, grassy carpeted habitats with intermittened shola forests. Rice (1986, 1988 a, b) studied various aspects of ecology, ranging pattern and behaviour of this species in Eravikulam National Park. The black buck (*Antelope cervicapra* Linnaeus, 1758), the four-horned antelope or chowsingha (*Tetracerus quadricornis* Blainville, 1816) and the blue bull or nilgai (*Boselaphus tragocamelus* Pallas, 1766) are the antelopes found in the dry-deciduous and desert regions of India. Black buck is now restricted in their distribution only in protected areas. However, in the deserts in Rajasthan, large herds of up to 500 individuals are seen around villages. The Vishnois (one of the local communities) hold the black buck in great reverence, resist their killing, and thus, protect the species in their surroundings. The black buck, unlike chinkara (*Gazella bennettii* Sykes, 1831), prefers to stay around villages. One of their good natural populations was seen in Point Calimer Sanctuary in Tamil Nadu. But at present, their number is drastically reduced in this location. They inhabit the dry-deciduous plains and chiefly subsist upon short grasses and herbs.

5.2.1. Project Tiger

Tiger is one among the larger species of cats. This magnificent animal was seen widely across Asia from Turkey in the West to the eastern ranges of Russia, and the Indonesian Islands of Java, Bali and Sumatra. They were reported in good numbers from most of the forest areas in the Indian subcontinent. However, most of its former ranges are now devoid of tiger. Habitat modification, fragmentation and poaching are responsible for the depletion of tiger populations all over its former ranges. Tiger is currently seen in parts of India, Siberia, China and certain areas of south-east Asia including Sumatra.

A tiger can be easily distinguished from other similar-sized cat species by its characteristic pattern of dark vertical stripes on the body. Its body is covered with reddish-orange fur with lighter underparts. Among the living cat species, the tiger has the longest canines. An adult tiger weighs around 300 kg with a total body length of over 3 m.

Tiger is a solitary and territorial animal seen in pairs only during the breeding period. A tigress may be seen with its cubs. They scent-mark their territorial sign-posts using a special pattern of urine spray at specific heights and at different localities all over the area of its territory. Being one of the tertiary carnivores, the tiger moves through larger home ranges.

The extant tigers are classified among the following six subspecies: *Panthera t. tigris*, *P.t. corbetti, P.t. jacksoni, P.t. sumatrae, P.t. altaica.* and *P.t. amoyensis.* The current global population of tigers is around 5000 individuals. Unfortunately, most of the remaining populations are in fragmented habitats, and isolated from each other. Occasionally, some individuals of tigers come outside their forest habitats and enter human settlements. They lift cattle and develop conflicts with villagers in such areas. The IUCN threatened category of tiger is 'Endangered'.

The tiger conservation programme launched in 1973 by the Government of India with the help of World Wide Fund for Nature (WWF) has 53 tiger reserves comprising over 53,547 km^2 of forest habitats. It is one of the most successful endangered animal conservation programmes in Asia. As a result of better habitat management and conservation, population of this magnificent carnivore has increased from about 2,000 heads in 1972 to over 5,000 in the recent times. Even though not much emphasis was put on research under this project, populations have increased, especially because human activities in tiger habitats were controlled and poaching for tiger skin was reduced. The problem of man-eating and livestock-predating tigers is common in some areas in India. However, conservation measures adopted in areas such as the Sundarbans Tiger Reserve have reduced man-tiger conflicts (Saha 1987). At present, there are only around 5,000–7,000 tigers left in the wild in all their natural habitats in the Russian Far-East, Malaysia, Nepal, Bangladesh, Sumatra (Indonesia) and India.

5.2.2. Gir Lion Project

During the first half of the last century, the Asiatic lion was abundant in India throughout the open forests in the north of the river Narmada. However, at present, the Gir forest in Gujarat is the only habitat in India where this large predator occurs naturally. The population of lions is only about 400. It is estimated that an adult lion kills prey animals at a rate of about 700 kg per week. The requirement of such large number of prey animals necessitates large extent of natural habitats for lions to survive.

Ecological studies in Gir forests in relation to the Asiatic lion, its habitat and prey were made under an Indo-US collaborative programme (Berwick 1976). Studies on predator–prey relationships and aspects of wildlife management programmes were initiated in India under this programme. As per the estimation of Joslin (1973), the population of domestic cattle from 20 villages and 135 settlements grazing in Gir sanctuary was over 20,000 heads. Further, his analysis of 480 lion scat samples collected from four areas showed that within the sanctuary, the occurrence of domestic livestock in lion scats varied from 75–83%. Livestock grazing is one of the major problems of conservation and management in this sanctuary.

5.2.3. Snow Leopard Project

The Government of India and the Wildlife Institute of India have initiated a snow leopard conservation project under an international collaborative research programme (Fox et al 1988). The habitat of snow leopard in the northern mountainous areas of India (Ladakh) was surveyed and the relative presence of snow leopard and their prey animals were assessed. Governments of India and Nepal have taken steps to conserve the endangered snow leopard in the Himalayan region. Both these nations have ratified the Convention on International Trade in Endangered Species (CITES) of wild fauna and flora. India has totally banned the export of wild animals and commercial transport of all wild animal products.

5.2.4. Asiatic Elephant

The status report on elephant populations in the Indian subcontinent prepared by the Asian Elephant Specialist Group of the IUCN Species Survival Commission stressed concern over the destruction of elephant habitats in India and adjoining regions due to the ever-increasing human pressure (Daniel 1980). Asiatic elephants are now confined to only few isolated areas where they are under stress due to increased density. Hence, their survival on the Indian subcontinent is threatened. A comparison of the past and present geographical distributions shows the deteriorating trend in the availability of environmental resources for their survival. Elephants range through almost all types of habitats in India including climax evergreen forests, dry thorn forests, swamps and grasslands. The optimal habitat is the moist-deciduous forest with large areas of grasslands around.

The Asiatic elephants in India have a total population of 17,000–22,000 heads. They exist in the following five disjoint populations:

1. North-west India in the forest divisions of Dehradun, Bijnor and Nainital districts of Uttar Pradesh
2. Southern population in the Western Ghats in the States of Karnataka, Kerala, Tamil Nadu, north of Palghat gap
3. Another southern population in the States of Kerala and Tamil Nadu, south of Palghat gap
4. Central Indian population discontinuously distributed in Odisha, South Bengal and extending up to southern Bihar
5. Eastern population discontinuously distributed in North Bengal, Assam and other States in eastern India.

The last one is continuous with the elephant populations in Myanmar and Bangladesh. Conservation of Asian elephant is strictly dependent on the management policies of the nations where they still exist in viable populations (Sukumar 1985, Easa 1989, Johnsing and Panwar 1990).

Behavioural ecology of the Asian elephants in Parambikulam Wildlife Sanctuary, South India was studied by Easa (1989). Feeding patterns, food preferences, activity patterns, movement and time budget, population ecology and dynamics of this magnificent animal were studied in detail (Sukumar 1985, Balakrishnan and Easa 1986, Easa 1989). In the recent past, elephants have become crop raiders in different parts of India and elsewhere (Sukumar 1985, 2003).

5.2.5. Indian Rhinoceros or the Greater One-horned Rhinoceros

The greater one-horned rhinoceros (*Rhinoceros unicornis* Linnaeus 1758), which was on the verge of extinction, was recovered considerably following conservation activities in the north-eastern parts of India. Poaching for its costly horn was a major problem for this magnificent animal to sustain for long (Martin et al 1987). Once they were present in larger extents of areas throughout the northern parts of the Indian subcontinent, along the Indus, Ganges and Brahmaputra river basins, from Pakistan to the Indian-Burmese border, including parts of Nepal, Bhutan, Myanmar, southern China and Indonesia. However, at present, they inhabit only in parts of north-eastern India, mainly in Assam, and in few protected areas in southern Nepal.

The Indian rhinoceros is a heavily built mammal with its height ranging from 163 to 193 cm in case of adult males and from 147 to 173 cm in case of adult females. Their respective mean weights are 2,075 kg and 1,600 kg. They have a single nasal horn in both sexes, which starts growing when they are around six years of age. The horn is slightly curved backwards with a base of about 18.5 x 12.0 cm, and narrow and pointed at the tip. Their skin is thick and folded in certain areas of body parts. Males have thick neck folds of the skin. The skin is silver-brown in colour, and pinkish near the folds. There are wart-like bumps on the upper legs and shoulders.

Rhinoceros are mainly grazers, and they prefer grasslands and wetlands where there is sufficient water for wallowing. Their habitats are depleted and are surrounded by areas occupied by villagers around secondary forests, adjacent to farmlands and livestock grazing areas.

The latest estimation of population of the Indian rhinoceros shows that there are only around 2,600 individuals, majority of which is recorded in protected areas in north-eastern India. Kaziranga National Park in Assam holds the largest population of them; around 2,000 individuals. Jaldapara Wildlife Sanctuary, Pobitora Wildlife Sanctuary, Katarniaghat Wildlife Sanctuary, Orang National Park, Dudhwa National Park and Manas National Park are the other locations in India where rhinos are present; although some of them are only in few numbers. The Royal Chitwan National Park (Nepal) and the surroundings hold the second best population of rhinoceros, numbering less than 400 individuals. Bardia National Park and Sukla Phanta Wildlife Reserve of Nepal also have small populations of this species.

Their food mainly consists of grass, but they eat leaves, branches, shrubs and fruits. They also eat aquatic plants. Rhinos are solitary animals, but they congregate in pools for wallowing. Pairs are seen together during breeding; and mothers with their calves. Most of the activities are during night and in the early morning hours. Mid-days are mostly spent in water (in rivers, lakes, and ponds), wallowing to cool down body temperature. They are also good swimmers. They maintain territories of the size 2–8 km^2, which overlap with neighbouring territories. They also form social groups. Mothers stay with their calves for up to four years of the age of the calves. Sub-adult males and females form groups of 2–3 on the edge of the home ranges of dominant males.

A variety of vocalizations form the major communication pattern among the rhinoceros. They also use olfactory cues. They have the special urine-spraying behaviour. They also maintain dung piles in their territories. They have pedal glands, which are also used for scent-marking at their latrine sites. Males constantly sniff the ground by lowering head for sensing olfactory signals laid by align individuals, particularly females in estrus. Sexual maturity is attained when the females are 4–6 years. Gestation period is around 15.7 months.

Sport hunting was the main threat for Indian rhinoceros. They were poached for their costly horn, and for the use in traditional Chinese medicine. Encroachment into their habitats also caused considerable damage on their populations. The Indian rhinoceros is listed in Appendix I of CITES. India and Nepal have been seriously involved in rhinoceros conservation. In the 1980s, a rhino translocation programme was initiated and the first pair of rhinos was reintroduced from the Terai of Nepal to Lal Suhanra National Park in Pakistan in 1982. The current IUCN threatened category status of the Indian rhinoceros is 'Vulnerable' (IUCN, 2011).

5.2.6. Threatened Wildlife of India

India has an excellent array of wildlife, although majority of which are now on the verge of extinction. In response to evolutionary processes, some species may become extinct and some may evolve with time. Further, population of any species may dwindle down after attaining a maximum growth rate for some time due to diverse reasons. Fluctuation of the population of a particular species can also affect the populations of other species co-habiting the same area and of those that are linked directly with the food web. However, as compared to the rapid rate of recent extermination of certain species of animals with that of evolutionary extinction and population fluctuations, the latter is very slow. The fact is that those species which are rapidly becoming extinct are those with costly horns, beautiful fur or feathers, tasty flesh and such useful properties. These properties attract human beings, and hence, it is believed that there are human involvements in the process of such extinctions. Apart from the direct involvement of human beings, there are also indirect causes for depletion of wildlife populations. Encroachment into natural habitats for agricultural expansion, intensive livestock grazing and multipurpose dam construction and related development activities also affect wildlife habitat and wildlife populations. It is estimated that at present, over 1,000 species of animals and 20,000 species of plants are threatened globally. It is also revealed that one species of animal per year faces extinction currently, which is 10 times greater than the comparable figures during the period from 1600 to 1950 AD. The present level of threat of many species is very high mainly due to unscientific and illegal activities of human beings with a view to achieve immediate benefits without any consideration for long-term benefits. As a result of such activities, the Indian cheetah (*Acinonyx jubatus* Griffith, 1821) is already extinct in natural habitats and the fate of larger animals such as the Asiatic elephant, Asiatic lion, rhinoceros, tiger and a number of antelopes and primates in addition to several species of reptiles and birds are threatened with extinction. Strict conservation efforts and management programmes implemented by the Government of India and State governments have been helpful to revive the populations of some of them including the tiger. Although the conditions are more or less similar in different parts of the world, the levels of deforestation and fragmentation of natural habitats and subsequent extermination of wildlife in the developing tropical nations are at higher levels as a result of poor socio-economic conditions of the local people. Hence, conservation of wildlife can be satisfactorily achieved only if the rural communities around forest areas are uplifted.

5.2.7. Threatened Mammals of India

India has over 350 species of mammals recorded in different forest habitats. Among them, over 80 species have been listed as threatened with extinction. Table 1 gives a list of Indian mammals that are currently under various categories of threat as per IUCN status report. Among the 19 species of primates in India, 12 are facing problems in maintaining their populations. They include the slender loris (*Loris tardigrandus* É. Geoffroy, 1758) of southern India and the slow loris (*Nycticebus coucang*

É. Geoffroy, 1812) of north-eastern India. The lion-tailed macaque, which is an endemic species of the southern Western Ghats, is a tropical evergreen dweller. Its total population is only around 1,000 individuals. A viable population of this species was reported from Silent Valley forest of Kerala (Vijayan and Balakrishnan 1977, Balakrishnan 1984). One of the considerations of the Government of India to conserve Silent Valley of Kerala as a National Park was to conserve populations of this species in Silent Valley. The golden langur (*Trachypithecus geei* Khajuria, 1956) of the deep forests of the north-east of India, especially in the forests of Assam, was not known to the scientific community until its first report in the 1960s. The populations of the only ape in India, the Hoolock gibbon (*Hoolock hoolock* Mootnick & Groves, 2005) of north-eastern India and the Nilgir langur of south India have dwindled as a result of habitat destruction in addition to large-scale poaching for their flesh.

The two species of pholidota seen in the Indian subcontinent are the Chinese pangolin (*Manis pentadactyla* Linnaeus, 1758), which is restricted in its distribution to the north-eastern regions and the Indian pangolin (*M. crassicaudata* Linnaeus, 1758) of the peninsular India. Both these species are threatened with extinction.

Among the 36 species of carnivores in India, 28 are facing extinction problems and their populations have dwindled during the recent past. Among the felids, wolf (*Canis lupus* Linnaeus, 1758), jackal (*C. aureus* Linnaeus, 1758), red fox (*Vulpes vulpes* Linnaeus, 1758), Indian fox or the Bengal fox (*V. bengalensis* Shaw, 1800) and wild dog were abundant in all forest habitats in India in the past. However, their populations have dwindled in the recent decades. Among the bears, the Himalayan brown bear (*Ursus arctos* Linnaeus, 1758), Malayan sun-bear (*Helarctos malayanus* Raffles, 1821) and the sloth bear (*Melursus ursinus* Shaw, 1791) are threatened with extinction as a result of loss of their natural habitats in addition to direct persecution by human beings. The red or lesser panda (*Ailurus fulgens* F. Cuvier, 1825), which is also known as the cat bear is threatened with extinction.

Among the 14 species of mustelids in India, the erminae (*Mustela ermine* Linnaeus, 1758), which is restricted to the forests of the western Himalayas is threatened with extinction. The honey badger or ratel (*Mellivora capensis* Storr, 1780) once enjoyed wide ranges of habitats extending through the whole forest areas in India. However, its populations are now dwindled. Among viverridae, the Malabar civet (*Viverra civettina* Blyth, 1862) reported mainly from the coastal areas of Malabar (Kerala); the spotted linsang or the tiger civet (*Prionodon pardicolor* Hodgson, 1842) and the binturong or the bear cat (*Arctictis binturong* Raffles, 1822), restricted to the eastern Nepal to Arunachal Pradesh, are threatened with extinction mainly due to habitat loss. Among hyena, population of the striped hyena (*Hyaena hyaena* Linnaeus, 1758) is considerably reduced.

Table 1. Threatened mammals of India*

Common name	Latin name	IUCN status
Asiatic golden cat	*Felis temmincki*	I
Asiatic lion	*Panthera leo persica*	EN
Asiatic wild dog or Dhole	*Cuon alpinus*	VU
Asiatic black bear	*Selenarctos thibetanus*	EN
Banteng	*Bos javanicus*	VU
Bengal fox	*Vulpus bengalensis*	I
Brow-antlered deer	*Cervus eldi eldi*	VU
Brown bear	*Ursus arctos*	VU
Brown palm civet	*Paradoxurus jerdoni*	VU
Clawless otter	*Anonyx cinerea*	K
Clouded leopard	*Neofelis nebulosa*	VU
Common otter	*Lutra lutra*	VU
Desert cat	*Felis libyca ornata*	EN
Dugong or Sea cow	*Dugong dugon*	VU
Ganges river dolphin	*Platanista gangetica*	VU
Gaur	*Bos gaurus*	VU
Golden langur	*Trachypithecus geei*	R
Goral	*Nemorhaedus goral*	VU
Great Indian rhinoceros	*Rhinoceros unicornis*	EN
Grey Indian wolf	*Canis lupus*	VU
Himalayan Whitetoothed shrew	*Crocidura attenuata*	VU
Himalayan marten	*Martes flavigula*	I
Himalayan musk deer	*Moschus chrysogaster*	VU
Himalayan shrew	*Soriculus nigrescens*	VU
Hispid hare	*Caprolagus hispidus*	EN
Hoolock gibbon	*Hylobates hoolock*	EN
Asian elephant	*Elephas maximus*	EN

Common name	Latin name	IUCN status
Indian wild ass	*Equus hemionus khur*	T
Jackal	*Canis aureus*	VU
Kashmir stag or Hangul	*Cervus elaphus hanglu*	EN
Leopard	*Panthera pardus*	T
Lesser panda	*Ailurus fulgens*	VU
Lion-tailed macaque	*Macaca silenus*	EN
Malabar civet	*Viverra civettina*	EN
Marbled cat	*Felis marmorata*	I
Markhor	*Capra falconeri*	EN
Nayan	*Ovis ammon hodgsoni*	EN
Nilgiri langur	*Trachypithecus johnii*	VU
Nilgiri marten	*Martes gwatkinsi*	I
Nilgiri tahr	*Hemitragus hylocrius*	VU
Phayre's langur	*Trachypithecus phayrei*	I
Pygmy hog	*Sus salvanius*	EN
Red fox	*Vulpes vulpes montana*	T
Rusty-spotted cat	*Prionailurus rubiginosa*	T
Serow	*Nemorhaedus sumatraensis*	EN
Sloth bear	*Melursus ursinus*	VU
Smooth Indian otter	*Lutra perspicillata*	K
Snow leopard	*Uncia uncia*	EN
Stump-tailed macaque	*Macaca arctoides*	VU
Barasingha or Swamp deer	*Cervus duvauceli duvauceli*	EN
Takin	*Budorcas taxicolor*	I
Tibetan wild ass	*Equus hemionus kiang*	VU
Tiger	*Panthera tigris*	EN

EN = Endangered, VU = Vulnerable, I = Indeterminate, R = Rare, K = Insufficiently known, T = Threatened

*Modified from Wildife Institute of India Report, 2010

Among the 15 species of cats seen in different forest habitats in India, population of the tiger was almost reduced to a meagre number of around 1,500 in the early 1970s. The wildlife authorities were worried about the alarming situation of this magnificent carnivore. Subsequently, the Government of India implemented the 'Project Tiger' to save the tiger population in India with the help of international conservation agencies. As a result of the serious efforts under this programme, the tiger population has increased to the present level of around 5,000 heads. The leopard has a wide distribution in the Indian subcontinent. Now and then, they are also reported from villages and even in townships as occasional visitors. When there are less prey animals in the natural forests, they are forced to roam around villages and townships, and lift domestic animals, particularly sheep and domestic dogs. They have been widely persecuted for their beautiful skin, which has high market in Europe. The snow leopard, which is restricted to the Himalayas, is also threatened with extinction. The desert cat (*Felis silvestris* Schreber, 1777), lynx (*Lynx lynx* Linnaeus, 1758), leopard cat (*Prionailurus bengalensis* Kerr, 1792), pallas cat (*Otocolobus manul* Pallas, 1776), rusty spotted cat (*Prionailurus rubiginosus* Geoffroy Saint-Hilaire, 1831), fishing cat (*Prionailurus viverrinus* Bennet, 1833*)*, Asian golden cat (*Pardofelis temminckii* Vogors and Horsfield, 1827), marbled cat *(Pardofelis marmorata* Martin, 1836) and the clouded leopard (*Neofelis nebulosa* Griffith, 1821) are threatened as a result of diverse human activities in their natural habitats. The magnificent Indian cheetah is already extinct in its natural habitats. They were reported to have roamed all over the geographic regions of India during the earlier periods of 1900s, but the population was subsequently reduced and the last sighting of this species in its natural habitat was in 1948. They were also hunted for their beautiful skin for illegal international trade. Scientists are now trying to use captive individuals of this species to breed and to reintroduce them into some of their original habitats in India.

The dugong (*Dugong dugon* Müller, 1776) was seen in good numbers in the Gulf of Cutch, Gulf of Mannar, coastal areas of Malabar, and in the Andaman Nicobar Islands, but now, they are rare and it is difficult to locate them in any of these areas.

The order Proboscidae is represented by only one species in India; i.e., the Asiatic elephant. Males of this species (tusker) have been hunted for tusks. Further, habitat destruction resulted in the reduction of availability of food, and elephant populations have been reduced all over its former ranges. Its African counterpart, the African elephant (*Loxodonta africana* Blumenbach, 1797) differs from the Asian elephant by having tusks in both males and females. Among the males of Asian elephants, there are also tuskless males.

Among Perissodactyla, the great Indian one-horned rhinoceros had a golden era in the first half of the last century in the Gangetic grasslands.

However, their number is considerably reduced to about 1,000 individuals in the recent past and the distribution is restricted to northern Bengal and in some low lands of Assam. They have been hunted for their costly horns, which are supposed to have aphrodisiac value. The lesser one-horned rhinoceros (*Rhinoceros sondaicus* Desmarest, 1822) and the Asiatic two-horned rhinoceros (*Didermoceros sumatrensis* Fischer, 1814) have already been extinct from their natural habitats. Distribution of the Indian wild ass (*Equus hemionus* Pallas, 1775) is restricted to the saline flats of Rann of Kutch, and is endangered. The Tibetan wild ass (*Equus kiang* Moorcroft, 1841) is restricted to Ladakh and Jammu and Kashmir in its distribution. However, its status is now improved and is considered as 'Least Concern'.

There are 32 species of deer in the forests of India, of which 20 are threatened with extinction. Kashmir stag or the hangul (*Cervus Canadensis* Erxleben, 1777), is now restricted to the northern regions of the Kashmir Valley with only about 160 heads, and is critically endangered. Swamp deer or the barasingha (*Rucervus duvaucelii* G. Cuvier, 1823) is seen in some forests of Madhya Pradesh, Uttar Pradesh and in Assam. They have a population of less than 2,500 individuals. Eld's deer or the brow-antlered deer (*Panolia eldi* M'Clelland, 1842) is strictly restricted to the reed forests and grasslands of Manipur and in the Loktak lake areas. This species was considered as extinct, but they were sighted again in some of its former ranges. Due to strict conservation measures, they are recovering from the threat of extinction, even though their population is less. The alpine musk deer (*Moschus chrysogaster* Hodgson, 1839) and the dwarf musk deer or the Chinese forest musk deer (*M. berezovskii* Flerov, 1929) are also listed as endangered. They have been hunted for their meet and musk. The population of the mouse deer (*Tragulus meminna* Milne-Edwards, 1864), which is the lone representative of Tragulidae in India, is also very low. Among the pigs, the Andaman wild pig (*Sus scrofa andamanensis* Linnaeus, 1758), the Nicobar wild pig (*S. scrofa nicobaricus* Linnaeus, 1758) are endemics to their respective islands. The pigmy hog (*Porcula salvania* Hodgson, 1847) is now restricted to Assam with its critically-endangered, small population of around 150 individuals.

Among bovidae, there are 21 species in India of which 14 are listed under threatened categories. These are as follows: Blackbuck (*Antelope cervicapra* Linnaeus, 1758), Tibetan antelope or chiru (*Pantholops hodgsonii* Abel, 1826), Tibetan gazelle (*Procapra picticaudata* Hodgson, 1846), four-horned antelope or chowsinga (*Tetracerus quadricornis* Blainville, 1816), gaur (*Bos gaurus* Smith 1827), wild yak (*Bos grunniens* Linnaues, 1766), wild water buffalo (*Bubalus arnee* Kerr, 1792), urial or shapu (*Ovis orientalis* Linnaeus 1758), merkhor (*Capra falconeri* Wagnet,

1839), Himalayan ibex (*C. ibex* Linnaeus, 1758), argali or the mountain sheep (*Ovis ammon* Linnaeus 1758), serow (*Capricornis atraensis* Ogilby, 1836), takin (*Budorcas taxicolor* Hodgson, 1850) and Nigiri tahr.

Among the four species of hares seen in the Indian subcontinent, the Assam hare or the hispid hare (*Caprolagus hispidus* Pearson, 1839) is endangered. They are seen only in the Brahmaputra Valleys, the Tarai and in Duars. Among the 95 species of rodents seen in India, the Travancore flying squirrel (*Petinomys fuscocapillus* Jerdon, 1847), woolly flying squirrel (*Eupetaurus cinereus* Thomas, 1888) and Hodgson's giant flying squirrel (*Petaurista magnificus* Hodgson, 1836) are threatened.

All the 25 species of Cetaceans represented in the Indian subcontinent are threatened with extinction. They include various species of whales and dolphins seen in the oceans and rivers.

5.2.8. Threatened Birds of India

India has a rich avifauna of over 1,200 species. In addition, several species of migratory birds make seasonal visits to various parts of India. Some of them are pests of agriculture, but as most of them feed on insects, they are equally good in controlling noxious pests. Therefore, the sparrow extermination programme launched by the Chinese Government in the 1960s was dispensed of. Though eagles, owls and crows appear to be harmful, they are, in fact, highly useful to human beings as they feed on human waste, and thus, act as environmental cleaners.

Rahmani (2012) gives the best account of the threatened bird of India with recommendations for their long-term conservation Table 2 gives the names of birds in India listed by IUCN in its threatened categories. The great Indian bustard (*Ardeotis nigriceps* Vigors, 1831) is one of the typical examples of birds facing extermination due to poaching. They weigh 15–20 kg and live in small flocks in certain parts of Rajasthan and Gujarat. They have been attracting sport hunters even from outside the country, who use trained falcons to locate and chase bustards. This species was once widespread with good populations in the dry plains of the northern regions of the Indian subcontinent, particularly in India and Pakistan. However, their population has depleted to about 250 individuals, and at present seen mainly in the central and western India in few numbers (Rahmani and Manakadan 1990). Hunting and loss of natural habitats were the main causes for the decline of its population. Agricultural development including canal irrigation, especially in the arid areas in Rajasthan resulted in the disappearance of this bird from such regions (Rahmani and Soni 1997).

Table 2. Threatened birds of India*

Common name	Latin name	IUCN status
Pink-headed duck	*Rhodonessa caryophyllacea*	CR
White-rumped vulture	*Gyps bengalensis*	CR
Long-billed vulture	*Gyps indicus*	CR
Himalayan quail	*Ophrysia superciliosa*	CR
Siberian crane**	*Grus leucogeranus*	CR
Jerdon's courser	*Rhinoptilus bitorquatus*	CR
Forest owlet	*Athene blewitti*	CR
White-bellied heron	*Ardea insignis*	EN
Oriental stork	*Ciconia boyciana*	EN
Greater adjutant	*Leptoptilos dubius*	EN
White-headed duck**	*Oxyura leucocephala*	EN
White-winged duck	*Cairina scutulata*	EN
Great Indian bustard	*Ardeotis nigriceps*	EN
Bengal florican	*Houbaropsis bengalensis*	EN
Lesser florican	*Sypheotides indica*	EN
Nordmann's greenshank**	*Tringa guttifer*	EN
Rufous-breasted laughing thrush	*Garrulax cachinnans*	EN
Spot-billed pelican	*Pelecanus philippensis*	VU
Lesser adjutant	*Leptoptilos javanicus*	VU
Lesser white-fronted goose	*Anser erythropus*	VU
Red-breasted goose	*Branta ruficollis*	VU
Baikal teal	*Anas formosa*	VU
Marbled teal	*Marmaronetta angustirostris*	VU
Baer's pochard**	*Aythya baeri*	VU
Palla's sea eagle	*Haliacetus leucoryphus*	VU
Greater spotted eagle**	*Aquila clanga*	VU
Imperial eagle**	*Aquila heliaca*	VU
Lesser kestrel**	*Falco naumanni*	VU
Nicobar scrubfowl	*Megapodius nicobariensis*	VU
Swamp francolin	*Francolinus gularis*	VU

Common name	Latin name	IUCN status
Manipur bush quail	*Perdicula manipurensis*	VU
Chestnut-breasted partridge	*Arborophila mandellii*	VU
Western tragopan	*Tragopan melanocephalus*	VU
Blyth's tragopan	*Tragopan blythii*	VU
Sclater's monal	*Lophophorus sclateri*	VU
Cheer pheasant	*Catreus wallichi*	VU
Hume's pheasant	*Syrmaticus humiae*	VU
Green peafowl	*Pavo muticus*	VU
Sarus crane	*Grus antigone*	VU
Black-necked crane	*Grus nigricollis*	VU
Hooded crane	*Grus monacha*	VU
Masked finfoot**	*Heliopais personata*	VU
Sociable lapwing**	*Vanellus gregarius*	VU
Wood snipe	*Gallinago nemoricola*	VU
Spoon-billed sandpiper**	*Eurynorhynchus pygmeus*	VU
Indian skimmer	*Rhynchops albicollis*	VU
Pale-backed pigeon**	*Columba eversmanni*	VU
Nilgiri wood pigeon	*Columba elphinstonii*	VU
Pale-capped pigeon	*Columba punicea*	VU
Dark-rumped swift	*Apus acuticauda*	VU
Rufous-necked hornbill	*Aceros nipalensis*	VU
Narcondam hornbill	*Aceros narcondami*	VU
Yellow-throated bulbul	*Pycnonotus xantholaemus*	VU
Nicobar bulbul	*Hypsipetes nicobriensis*	VU
Grey-sided thrush**	*Turdus feae*	VU
Rusty-bellied shortwing	*Brachypteryx hyperythra*	VU
White-bellied shortwing	*Brachypteryx major*	VU
White-browed bushchat	*Saxicola macrorhyncha*	VU
White-throated bushchat**	*Saxicola insignis*	VU
Marsh babbler	*Pellorneum palustre*	VU
Rusty-throated wren babbler	*Spelaeornis badeigularis*	VU

Common name	Latin name	IUCN status
Tawny-breasted wren babbler	*Spelaeornis longicaudatus*	VU
Snowy-throated babbler	*Stachyris oglei*	VU
Jerdon's babbler	*Chrysomma altirostre*	VU
Slender-billed babbler	*Turdoides longirostris*	VU
Black-breasted parrotbill	*Paradoxornis flavirostris*	VU
Grey-crowned prinia	*Prinia cinereocapilla*	VU
Bristled grass warbler	*Chaetornis striatus*	VU
Broad-tailed grass bird	*Schoenicola platyura*	VU
Kashmir flycatcher	*Ficedula subrubra*	VU
White-naped tit	*Parus nuchalis*	VU
Beautiful nuthatch	*Sitta formosa*	VU
Green avadavat	*Amandava formosa*	VU
Yellow weaver	*Ploceus megarhynchus*	VU
Dalmatian pelican**	*Pelecanus crispus*	VU
Persian shearwater**	*Puffinus persicus*	NT
Oriental darter	*Anhinga melanogaster*	NT
Painted stork	*Mycteria leucocephala*	NT
Black-necked stork	*Ephippiorhynchus asiaticus*	NT
Black-headed ibis	*Threskiornis melanocephalus*	NT
Lesser flamingo	*Phoenicopterus minor*	NT
Ferruginous duck	*Aythya nyroca*	NT
White-tailed eagle**	*Haliaeetus albicilla*	NT
Lesser fish eagle	*Ichthyophaga humilis*	NT
Grey-headed fish eagle	*Ichthyophaga ichthyaetus*	NT
Cinereous vulture	*Aegypius monachus*	NT
Red-headed vulture	*Sarcogypus calvus*	NT
Nicobar serpent eagle	*Spilornis minimus*	NT
Andaman serpent eagle	*Spilornis elgini*	NT
Pallid harrier	*Circus macrourus*	NT
Nicobar sparrow hawk	*Accipiter butleri*	NT
White-cheeked patridge	*Arborophila atrogularis*	NT

Common name	Latin name	IUCN status
Satyr tragopan	*Tragopan satyra*	NT
Tibetan eared-pheasant	*Crossoptilon harmani*	NT
Houbara bustard	*Chlamydotis undulata*	NT
Asian dowitcher	*Limnodromus semipalmatus*	NT
Beach thick-knee	*Esacus magnirostris*	NT
Black-bellied tern	*Sterna acuticauda*	NT
Andaman wood pigeon	*Columba palumbodies*	NT
Andaman cuckoo dove	*Macropygia rufipennis*	NT
Nicobar pigeon	*Caloenas nicobarica*	NT
Nicobar parakeet	*Psittacula caniceps*	NT
Long-tailed parakeet	*Psittacula longicauda*	NT
Andaman scops owl	*Otus balli*	NT
Andaman hawk owl	*Ninox affinis*	NT
Ward's trogon	*Harpactes wardi*	NT
Blyth's kingfisher	*Alcedo hercules*	NT
Brown-winged kingfisher	*Pelargopsis amauropterus*	NT
Malabar pied-hornbill	*Anthracoceros coronatus*	NT
Great Indian hornbill	*Buceros bicornis*	NT
Brown hornbill	*Anorrhinus tickelli*	NT
Yellow-rumped honeyguide	*Indicator xanthonotus*	NT
Andaman woodpecker	*Dryocopus hodgei*	NT
Nilgiri pipit	*Anthus nilgiriensis*	NT
Firethroat**	*Luscinic pectardens*	NT
Chest-backed laughing thrush	*Garrulax nuchalis*	NT
Grey-breasted laughing thrush	*Garrulax jerdoni*	NT
Rufous-throated wren warbler	*Spelaeornis caudatus*	NT
Wedge-billed wren babbler	*Spenocichla humei*	NT
Giant babax	*Babax waddelli*	NT
Rufous-vented prinia	*Prinia burnesii*	NT
Long-billed bush warbler	*Bradypterus major*	NT
Rufous-rumped grassbird	*Graminicola bengalensis*	NT

Common name	Latin name	IUCN status
Black and rufous flycatcher	*Ficedula nigrorufa*	NT
Nilgiri flycatcher	*Eumyias albicaudata*	NT
Andaman drongo	*Dicrurus andamanensis*	NT
Andaman treepie	*Dendrocitta bayleyi*	NT
Andaman crake	*Rallina canningi*	DD
Nicobar scops owl	*Otus alius*	DD

CR = Critically endangered, EN = Endangered, VU = Vulnerable,
NT = Near Threatened, DD = Data Deficient
**Non-breeding visitor
*Modified from Wildlife Institute of India Report (2010)

The tragopans of the Himalayas are another endangered group of birds in India. Among the four species of them in India, *Tragopan blythii* Cuvier, 1829 is the most endangered species. They were poached for their flesh. To save this bird from extinction, the Kohima Zoo initiated a breeding programme in 1973. They live in single, pairs or even in small family flocks. Wide use of pesticides in agricultural practices has also affected them adversely. Several species of ducks such as white-winged wood duck (*Asarcornis scutulata* Müller, 1842), pink-headed duck (*Rhodonessa caryophyllacea* Latham, 1790) are also threatened. The Jerdon's courser (*Rhinoptilus bitorquatus* Blyth, 1848), which was recorded in small flocks in Pennar Valley and Godavari basin, is also endangered.

Shooting and habitat destruction are the major factors causing depletion of populations of the Cheer pheasant (*Catreus wallichii* Hardwicke, 1827) of the northern India. Even though not threatened, the status of the national bird of India, the peacock (*Pavo cristatus* Linnaeus, 1758) is also not promising due to habitat loss. At present, they are safe only in certain parts of Gujarat and Rajasthan where they are left undisturbed. In the southern States of India, including Kerala, their population has come down mainly due to poaching and habitat destruction. They can be sighted in Bandipur and Parambikulam Wildlife Sanctuaries.

The great Indian hornbill (*Buceros bicorni* Linnaeus, 1758) (Fig 48) of the Western Ghats and the eastern Himalayas is a 'near threatened' species. A good population of them is still seen in Silent Valley forests of Kerala, which is one of the comparatively less-disturbed tropical evergreen habitats in the whole of South India. They are also seen in Periyar Tiger Reserve in Kerala.

Fig 48. The great Indian hornbill is seen feeding on a *Ficus* tree (Photo: S Anilkumar)

5.2.9. Threatened Reptiles of India

Many species of reptiles found in India are also on the verge of extinction. Crocodiles, monitor lizards, pythons and a few species of snakes are listed in the IUCN Red Data Book of animals (Table 3). They deserve special attention for their survival. Crocodile populations are now strictly protected all over India. The Crocodile Breeding and Training Centre in Hyderabad, and the Madras Crocodile Bank associated with the Madras Snake Park are well-known programmes associated with crocodile conservation in India. Crocodiles are seen in some of the reservoirs and natural forest areas, where the Forest and Wildlife Department of the concerned State is responsible for the conservation activities.

Crocodiles were killed in large numbers to sustain the skin trade, and for the export of their lucrative hard coat. The population of estuarine crocodile or saltwater crocodile (*Crocodylus porosus* Schneider, 1801), which was once abundant all over India, has been depleted. This is the largest of all living reptiles as well as the largest terrestrial and riparian predator in the world. However, conservation efforts have helped to improve its population status, and became one of the 'Least Concern' species as per IUCN status. The marsh crocodile or the mugger crocodile (*Crocodylus palustris* Lesson, 1831) and the gharial (*Gavialis gangeticus* Gmelin, 1789) are threatened crocodile species in India.

Table 3. Threatened reptiles of India*

Common name	Latin name	IUCN status
Asian giant tortoise	*Manouria emys*	VU
Assam roofed turtle	*Pangshura sylhetensis*	EN
Black pond turtle	*Geoclemys hamiltonii*	I
Central Asian/Oxus cobra	*Naja naja oxiana*	K
Cochin forest cane turtle	*Geoemyda silvatica*	VU
Elongated tortoise	*Indotestudo elongata*	K
Estuarine crocodile	*Crocodylus porosus*	VU
Gharial (Long-snouted crocodile)	*Gavialis gangeticus*	EN
Green sea turtle	*Chelonia mydas*	EN
Hawksbill turtle	*Eretmochelys imbricata*	EN
Indian egg-eating snake	*Elachistodon westerwanii*	R
Indian python	*Python molurus molurus*	VU
Leathery turtle	*Dermochelys coriacea*	EN
Loggerhead	*Sternotherus minor*	VU
Mugger (Marsh crocodile)	*Crocodylus palustris*	VU
Olive Ridley turtle	*Lepidochelys olivacea*	EN
Red-crowned roofed turtle	*Kachuga kachuga*	I
Three-keeled land tortoise	*Melanochelys tricarinata*	I
Travancore tortoise	*Indotestudo travancorica*	VU
Yellow monitor lizard	*Varanus flavescens*	I
Nikhili's Kukri snake	*Oligodon nikhili*	DD

DD = Data Deficient, EN = Endangered, I = Indeterminate,
K = Insufficiently known, R = Rare, T = Threatened, VU = Vulnerable
*Modified from Wildlife Institute of India Report (2010)

There are around 70 species of the Monitor lizards or varanus (*Varanus* species Merrem, 1820) in the world. In general, their populations are dwindling fastly. They have been killed for their flesh, which is considered to have medicinal properties. The python (*Python moluru* Linnaeus, 1758) is yet another species of reptile, which is facing problems for its survival. Monitor lizards are also killed for flesh and fat. Habitat destruction is another threat for the survival of both varanus and python in India. Several species of snakes are also killed for their skin in India.

Box 5.1 (Figs 49 to 54) gives an account of wildlife research activities of Wildlife Institute of India and Box 5.2 (Figs 55 to 56) gives those of Bombay Natural History Society. These two boxes cover recent major wildlife research projects in India.

Box 5.1

Wildlife Research in India in Two and Half Decades by the Wildlife Institute of India – Achievements at a Glance

K Sankar

Wildlife Institute of India, P.O. Box No. 18, Chandrabani,
Dehradun – 248 001, Uttarakhand, India
(E-mail: sankark@wii.gov.in)

One of the primary objectives of Wildlife Institute of India (WII) is to conduct and coordinate wildlife research in selected priority areas in the country and evolve appropriate management strategies for conservation of fauna and flora. Research activities are the means by which the faculty of the institute keeps abreast of current field situations, management needs and research trends. This ultimately helps in strengthening teaching and training programmes of the institute. Research activities of WII are guided by the Training, Research and Academic Council (TRAC) represented by eminent naturalists, scientists, wildlife managers and decision makers.

In the last two and half decades, the institute has executed about 190 research projects covering ecological studies, inventory surveys, management-specific studies and landscape-level studies. About 10% of these projects were on biology/ecology of species, 30% on floral and faunal inventories, 50% on applied research and 10% on human aspects. Wildlife Institute of India has published 190 reports, training manuals and books, and more than 900 peer-reviewed scientific papers. It has produced more than 400 professionals in wildlife science. Currently, the institute is conducting 48 research projects in all major biogeographic zones of the country.

Fig 49. Tiger on a chital kill

To date, several surveys covering highly-endangered species, taxonomic groups and their distribution range in different biogeographic zones have been undertaken. These surveys have provided valuable information on the conservation status and distribution of species, use of habitat, potential threats, identification of important breeding sites, range extensions and discovery of new species and subspecies. Results of these surveys have ultimately helped in effective planning and prioritizing conservation measures. For instance, the survey on snow leopards in the Trans-Himalayan region has identified potential sites for conservation, generated baseline information and standardized methods to assess indirect evidence. The other important surveys are – the surveys for takin in Arunachal Pradesh; Malabar civet and herpetofauna in Western Ghats; clouded leopard and pheasants in north-east India; orchids in Uttaranchal and Bihar; and the country-wide survey for the sarus cranes. The herpetofaunal survey in the Western Ghats described a new species of flying frog *Rhacophorus pseudomalabaricus,* a new species of burrowing frog *Nasikabatrachus sahyadrensis* and a new species of skink, the barred tree skink *Dasia johnsinghi* (Harikrishnan et al 2012). In addition, a new species of agamid lizard or short-tailed Bay Island forest lizard (*Coryphophylax brevicaudus*) in Andaman and Nicobar Islands has also been described (Harikrishnan et al 2012). The pheasant survey in Arunachal Pradesh described a new subspecies of Sclater's monal pheasant (*Lophophorus sclateri arunachalensis*) (Suresh Kumar and Singh 2004). The survey of potential sites for re-introduction of Asiatic lion in Western and Central India led to identification of a second home for the Asiatic lion in Kuno-Palpur Wildlife Sanctuary, Madhya Pradesh. The outcome of the survey on nesting sites of Olive Ridley sea turtle along the Odisha coast facilitated planning for long-term intensive study. The status survey of elephants in Garo Hills, Meghalaya led to the identification of three crucial corridors for wild elephants and also provided insight on the potential threats of developmental projects.

Species-specific or habitat-specific studies on tiger, lion, leopard, snow leopard, caracal, jungle cat, wolf, dhole, hyaena, sloth bear, elephant, rhino, gaur, black buck, wild buffalo, wild ass, Nilgiri and Himalayan tahr, ibex, musk deer, swamp deer, sangai, wild pig, hoolock gibbon, lionmtailed macaque, Phayre's leaf monkey, Nilgiri langur, giant squirrel, otter, Sarus crane, pheasant, hornbill, megapode, freshwater turtle, tortoise, spider and tiger beetle are being persued or have been completed. The ecological study on the endangered grizzled giant squirrel in Tamil Nadu helped in the creation of Srivilliputtur Grizzled Giant Squirrel Sanctuary.

Fig 50. Radio-tracking

Studies on Asiatic elephants covering ecology, ranging pattern and conservation were addressed by seven different projects in north-west Uttar Pradesh, Bihar, West Bengal and Meghalaya. Findings of these studies have significantly contributed to the conservation of the endangered elephants by providing insight into crucial corridors, elephant–human conflicts and impact of developmental activities on elephants and their habitats. The studies on snow leopard, ibex, musk deer, blue sheep, hangul, wild yak, Tibetan gazelle and Himalayan tahr of high-altitude ecosystems (Ladakh, Nanda Devi, Pin Valley, Dachigam) have covered different aspects such as distribution, status, biology, habitat use, feeding and interactions with other sympatric species. They have provided insights into biology, ecology and management needs of these critically endangered species, and the findings of these studies have been incorporated into the management plans of respective Protected Areas (PAs).

The long-term research projects addressed conservation and management of endangered species, critical habitats and fragile ecosystems. Five successive studies in Gir ecosystem on ecology of lion, ungulate–habitat interactions, impact of management practices, and survey of potential sites for re-introduction of the lion resulted in understanding of changing trends in the lion population, prey species and habitat dynamics. These findings have been incorporated into the management plan for the Gir Conservation Area, and in identifying a second home and setting up a framework for re-introduction and monitoring of the population. Research on rhinoceros included monitoring of re-introduced population in Dudhwa Tiger Reserve, Uttar Pradesh, and management of rhinos in West Bengal and Assam. Studies on their population structure, dynamics, habitat use, effect of tourism and population genetics contributed to conservation of this critically endangered species.

Fig 51a. Caracal

Fig 51b. Asiatic lion

Fig 51c. Nilgiri tahr

Five projects addressed different aspects of wetland ecology. These projects involved studies on mugger crocodiles in Andhra Pradesh and otter ecology in river Chambal, monitoring gharial, survey of sarus and common cranes and associated fauna, and resulted in valuable contribution to improve our knowledge on the ecology of the concerned species, and in general about wetlands. Relevant management recommendations are being absorbed in management strategies. Wildlife Institute of India also worked on assessment of current status of trading threatened and protected marine flora and fauna in India, preparation of a report on the status of Indian Coastal and Marine Environment and a network of Marine Protected Areas, and preparation of Management Plan for Gulf of Mannar Biosphere Reserve, Tamil Nadu.

Fig 52a. Asiatic wild ass

Fig 52b. Olive Ridley sea turtle with satellite transmitter

Fig 52c. Asiatic wild buffalo

A 5-year WII-USFS collaborative project was carried out in four field sites across the country – Anamalai Conservation Area, Garo Hill Conservation Area, Satpura Conservation Area and Terai Conservation Area – to evolve pointers for integrated land-use management strategies in support of the role of PAs in maintaining biological diversity. This was a very significant initiative in evolving a system of integrated land-use management to support conservation of biological diversity. It helped to establish 'hierarchical scales of assessments and management' by considering site-level to regional-level patterns and issues. Another 5-year WII-USFWS project executed by the WII with an objective to develop an 'Indian Wildlife Health Co-operative Programme' was a major initiative to boost the wildlife health care discipline. Five wildlife health co-operatives (Anand, Chennai, Hissar, Guwahati and Jabalpur) have been established in different zones of the country. Each of these centres has a veterinarian – wildlife expert trained at WII. These centres develop and impart training for veterinarians and extend help within the region to wildlife managers on matters of wildlife health issues. A WII-USFWS collaborative project of national and regional significance is related to 'Development of a Wildlife Forensic Capacity at WII'. Between 1987 and 2011, more than 2,000 offence cases were processed with more than 2,500 samples received from various enforcement agencies.

The discipline of 'conservation genetics' is gradually gaining attention in India. Wildlife Institute of India has made a modest beginning in using DNA technology for addressing some pertinent questions relevant to conservation. The application of DNA technology has been successfully demonstrated as an integral component of major research studies on the Asiatic wild buffalo, rhinoceros and wolf. A recently-concluded collaborative study with the National Institute of Immunology (NII), New Delhi was able to establish the pattern and extent of genetic variation at the DNA-sequence level among the rhino population in Jaldapara Wildlife Sanctuary. This was done using conventional DNA profiling and P-PCR mediated approaches. The findings revealed a high level of genetic homogeneity. The development of a cloned probe pSS(R)2 is the most significant achievement of this study. The total sequence of the clone has been deposited in the gene bank. The probe is species-specific and can be used successfully for ascertaining the origin of tissue, blood, horn, hoof or skin samples from the rhino genome. The DNA studies conducted on Indian wolf revealed that the Indian wolves are indeed the most ancient lineage from where the other species of wolves and dogs in the world branched off (Dinesh et al 2004). The DNA studies on sea turtles, tigers and herpetofauna are currently being carried out by the WII in collaboration with the Centre for Cellular and Molecular Biology (CCMB), Hyderabad.

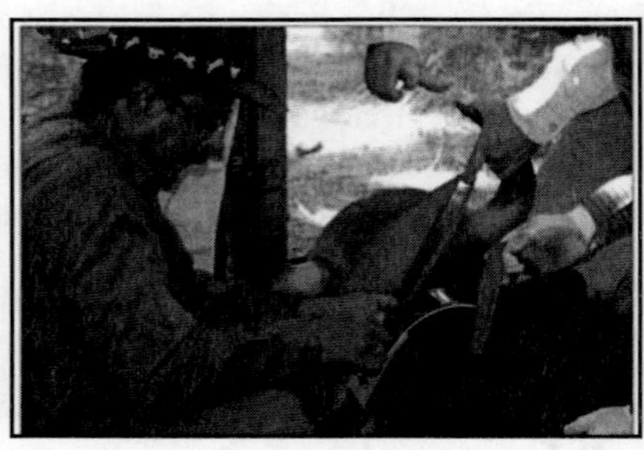

Fig 53a. Radio-collaring a gaur

Fig 53b. Indian wolf

Fig 53c. Asiatic one-horned Rhinoceros

The application of Remote Sensing and GIS technology in research projects helped in identification and assessment of crucial corridors, resource mapping, habitat characterization, landscape dynamics, developing spatial database, monitoring and predictive modelling and decision making. The application of chemical immobilization and radio-telemetry techniques are becoming an integral part of major researches. These techniques have been successfully demonstrated on over 40 different species such as lion, tiger, leopard, snow leopard, jungle cat, wolf, civet, dhole, hyaena, sloth bear, elephant, gaur, chital, sambar, nilgai, wild ass, ibex, Olive Ridley sea turtle, otter and others.

The institute was engaged in population estimation of 'tiger, co-predators, prey and habitat monitoring in all the States in the country having tiger habitats for two consecutive terms (2007 and 2011) (Jhala et al 2011). This exercise is first of its kind in the world involving such a huge landscape and manpower. The methodology for doing this has been developed by WII in association with National Tiger Conservation Authority (NTCA). The new methodology is a path-breaking endeavour, which has provided the Central and State governments a baseline data for monitoring ecological changes and formulation of appropriate policies and informed decision making.

A gradual shift from 'species-oriented research' to multidisciplinary and integrated ecosystem research was evident in the last fifteen years. These studies were being demonstrated in the Great Himalayan National Park (GHNP), Kalakad–Mundanthurai Tiger Reserve (KMTR), Sariska Tiger Reserve, Gir forest and Bhal ecosystem, Gujarat and Pench Tiger Reserve, Madhya Pradesh. Simultaneous researches on featured species and different ecological aspects have started reflecting a high level of scientific information during the initial period of three years. There is an urgent need to replicate the above coordinated research programmes in other prominent regions in the country. More emphasis has been given on research related to wildlife–human conflicts, *ex-situ* conservation, ecotourism, interpretation and environmental impact assessment during the last ten years.

Fig 54. Asiatic wild dogs

Considerable progress and significant contributions in wildlife research in a short span of 25 years are evident. The research projects of WII have tried to cover different aspects of biology and ecology of many endangered species in the country. However, there are still large gaps considering country's vast diversity of ecosystems and species. There is a need to further strengthen the new initiatives; viz., impact of climate change, landscape ecology and dynamics, community ecology, nutritional ecology, wildlife forensics, conservation genetics, wildlife health, human–wildlife conflicts, socio-economic studies relevant to wildlife conservation, PA management and eco-development planning. Islands, coastal and marine ecosystems need exploration in view of their value as a biodiversity treasure. To strengthen field conservations, experimental studies on management of grassland, forest management, use of fire as a management tool, etc. need encouragement and support through greater involvement of PA officials. There is a need for representation of scientific community in policy and decision-making bodies at all levels, periodic wildlife research seminars and career opportunities for wildlife researchers and other human resources development issues.

More multidisciplinary, coordinated and long-term research in all the prominent Protected Areas needs to be conducted. Selected PAs in the country were able to attract national and international researchers even 2–3 decades ago, and since then, the tempo for research activities in these PAs have been maintained. These PAs are serving as 'learning centres'. There is a need to recognize these PAs and replicate several such models. Wildlife research is difficult, expensive, time consuming and full of risk. It calls for adequate administrative support, encouragement at all levels and greater budgetary support. There is an urgent need to use modern technologies and tools for promoting field conservation, to build field capacity for monitoring and research through participatory research, training and technology transfer, to promote active research networks with institutions or individuals globally and to assist the Ministry of Environment and Forests, Government of India and the Indian States in formulating research and monitoring protocols and policies. The recent tiger re-introduction in Sariska (Sankar et al 2010) and Panna Tiger Reserves and gaur re-introduction in Bandhavgarh Tiger Reserve (Sankar et al 2013) demonstrated the capabilities of WII in successful translocation of large cats and in mass capture of mega herbivores, and developed protocols for future endeavors.

References

Sharma DK, Maldonado JE, Jhala YV and Fleischer RC 2004 Ancient wolf lineages in India, *Proceedings of the Royal Society London B Supplement* **271** S1–S4

Harikrishnan S, Vasudevan K, Chandramouli SR, Choudhury BC, Dutta SK and Das I 2012 A new species of *Coryphophylax* Fitzinger Steindachner, 1867 (Sauria: Iguania: Agamidae) from the Andaman Islands, India, *Zootaxa* 3451 31–45

Harikrishnan S, Vasudevan K, Silva AD, Deepak V, Kar NB, Nanawadekar R Lalremruats A, Prasoona KR and Aggarwal RK 2012 Phylogeography of

Dasia Gray, 1830 (Reptilia: Scincidae), with the description of a new species from southern India, *Zootaxa* 3233 37–51

Jhala YV, Qureshi Q, Gopal R, and Sinha PR 2011 *Status of Tigers, Co-Predators, Prey in India*, New Delhi: Government of India, National Tiger Conservation Authority, Dehra Dun: Wildlife Institute of India

Sankar K, Qureshi O, Nigam P, Malik PK, Sinha PR, Mehrotra RN, Gopal R, Bhattacharjee S, Mondal K and Gupta S 2010 Monitoring of reintroduced tigers in Sariska Tiger Reserve, western India: Preliminary findings on home range, prey selection and food habits, *Tropical Conservation Science* **3** 301–318

Sankar K, Pabla HS, Patil CK, Nigam P, Qureshi Q, Navaneethan B, Manjreakar M, Virkar PS and Mondal K 2013 Home range, habitat use and food habits of re-introduced gaur (*Bos gaurus gaurus*) in Bandhavgarh Tiger Reserve, Central India, *Tropical Conservation Science* **6** 50–69

Kumar SR and Singh P 2004 A new subspecies of Sclater's monal *Lophophorus sclateri* from western Arunachal Pradesh, India, *Bulletin of the British Ornithologists Club* **124** 16–2

Box 5.2

Bombay Natural History Society – Conservation Activities and Recent Achievements

AR Rahmani

Bombay Natural History Society, Hornbill House,
Shaheed Bhagat Singh Marg, Mumbai 400 001
(E-mail: rahmani.asad@gmail.com)

Bombay Natural History Society (BNHS) has been the pioneer organization in wildlife research in India, since its inception in 1883. Its Mission Statement "Conservation of Nature, primarily Biological Diversity, through Actions based on Research, Education and Public Awareness" conveys the focus areas appropriately. Research at BNHS contributes towards identifying, monitoring and mitigating the adverse impacts of unscientifically planned and unsustainable developmental activities on natural environment. Its guiding principle is that development activities and conservation initiatives must be based on scientific truth. Towards this end, it is also engaged in networking, surveying and implementing conservation projects, mitigating adverse impacts of development projects and creating public awareness.

From eight amateurs and hunters who set up BNHS, it graduated into a members' organization conducting natural history research, and finally into a premier wildlife research organization in the country with talented professional scientists. The tradition of research at BNHS includes many significant past projects including mammal survey of India, ornithological surveys, Vernay scientific survey of eastern ghats and study of Asiatic lion in Gujarat. Research picked up further after the studies of Bharatpur and Point Calimere Sanctuaries in association with US Fish and Wildlife Service in the 1980s. Some of the major ongoing research and conservation projects of BNHS are the following:

Vulture Conservation Breeding Programme

It is a partnership between BNHS, UK-based Royal Society for the Protection of Birds, several State Governments in India and other research organizations. It includes three Vulture Conservation Breeding Centres in Haryana, West Bengal and Assam, where over 250 critically endangered white-backed, long-billed and slender-billed vultures are being taken care off. This includes hatching of chicks and raising them successfully. The decline in the number of vultures in South Asia by nearly 99% over the past decade was caused by veterinary painkiller diclofenac, which when ingested by vultures through treated cattle cadavers, caused death due to kidney failure. Efforts of BNHS resulted in ban on veterinary diclofenac. The programme also includes advocacy with governments, traders and cattle owners and identifying 'Vulture Safe Zones' for future re-introduction. The final challenge is to prevent other potentially-hazardous painkillers from entering the food chain and ensuring that diclofenac is totally eradicated.

Fig 55a. Vulture
(Photo: Abrar Ahmed)

Fig 55b. Great Indian Bustard
(Photo: Ashoka)

Migratory Birds

India, with its unique location and great biodiversity, is a favoured destination of many species of migratory birds. Bombay Natural History Society has a full-fledged 'Migratory Birds Study Centre' at Point Calimere Sanctuary in Tamil Nadu, dedicated to research of migratory birds that arrive there in large numbers. This includes bird ringing, fitting of satellite transmitters and monitoring movements of birds. Such studies, in association with local organizations, are also regularly carried out in States such as Odisha, Himachal Pradesh, Uttar Pradesh, West Bengal, Assam, Manipur and Punjab. The BNHS team has also visited Bangladesh to train local researchers in studying migratory birds. Ducks, geese and flamingos have been studied. Several interesting findings pertaining to migration patterns, migration routes, choice of resting places and feeding destinations and threats to birds have been documented till now.

Fig 56. Forest owlet (Photo: Girish Jathar)

Important Bird Area (IBA) Programme

It is a partnership between BNHS, India and UK-based Birdlife International. Till now, 466 IBAs have been identified across India and several more have been proposed. Establishment of Indian Bird Conservation Network (IBCN) – a network of individuals, organizations and governments – has enabled action at local, state and national levels. Indian Bird Conservation Network has conducted research with emphasis on conservation for several species of birds such as Forest Owlet, Himalayan Quail, Manipur Bush-Quail, Great Indian Bustard, Lesser Florican and Sarus Crane. In all States of India, coordinators have been appointed for monitoring and safeguarding IBAs. It is envisaged to have nodal persons for all IBAs. Till now, 250 nodal persons have been identified. Five Marine IBAs have also been identified; *viz.,* Pitti Island, Beleapani Reef, Gulf of Mannar/Palk Bay, Point Calimere and Vengurla Rocks. Several books have been published including 'Important Bird Areas of India', 'Potential and Existing Ramsar Sites in India' and 'Threatened Birds of India'.

Jerdon's Courser Project

Jerdon's Courser is a critically endangered land-bird known to survive only in a couple of districts of Andhra Pradesh in India. Confirmed records have been obtained only from a few sites in Sri Lankamaleswara Sanctuary in Kadapa District. Considered to be extinct for nearly a century, it was rediscovered in 1986. Endemism and loss of habitat have been threatening this species of bird. Bombay Natural History Society has been working on this project since 2000 in collaboration with other research and educational institutions and Andhra Pradesh Forest Department and has succeeded in diverting the Telugu-Ganga irrigation canal that would have cut right through its habitat; but other threats such as conversion of habitat for human use and plantation of non-native species continue to plague this highly elusive bird species. Bombay Natural History Society has been doing extensive field studies using camera traps and other modern tools to get first-hand evidence of the existence of this bird species in the region. Also, BNHS is doing advocacy and creating public awareness among the locals regarding conservation of this species. In November 2010, Andhra Pradesh Forest Department submitted a 'Species Recovery Plan' for Jerdon's Courser to the Ministry of Environment and Forests, Government of India, based on BNHS studies.

Satpuda Landscape Tiger Programme

Tiger, the national animal of India, has been the flagship species for conservation action in India. The central Indian landscape, which comprises Satpuda mountain range, Vidarbh region of Maharashtra and parts of Madhya Pradesh, has some of the most continuous and viable tiger habitats in the country. Bombay Natural History Society has been engaged in tiger conservation through Satpuda Landscape Tiger Programme (SLTP) in association with UK-based Born Free Foundation, several local and international organizations and corporates. This programme covers the tiger reserves of Melghat, Tadoba-Andhari and Pench. The scope of SLTP includes survey and field research of tigers and their habitats, their conservation, corridor development, education and public awareness, community participation, sustainable livelihoods, energy efficiency and networking. The programme also runs a mobile education unit and mobile health unit for the surrounding villages. It revives gobar gas units and advocates initiatives with the government and other entities.

Project Mangrove

Mangroves play a key role in controlling coastal erosion and in reducing intensity of cyclone and tsunami. They also act as breeding areas for fish and habitats for birds. With rapid urbanization and development along the coast, mangroves are under serious threat. Bombay Natural History Society has been working for mangrove conservation in Maharashtra and Gujarat. Eco-restoration of over 100 hectares of degraded mangroves in the Gandhar area near Surat, with corporate support, has been a success story. It has provided employment to nearly 100 locals. Apart from Gujarat, public awareness campaigns have been going on in the coastal districts of Maharashtra, reaching out to coastal communities, school and college students and government officers. Research is being undertaken in Ratnagiri district, which includes species composition, threats such as dumping wastes, bund building and industrial activities, and ways and means of mitigating the impacts. Recently, coastal plateaus that are rich in biodiversity are also being studied in the region.

Some of the other ongoing projects include ENVIS Centre for Avian Ecology, Conservation of Giant Clams (Lakshadweep), Kondhana Rat Study (Maharashtra) and Molluscan Taxonomy Studies (all along Indian coastline). Some of the recently concluded projects of BNHS include Caecilian Study Project (Maharashtra) and Elephant Migration Studies (Tamil Nadu and Karnataka).

5.3. NEPAL

A wildlife conservation-oriented management programme was initiated in Nepal in 1973 with joint financial and technical assistance of the United States of America and Food and Agricultural Organization. This programme was responsible to establish a separate Department of National Parks and Wildlife Conservation (within the Ministry of Forest) and a new National Parks and Wildlife Conservation Act in Nepal. Currently, in addition to the three Ramsar sites, Nepal has 17 protected areas including 10 national parks and one hunting reserve. There are over 1,120 species of non-flowering plants, 5,160 species of flowering plants, 181 species of mammals, 844 species of birds, 100 species of reptiles and 143 species of amphibians reported from Nepal. Two non-governmental organizations are actively engaged in nature conservation, *viz.*, King Mahendra Trust for Nature Conservation and International Centre for Integrated Mountain Development. Long-term research projects on lowland large mammals such as tiger, ungulates and rhinoceros were also implemented. As a result of these developments, conservation and management of wildlife in Nepal has become increasingly integrated with other land-use development programmes. A pioneering effort in this direction was the Annapurna Conservation Area Project, which encompasses all natural resources and human activities within a 4,000 km^2 area in north-central part of the country. Initiated and implemented by the King Mahendra Trust of Nature Conservation, this approach has been serving as a model for management of wildlife and other natural resources in Nepal (Mishra and Sherpa 1987, Rowell 1989).

5.3.1. Lowland Projects

The Royal Chitawan National Park in the Terai belt near India has been the main site of research on lowland mammals in Nepal. Outstanding data on the population ecology of tiger and its prey species in the region; *viz*., chital, sambar and Indian hog deer (*Hyelaphus porcinus* Zimmermann, 1780) were gathered by a number of researchers (Seidensticker 1976, McDougal 1977, Dinerstein 1979 a, b, c, Sunquist 1981, Tamang 1982, Mishra 1982, Smith 1984, Smith et al 1987, 1989). The population ecology of the endangered rhinoceros was studied by Laurie (1982) and later by Dinerstein and Wemmer (1988) and Dinerstein et al (1988). The problem of crop depredation by rhinos in Nepal was studied by Jnawali (1989).

The Royal Chitawan National Park has been successful in protecting the depleting populations of tiger and rhinoceros in Nepal's Terai region, but in achieving this success, there have been conflicts between the Park management and the local communities (Jnawali 1989). Restrictions imposed by the Park regulations and loss of livestock, crops and human lives due to wildlife were the principal causes of these conflicts. To resolve the conflicts in the Park, authorities have experimented by permitting local people to have access to the Park resources on a controlled basis, mainly allowing collection of thatching grass and firewood. The increased population of the rhinoceros in the area as a result of strict conservation measures was translocated to Bardia National Park in an attempt to build up another rhino population in the country, and also to safeguard against epidemic diseases that might affect the Chitwan population and to alleviate the serious problem of crop damage in the area (Jnawali 1989).

Dinerstein (1979 a, b) studied the ecological separation of several large herbivores in Karnali – Bardia in west-central Terai, with emphasis on the vegetation-exploitation patterns by chital, sambar deer and nilgai. Different feeding and anti-predator strategies could explain most of their separation. Chital and nilgai constituted 88% of the total herbivore biomass in the area. During the hot dry season, both the wild and domestic herbivores compete for the same forage despite having different feeding habits. Following improved protection after the establishment of the Royal Bindra National Park, the small local population of barasingha has increased rapidly, exerting a competitive pressure on other herbivores in the area.

The black buck is almost extinct in Nepal and only exists in a remnant herd of about 200 animals on mostly cultivated land adjacent to the Indian border (Bauer and Ellenberg 1989). Due to the increasing human pressure, attempts were made to translocate them and establish a viable population inside Bardia National Park. However, there was high predation pressure on them from tigers and possibly from wild dogs.

Schaaf (1978) studied habitat ecology of the main population of barasingha in Nepal in Suklaphanta Wildlife Reserve situated at the south-western corner of the country. The management programme was directed at maintaining open grasslands, a critical habitat element for this species, through a controlled burning programme. The traditional practice of grass cutting by local villagers was a disputed issue, but might be compatible with the habitat requirements of this species.

5.3.2. Mountain Research

The population ecology of the naur (*Pseudois nayaur* Hodgson, 1833) was studied in the west-central mountains of Nepal (Dhorpatan), with the objective of obtaining background data for a trophy-hunting programme. In high-quality habitats, yearlings were found to conceive, but annual lamb crops were often severely reduced due to unpredictable late spring, snow storms and predation (Wegge 1976, 1979, Wilson 1981). Dhorpatan Hunting Reserve yields substantial revenues from hunting fees from harvests kept within the sustainable yield of the different sub-populations of naur. Although a large number of domestic livestock, mainly sheep and goats also graze in this reserve, habitats were not adversely affected, except possible conflicts to wild stock from spring burning near lambing areas. Naur is widely distributed in the western mountains of Nepal where they constitute an important prey base for the endangered predator, snow leopard.

A radio-telemetry study on snow leopard was conducted adjacent to Shey Wildife Reserve in north-west Nepal, which revealed information on movement pattern and prey base (Jackson and Hillard 1986). Due to the populations of snow leopard and naur prey, the reserve was enlarged and developed as a National Park, including the Upper Langu area. Based on the findings of this investigation, a general habitat suitability model for snow leopard was developed.

Green (1986) evaluated the musk deer population and conservation status of the species throughout its ranges in the Himalayas, which included eastern Afghanistan, Pakistan, northern India, Nepal, Bhutan and Myanmar. He estimated that about 4,000 males of them were killed every year illegally, to extract musk. Habitat destruction has partly resulted from the mounting human population in the hilly regions. There is a potential of 56,000 km^2 area available for the musk deer in this region. There are around 30,000 musk deer in the whole of its ranges. They are hunted heavily for the musk, and hence, their population has been declining. The price of the musk would continue to be so for long as this natural product is used in pharmaceutics. Sustainable exploitation of this species may be possible, as the animal need not be killed for the extraction of the musk. Musk can be extracted from the musk gland by inserting a tube into the aperture, and a stream can be squeezed out by slight pressure on the gland. The wild populations can be live-trapped and released in the same place immediately after the extraction of the musk. This can be practiced with the help of local people so as to form a rural development programme. Even though a ban on the trade of musk was suggested, it would only reduce the value of the musk deer (Green 1986).

5.4. SRI LANKA

Among the nations in the Indian subcontinent, Sri Lanka has been in the forefront in achieving conservation objectives. There are over 500 protected areas in Sri Lanka comprising an extent of 17,67,000 ha, which is about 26.5% of the total area of the nation. When compared to other nations in Asia and elsewhere, Sri Lanka has the highest percentage of area under protective coverage. The Wildlife and Nature Protection Society of Sri Lanka and the Ceylon Bird Club have been actively engaged in conservation activities. There are over 90 species of mammals in Sri Lanka, of which 41 are threatened. Among the mammals in Sri Lanka, 16 are endemics. Among the biological diversity of Sri Lanka, 16% of the fauna and 23% of the flora are endemics.

Erdelen (1988) studied the problems related to conservation of natural forests and associated faunal communities in Sri Lanka and remarked that the problems related to the growth of human population and associated changes in the land-use pattern were serious threats to the natural communities and wildlife of the nation. The utilization of woody plants by elephants was analyzed in relation to the type of damage, woody plant size-classes and species (Ishwaran 1983). The percentage of damaged woody plants and the preference ratios for size-classes did not differ significantly between the two forest sites studied. Elephant damage on woody plants could become an important factor in effecting changes in the landscape, particularly in habitats where population densities are high. Eisenberg (1980) reviewed the literature on Asian elephants in Sri Lanka, and the status and distribution of the species in different areas of Sri Lanka were reported (Ishwaran 1984). It was stressed that effective conservation measures are essential for the future protection of this species in Sri Lanka. Investigations with particular reference to the eco-requirements, feeding habits, habitat analyses and reproductive behaviour in various habitats in Sri Lanka have also been made by Eisenberg and Lockhart (1972), Vancuylenberg (1977), and Santiapillai et al (1984).

The water buffalo is one of the eight most utilized species of animals that have been domesticated, but little studied in the wild. In India, wild populations of this species are confined to small reserves in the north-east of the country, such as Kaziranga and Manas National Parks in Assam, where they are at risk of decimation. Nepal has a small population of this species in the Kosi Tappu Wildife Reserve. The free-living populations of them in the Philippines are feral. In Myanmar and in Thailand, they are truly wild. Sri Lanka has the largest free-living populations of water buffalos. Ashby and Santiapillai (1983) studied its ecology, movement pattern, habitat utilization and social behaviour in the Ruhuna National Park and suggested that the wild water buffalo could be sustainably utilized outside the national parks for human needs, if their populations are sustainably managed.

5.5. PAKISTAN

Pakistan has 157 protected areas including one biosphere reserve. Kirthar National Park is one of the largest among the conservation areas in Pakistan with an area about 3,000 km^2. Others are the semi-natural wetlands of Lal Suhanra in Bahawalpur, Khunjerab in the high mountains, north of Hunza along the Pakistan-China border, Chitral Gol in the north-west mountains near Afganistan, the Margalla Hills near Islamabad, and Chiltan National Park near Quetta. The hill ranges of Kirthar National Park, most of which lie below 2,000 feet asl, provide the last refuge for some of the larger and rare wild animals, which once roamed in large numbers throughout the hilly dry tracts of Sind and Baluchistan. Rare mammals such as the Sind ibex (*Capra aegagrus* Erxleben, 1777), urial (*Ovis aries* Linnaeus, 1758), chinkara gazelle (*Gazella bennettii* Sykes, 1831) and desert wolf or grey wolf (*Canis lupus* Linnaeus, 1758) are found in this area. The desert wolf, a geographical subspecies, is in serious threat of extinction throughout its ranges. Population of the Sindh ibex (*Capra aegagrus* Hune, 1875) in 1971 was only about 1,200 heads, but conservation measures have allowed them to increase to about 4,000 animals. Populations of urial and chinkara have also increased to viable levels.

Following the wildlife survey of Schaller (1976 a, b), Khunjerab National Park was established in Upper Hunza in the northern areas. This rugged and extensive park adjoining the Taxkorgan Widlife Reserve in

China contains a viable population of the endangered snow leopard, besides healthy populations of Himalayan ibex (*Capre ibex*) and naur, the latter representing the western-most extension of its wide geographical range in Asia (Wegge 1988). The Marcopolo sheep (*Ovis ammon polii* Blyth, 1841), formerly abundant near the Pak–China pass, has decreased to an alarming population level, partly as a result of poaching during the construction of the Karakoram highway. Strict conservation measures were proposed, but inadequately implemented, with the result that the population is nearly extinct within the Park. Reportedly, a more viable population of this species exists outside the Park in a small Game Reserve near Mintaka pass, which seasonally migrate to China.

A small population of the Tibetan wild ass, not found elsewhere in Pakistan, was reported to occur in the south-eastern part of Khunjerab National Park. With proper conservation measures and in co-operation with the Chinese authorities, the remnant populations of both Marcopolo sheep and wild ass may now be saved through concerted efforts by international and national organizations. As Khunjerab National Park contains vital summer pastures for traditional livestock grazers living at the periphery of the Park, management of the area is to be designed to incorporate other land-use patterns in an expanded multipurpose conservation area (Wegge 1988).

One of the subspecies of markhor (*Capra falconeri* Wagner, 1839), geographically separated into Kashmir and Astor markhors, has drastically reduced in number. A small population of the former is protected in the Chitral Gol Sanctuary. Similarly, the situation is crucial for the Ladakh urial, which used to have a wide range of distribution in the north-eastern parts of Pakistan. The combination of increased livestock grazing in key wintering habitats and hunter access through roads have put many of Pakistan's unique mountain ungulates under heavy pressure (Schaller 1976 b). Improved conservation and management measures are envisaged through the National Council for Conservation of Wildlife, and hope to overcome conservation problems to revert the status of Pakistan's wildlife resources.

5.6. FUTURE PROSPECTS

The Federal Governments of the Indian subcontinent have already realized the need for conservation of wildlife and the wild habitats. India has already adopted the National Wildlife Action Plan (Panwar and Sale 1986), which envisages strengthening wildlife research in the country. To achieve this objective, the WII was established as a major institution to undertake research and training in all aspects of wildlife. Pakistan also established a similar institute for advanced studies on wildlife with emphasis on conservation measures. Nepal is a leading country in the subcontinent where pioneering research on wildlife has directed through international collaboration through the Terai Ecology Project. This project was terminated in 1989, but a new research station, NECTARI was established to continue research programmes in lowland Terai region. Further, the King Mahendra Trust for Nature Conservation has given Nepal a strong institutional support for implementing effective conservation activities in the country.

Adequate representation of protected areas with respect to biological diversity, improved management practice, wildlife protection and conservation, rehabilitation of critically endangered species, advanced research, public awareness programmes, incorporation of stake holders in forest and wildlife conservation-related committees and involvement of voluntary non-governmental organizations and local people in conservation programmes are the major policies now being adopted by the Governments of the Indian subcontinent for effective management and conservation of wildlife resources in this region. Besides focusing on conservation of natural habitats and endemic fauna and flora, there is an increasing awareness of the necessity of integrating conservation with the needs of local people, both directly in the form of access to natural resources and indirectly through benefits accruing from local conservation programmes. Further, in Nepal, utilization of wildlife populations through controlled, sustained harvesting has long been promoted as a viable option of wildlife conservation. A large number of research programmes in diverse aspect of wildlife have already been conducted and those projects in progress in the Indian subcontinent have already gathered useful information on all aspects of natural biodiversity in the region.

Questions for Review/Exercise

1. Have you understood the objectives and prospects of the following conservation projects?

   a. Project Tiger — b. Gir Lion Project

   c. Snow Leopard Project — d. Asian Elephant Project

2. What are your perceptions on the ecology, population, distribution and conservation status of the species of rhinoceros found in your country or in one of the nearby countries?
3. Have you seen any of the threatened mammals in your area? What do you think about the reasons for such threats?

4. What are the reptiles seen in the natural habitats in your region? Have you seen any of them in the wild? Have you seen varanus? How is it different from lizards?
5. How many species of birds in your locality can be identified by yourself? Write short morphological descriptions and ecological adaptations of each of them (at least 10 species).
6. How many species of mammals have you seen in the forest you have visited? Write short notes on them (at least 10 species).
7. What are the small mammals seen in your locality? How many species of rodents have you seen? Can you identify any of them?
8. Make a field survey of amphibians in your village. Write a brief survey report of your field study.
9. See Tables 1 and 2. Make similar tables showing mammal and avian species in your locality. Do not forget to mark the IUCN threatened category status for each of the species you have recorded.
10. What are your perceptions on the future of wildlife of your country? (This should be based on the understanding you have had from your locality and the various recent developments in the contexts of wildlife conservation you have seen in your country).
11. What have you understood from Boxed items 5.1–5.2? Briefly write on the recent advancements in wildlife-related studies in your country.

Human–Wildlife Conflict 6

Human–wildlife conflict, cattle lifting and crop raiding by wildlife, case studies, wildlife-related conflicts in African Wildlife Foundation Heartlands, measures to mitigate human–wildlife conflicts, physical barriers, visual, auditory and olfactory stimuli to scare wild animals, hot chilly smoke, beehive fence, guarding crops, creating water points, landscape modification and management, translocation of problem animals, compensation and benefit sharing, highway traffic through forests

---

6.1. HUMAN–WILDLIFE CONFLICT

There was a time when humans were free to involve in any activity in forest areas. There were no restrictions on encroachments, killing wild animals and extraction of forest and animal products. Even after establishment of legal regulations on such practices, the elites and influential people with powers, such as the officials holding high positions including police, forest and defence establishments were regularly involved in poaching activities. A much more intensive human persecution of wildlife began with the large-scale encroachment into forest areas as a result of which the extent of forest got dwindled and the remaining forest lands became fragmented. During this process, large extents of forest lands were brought under human control; mainly converted for farming activities and for human settlements. Most of such events took place during the middle of the 20th century with incentives from governments under various programmes aimed at food security. As a result, wildlife populations have dwindled all over the world. However, conservation initiatives in the past few decades helped in the process of re-establishment of wildlife habitats and populations of dwindled wildlife in several of its former ranges. There are also regional differences in the States of India in this matter. Effective conservation is seen in the southern Indian States whereas in some of the northern and north-eastern States of India, conservation status is still poor. In areas where wildlife population density is high, there is a tendency among wildlife to come out from forest and forage in the neighbouring villages. Rural people who depend solely on their farming activities for livelihood are severely affected by crop raiding and cattle lifting by wild animals. Eventually, rural people develop hatred towards wildlife and poison or kill them directly or by indirect means. Accidental human casualties by wild animals are also not uncommon in some of the rural areas. Rodents have been a major menace in agricultural areas, and are responsible for destruction of over 30% of standing crops and stored products. This problem is not restricted only to the developing countries, but is seen all over the world.

Thus, Human–Wildlife Conflict (HWC) is developed mainly in villages around forest areas. Human–Wildlife Conflict is related to any of the events in which wild animals destroy or damage crops or property, injure or kill human beings, injure or kill livestock and domestic animals, or human beings injure, trap or kill wild animals. Such conflicts occur when an animal or a human crosses a perceived borderline between nature and man-made areas, and involve in activities that may adversely affect each other.

Successful and sustainable development requires harmonization of both environmental and human-development goals. Resolving HWC is a major issue in this regard (Jones and Barnes 2006). Human–wildlife conflict is a phenomenon where a conflicting situation between people and wildlife takes place in the form of crop raiding, livestock depredation, predation on a managed wild animal species or killing of people. Such conflicts are not restricted to any special region or country, but have been occurring all over the world, on land as well as in waters, and in cities as well as in the countryside. Even though, such conflicts occur more frequently in rural areas and in villages bordering forests, it is common even in suburbs of large cities. For example, incursion of elephants in the suburbs of Bangalore (India) and in Mysore (India) has been reported recently, which is an example of HWC. Major categories of HWCs are distinguished as attacks on people, attacks on livestock, crop raiding, competition for forage with livestock and wildlife, and infestation of rural houses. There were few incidents of tiger depredation of livestock and attacking humans including government officials during December 2012, January 2013 and in January 2015 in certain areas in Wayanad in the State of Kerala and in areas bordering Kerala and Tamil Nadu States (India).

In Africa, a variety of herbivorous and carnivorous larger mammals, including several hundred thousands of wild elephants and more than 20,000 lions, still roam freely in rangeland areas. Pastoralists, agropastoralists, and other small and large-scale farmers, who live in these areas and depend upon resources around, have to face the impacts of activities of the free-roaming wildlife in their surroundings such as destruction of crops, livestock predation, competition for grazing and water, increased risk of some livestock diseases, inconveniences such as loss of sleep due to protecting crops at night, and even direct threats to human life (Muruthi 2005). All these events and consequences lead to HWCs in the region. Muruthi (2005) argues that it is the responsibility of the Wildlife Management Authorities to put in place policies and measures that at least reduce the threats posed by wildlife and enable local people to reap benefits out of the wildlife resources in their region, such as revenues from wildlife-based tourism enterprises. In the absence of clearly established policies, local people frequently take actions to defend their interests, and their own lives and farms by harassing and even killing wild animals. Unfortunately, some of such problem-species are among the threatened categories, and others are keystone species, and hence, the repercussions of any such action of local people can be felt at national and international levels. Such conflicts between local people and wildlife are amongst the main threats to conservation. Elephants have been killed, and even today they are being killed in large numbers to extract tusk as well as revenge against crop depredation (Figs 57, 58). During the period from May 2013 to August 2013, poachers killed 69 elephants in Zimbabwe. There are also news reports that 41 elephants were killed by poisoning drinking water using cyanide in Hwange National Park, Zimbabwe. Such activities have far-reaching effects as it would also kill large number of other animals as they all use the same water resource.

Fig 57 (above) and **58** (below). African elephants killed by local people in Babile Elephant Sanctuary, Ethiopia. Just in five months (June–October 2012) as many as 34 elephants were killed in this Sanctuary (Photo: Synthayu Workneh)

Most of the wildlife attacks on humans occur accidentally. This happens when people move through forests to collect forest resources such as firewood, honey, minor forest resources including medicinal plants, fruits and seeds of various plants. Very rarely, wild animals attack human beings in villages or in settlements. In case of elephants, it happens when the members of the herd sense the presence of humans in their immediate surroundings, and they flee away spreading all over the area. Attacks on livestock by larger predators occur in forest areas when cattle are let free to graze in wildlife habitats. Occasionally, animals such as leopards wander through villages, and lift sheep and domestic dogs. Such events may occur when there is shortage of food in forest areas and it becomes easy for such predators to find food in villages. Lifting of hen by fox, jackal, toddy cat, jungle cat and mongoose is common in villages near forests. There is nothing new in this as such smaller carnivores have been quite frequent in villages since long. However, their number is considerably decreased in the recent past and the frequency of such events has reduced.

Crop raiding by wild animals in farmlands around forest is common in all nations where effective fencing is not erected to prevent the movement of wild animals from forest to human habitats. A variety of animals such as elephant, rhinoceros, wild boar, deer, hippopotamus and primate are known as major crop raiders that affect the livelihood of rural people. Crop damage by different species such as nilgai, blackbuck, wild boar, Indian gazelle, Indian porcupine and Asian elephant has been reported from different parts of India (Prater 1980, Rajpurohit and Mohnot 1988).

Elephants forage through large areas in a single day and destroy all crops through which they move (Fig 59 and 60). Elephants occasionally raid farmlands in rural areas, and during this process they also destroy houses in the village, causing hazards to the occupants. Different species of bovids are pests of a variety of crops. Primates are of major menace in agricultural lands as they eat all types of crops throughout the year. In southern Indian States, the bonnet macaques are the most notorious among the primates, which regularly visit farmlands in large troupes and eat away the crops (Fig 61). It is difficult to chase them away from farmlands as they are not afraid of the presence of human beings. When livestock is permitted to graze in forest areas, there develops competition for food between the wildlife and livestock, as both of them have more or less the same food preferences. In such a situation, the competition is more severe during the summer months when there is shortage of fodder.

Domestic animals are carriers of diseases such as rinder pest, and foot and mouth disease, which they spread among wild animals as a consequence of livestock grazing in forests. In certain areas, even in townships, traditional wooden-roofed houses are occupied by toddy cats. They defecate and urinate from the hidings, mostly in the wooden ceilings, which drips down on the floor, and spreads the smell of urine and faeces inside the house.

Fig 59. Elephant attack in a wire-mesh fenced coffee plantation (Photo: Sharon M Balakrishnan)

Fig 60. Elephant attack on solar-powered fencing (Photo: Sharon M Balakrishnan)

Fig 61. Bonnet macaque, the most notorious crop-raiding primate in a farm in Wayanad, South India (Photo: Jaya Tawde)

There has been an ever-increasing human population since long, with direct consequences on wildlife habitats all over the tropics. As a result of this, competition between humans and wildlife has been on the increase, especially for living space and food. Those species of animals such as larger cats (tiger, leopard and lion) need larger extents of natural habitats as home range. However, in the context of increase in human population, the extents of forest as well as the population of wild prey are declining. When there is shortage of food in the natural habitats, such large predators have no choice other than moving into the marginal areas and in villages in search of suitable prey, and they find the domestic livestock as an easy prey in villages. As rural people heavily depend on their livestock for livelihood, and when they lose their livestock as a result of predator attacks, they adopt several methods to kill such predators. As direct shooting and killing may provoke legal problems, they usually poison to kill such predators. Poisoning will also kill other animals as they too may locate and eat the poison-bait. Such intentional persecution of wild predators has become a major threat for the survival of larger cats in Asia and elsewhere in forests around the villages.

Tigers are killed by shooting, trapping and poisoning, in retaliation against livestock lifting and attack on humans. Leopards are also losing their habitat and prey species. The growing human and livestock populations are putting increasing pressure on the habitats of leopards including that of the snow leopard. For example, on an average, the population of livestock increases around one million per year in some of the south-east Asian countries. This creates pressure on food sources for wild herbivores leading to reduced wild herbivore populations, which results in livestock predation by leopards and other larger predators. Between 1996 and 2002, at least 16 snow leopards were reported as killed in Zaskar in northern India, eight in one village alone (Spearing 2002 as cited in Theile 2003).

It is essential to find out possible means for controlling HWCs. The most appreciated programme in this context is the participation of stakeholders involving the communities affected by wildlife depredation and crop raiding. In order to have better results, programmes are oriented around protected areas (e.g., compensation schemes around the Corbett and Dudhwa Tiger Reserves in India), or in the corridors between protected areas or key habitats (e.g., snow leopard programmes in the Altan Khokhii Mountain range in Mongolia, which links existing protected areas, and is a migration corridor for snow leopards and their prey). As a part of mitigation programmes, improving the livelihoods of affected communities and providing incentives to support conservation activities to sustain with wildlife are being undertaken. Development of awareness among the local communities, particularly those who are affected by wildlife-related problems; programmes to improve the habitat quality and the extent of contiguous habitats for wildlife (both predators and their prey) are found to be better mitigative measures to control HWCs.

6.2. CATTLE LIFTING AND CROP RAIDING BY WILDLIFE

Effect of wildlife in rural areas is an important socio-economic issue of the villagers around forest areas. There are instances such as cattle lifting by large carnivores (e.g., tiger and leopard) and crop raid by elephants in agricultural areas. Poor farmers cannot withstand the effects of such acts of wildlife; rather they develop hatred against wildlife conservation if such instances continue. When there is shortage of food for wildlife in natural habitats, they may be forced to forage in villages. Effective fencing (electric fencing using solar power is being used in several places) is a means by which wildlife movement through villages can be controlled. In case of crop raid and cattle lifting, proper compensation should be given to the villagers affected by wildlife.

6.3. CASE STUDIES

Crop yield losses and livestock depredation were studied in buffer-zone villages of Nanda Devi Biosphere Reserve by Rao et al (2002). This study was carried out in 10 villages situated in the buffer zone of the Reserve in Chamoli district of Uttaranchal State (India) with the help of a questionnaire survey. The survey included 419 households with total population of 2,253 people. An extent of 273 ha area was under cultivation during this survey. Estimates of crop yield losses were made in four representative villages for each crop variety. The magnitude of crop yield losses varied significantly with the distance of agricultural areas from the forest boundary. The total crop yield losses were high for wheat and potato in all the villages studied. Losses from areas near forest contributed to more than 50% of the total losses for each crop in all villages. In monetary terms, potato alone represented 43% of total crop yield loss due to wildlife in Dronagiri village. Among the crops, the monetary value of yield losses was least for amaranth and highest for kidney bean. The projected total value of crop yield losses due to wildlife damage for buffer-zone villages located in Garhwal Himalaya was about Rs. 5,38,000 (over USD 11,000). Horticultural crops were also affected by wildlife. Wild boar, bear, porcupine, monkey, musk deer and partridge were responsible for major crop losses in the area. Primates and wild boar alone accounted for about 50%–60% of total crop damage. Goat and sheep were the major livestock depredated and leopard was the main culprit for this loss. The total value of livestock losses was around Rs. 1,02,4000 (Over USD 22,000). As a result of these losses, local people have negative attitude towards wildlife conservation establishments in the study area.

Crop damage and livestock depredation by wild animals were studied in villages in and around the Sariska Tiger Reserve, Rajasthan (India) by Sekhar (1998). Data were collected using a semi-structured questionnaire from 180 households in 37 villages, followed by quadrat sampling and focus group discussions. Crop loss and livestock depredation were evident in almost half of the households in villages adjacent to the Reserve. As per the claims of the villagers, crops were vulnerable from mid-harvest to harvest stages, which was also confirmed by field data. However, the losses to the crops in the late phase were more serious. Losses of maize and gram were highest, while bajra and wheat crops were the least damaged by wildlife. Nilgai was the major cause for crop damages, followed by wild boar, sambar, chital, common langur, rhesus monkey, Indian porcupine, parakeet and peafowl. Nilgai and wild boar were responsible for at least half of the total crop loss, and tiger and leopard were the main livestock predators in the area. Tigers preyed mostly on larger livestock such as cattle and buffaloes, whereas leopards preyed on goats and sheep. Distance of the village from the border of the Reserve was a factor in the intensity of the damage caused by wildlife. Measures adopted by villagers to protect their farm from wildlife had influence on the level of damage. More than two-thirds of the villagers in the study area spent time and money guarding crops and protecting livestock. Guarding was adopted by most of the villagers, followed by protection by the fences erected around the farming areas. Different methods were used to protect farmlands from the attack of wild animals such as guarding and providing physical barriers such as stone wall, mud wall and brushwood, all using locally available materials. Around 80% of the farmers are engaged in various kinds of protection for their crops. A combination of more than one measure was used to protect crops. Guarding crops during nights combined with fencing was the most popular measure. The number of nights spent guarding on the farm varied from 50 to 240 in a year in the study area. On an average, a household spent 184 nights in guarding crops from wildlife. Scarecrows, lighting, fire, crackers and gun shots were also in practice to drive away wild animals from entering farms.

In contrast to the case study reported by Rao et al (2002) in Nanda Devi Biosphere Reserve, in and around Sariska Tiger Reserve, local people had a positive attitude towards wildlife conservation, because of benefits they have been deriving from the Reserve such as fodder and fuel wood. The local people in this study area are religious, and they culturally maintain positive attitude towards wild animals. Majority of the people in this locality worship plants and animals, which was also responsible for their positive perception of wildlife.

Human–tiger conflict in Sumatra (Indonesia) was analyzed by Nyhus and Tilson (2004). Human–carnivore (large) conflicts occur mainly when carnivores lift cattle and other domestic animals. But in Indonesia, during the period 1978–1997, in addition to killing of over 870 livestock, Sumatran tigers killed 146 people and injured 30. The conflict was more common in intermediate disturbance areas such as multiple-use forests where tigers and people co-existed. Most of the victims were middle-aged males working in agricultural fields near forest edges during the daytime. However, there were also children of around 6 years and old people of 70 years of age as victims. Based on the available information, several attacks were attributed to single tiger, whereas in some cases more than one took part, including a tigress and her two cubs. Out of the 15 cases of known identity of sex of the tiger, 11 were males and four were females. There were accounts of over 265 tigers poisoned or shot-killed in retaliation or by accident, and 97 were trapped alive. Conservation authorities should be able to effectively control such killing of tigers as well as their prey species in tiger conservation areas to enhance the possibilities of co-existence of humans and tigers in Sumatra and elsewhere. It is also essential to reduce fragmentation of tiger habitats and to evict people forcefully by providing alternate relocation facilities and financial incentives to reduce chances of human–tiger interactions. Forest-edge villagers should also be educated on methods to reduce such conflicts. When proper incentives are provided, the forest and wildlife administration can also achieve the co-operation of villagers in conservation activities.

Sillero-Zubiri and Switzer (2001) reported on wildlife-related crop raiding in some of the regions in Tanzania with special reference to primates and discussed on the possible models to resolve HWCs. Encroachment of wildlife habitats by subsistence farmers in Africa has resulted in increased interactions between people and wildlife, which leads to the conflicts between them. Baboons, elephants and other herbivores cause serious crop damage in the study areas and hence, farmers retaliate. However, both the governments and the farmers fail to address long-term solutions to such problems and conflicts. Regular access to water is crucial for many of the wildlife species, and in their attempt to secure such essential resources, they may have to cross human habitats and come in conflict with people, who also depend on the same water source for livestock and for irrigation, especially in arid regions. Expansion of agriculture into forest habitats, formerly occupied by wildlife, is one of the major reasons for conflicts between people and wildlife, and such conflicts have become frequent in the recent past, especially in areas close to forests and conservation areas. Intensive agricultural practices with monocultures, which are attractive to wildlife and easy to raid, form good sources of food with less foraging efforts. For example, seed-eating birds can be easily attracted to such cereal farms when the crop matures, and farms of tuber crops and a variety of related crops are attractive to wild boars and primates. Some animals such as baboons are attracted to a number of crops as they are omnivorous. In addition, positive effects of conservation activities in the recent decades have helped several of the wildlife species to increase their population. Related increase in wildlife density in areas close to farmlands also results in developing crop-raiding behaviour among wildlife.

The most problematic wild animals, as revealed by respondents were primates (51.7%), bushpigs (13.3%), rodents (10.6%) and elephants. Among primates, the olive baboon (*Papio anubis* Lesson, 1827*)*, yellow baboon (*Papio cynocephalus* Linnaeus, 1766), vervet monkey (*Chlorocebus pygerythrus* F. Cuviet, 1821), black and white colobus

(*Colobus guereza* Illiget, 1811), patas monkey (*Erythrocebcus patas* Schreber, 1775) and chimpanzee (*Pan troglodytes* Blumenbach, 1775) were identified as crop raiders during different growth stages of the crops, even though it was not strictly restricted to any of the seasons. Primates constituted one of the major groups of animals responsible for crop raid and related conflicts with humans in Africa due to their intelligence, adaptability, complex social organization and opportunistic tendencies. Most of the non-human primates are adapted to a broad range of feeding strategies, and they eat a wide range of food items including items from natural habitats and from human dwellings. They have the potential to be involved in conflict through a variety of situations such as by raiding crops, houses, gardens and chicken coups, by killing young or small livestock, and by raiding of guest-rooms and halls in tourist lodges. In India, rhesus macaque (*Macaca mulata* Zimmermann, 1780) and bonnet macaque (*Macaca radi*ata É. Geoffroy, 1812) are considered sacred, and they often live alongside villages and even in townships, and raid crops in their surroundings. The Japanese macaque (*Macaca fuscata* Blyth, 1875) is reported as being serious crop raider across their ranges in south-east Asia and in Japan.

A wide variety of rodents feed on crops. These rodents include several mouse and rat species (Muridae), cane-rats and porcupines. Elephants are also responsible for large amounts of crop damage in African and Asian countries. Studies in Uganda have revealed that animals such as bushbuck (*Tragelaphus scriptus* Pallas, 1776), duikers (*Cephalophus* species), striped ground-squirrel (*Xerus erythropus* Desmarest 1817) and African civet (*Civettictis civetta* Schreber, 1776) are crop raiders. As many as 23 species of vertebrates have been reported to raid crops around the Budongo Forest Reserve in Western Uganda. Certain types of crops are more prone to be affected by wildlife than others. On the other hand, all crops are not seriously damaged by wild animals. In the Budongo area, maize, cassava, finger millet, and sweet potato are, in that order, the four most commonly grown crops rich in carbohydrates. Among these, cassava (25%) and maize (19%) are more raided by wildlife. In general, crop raiding is more severe during the harvesting period. This shows that wild animals are more attracted to crops when the crops mature (Workneh et al 2004).

Jones and Barnes (2006) have made a detailed analysis of various aspects of HWCs in Namibia, southern Africa. They studied the dynamics of HWCs and the root causes of such conflicts, and tried to identify opportunities and models for preventing and mitigating HWCs. They focused on three levels such as the macro level (policies, conventions and agreements), the meso level (regional institutions, development activities and land-use planning) and the local level (communities, farmers and non-governmental organizations). Namibia is one of the nations where wildlife populations have increased in the recent past as a result of conservation and participatory management programmes. Even populations of elephants and black rhino are on the increase in Namibia, whereas in a number of other African nations, their populations have declined considerably. Since the late 1960s, wildlife numbers on different land-use types have increased by 40%–70%, and diversity of large mammal species has increased by 44%. Approximately 80% of the larger game mammal species are found on privately-owned commercial farms, and free-hold farmlands in Namibia hosts the largest cheetah population. The elephant population in Namibia has almost doubled during 1984 and 2003 to an estimated population of 11,262 heads and a further increase to an estimated 16,397 was noticed by 2005 (Martin 2005). It was also revealed that there were only 15 lions in the Palmwag Tourism Concession in a range of around 4,000 km^2, but by 2006, there were 85 lions in an increased range of 24,000 km^2 area (Stander 2006).

The Community-Based Natural Resource Management (CBNRM) programme is praised for such increase in wildlife populations in Namibia. Under the CBNRM programme, the rural communities form local Natural Resource Management Institutions called Conservancies and get the rights from the government to manage wildlife and eco-tourism. This has become a good income-generating activity and the rural people have developed a positive attitude towards wildlife. As a result of this positive attitude towards wildlife, poaching has reduced considerably and wildlife populations have sustained and increased, leading to HWCs in some of the ranges. Immigration from the surrounding areas of harsh conditions into better areas with resources and protection also enhanced the populations of wildlife in the region. In most of the ranges in southern Africa, the elephant populations are stable as an effect of CITES and in some of its ranges, their population density has been on the increase. This forces them to migrate into the north-eastern parts of Namibia from Botswana. Hence, transboundary elephant management is required to reduce the conflict by successfully moving the elephants to less densely populated and safe areas in the region.

Analyses on HWC in Namibia have shown that elephants and large carnivores were responsible for most of the incidents, and these species live permanently on communal and freehold lands where they regularly come into contact with local communities. According to NACSO (2006), a total of 3,194 HWC incidents were reported in Namibia in conservancies during 2005. The species involved and percentage of incidents were:

elephants (23%), hyena (17%), jackal (10%), leopard (10%), cheetah (9%), bushpig (6%), hippopotamus (5%), crocodile (5%), antelopes (5%), lion (4%), baboon (2%), porcupine (2%) and caracal (1%). Data on HWCs in 12 Conservancies in northern Kunene Region shows that among the predators, leopard (23%), hyena (23%) and cheetah (22%) were the problem animals. In this region, elephants were responsible for only 9% of the incidents. In some of the conservancies such as Kwandu and Mayuni, incidences of HWCs have increased when settlements were across well-used elephant paths to and from the Kwando River. This leads to more crop damage by elephants.

There should be an integrated coordination between macro, meso and micro levels for reducing HWCs. Better planning and effective implementation of coordinated activities involving local people are essential for controlling HWCs, and to have a positive consideration for wildlife conservation among the public. It was also suggested that the costs of HWCs to communities and governments can be outweighed by the economic benefits generated by CBNRM and its sustainable use to local communities. Eventhough the financial benefits to local households have been only limited, it is envisaged to increase through various means to compensate losses caused by HWCs. The CBNRM provides a number of other benefits that help to increase the tolerance of wildlife that causes problems such as elephants and predators. While planning mitigative measures, it is also essential to visualize the difference in HWCs in relation to different areas, and different species of animals involved, as different approaches may be required for each area as well as for each species of wild animal.

6.4. WILDLIFE-RELATED CONFLICTS IN AFRICAN WILDLIFE FOUNDATION (AWF) HEARTLANDS

Human–wildlife conflicts have been extensively studied in some of the AWF's Heartlands (AWF Heartlands are Lopori Wamba, Kazungula, Kilimanjaro, Limpopo, Maasai Steppe, Samburu, Virunga and Zambezi) such as Samburu, Kilimanjaro, Maasai Steppe and in Virunga, and reviewed by Muruthi (2005). In some areas, crop damage by wildlife is perceived as a major problem for farmers. In Zimbabwe (portion of the Zambezi Heartland), elephants were responsible for up to three-quarters of the total crop damage caused by wildlife. In Kilimanjaro Heartland, in 1996 and 1997, at least 15 elephants representing three-quarters of the local population were killed by local people. Between 1974 and 1990, 141 out of 437 elephant deaths in the Amboseli ecosystem were caused by people through spearing. Main problems in Kilimanjaro Heartland were crop damage, competition for water and grazing, killing of livestock, disease transmission and human fatalities. In semi-arid areas, where livestock production constitutes a major part of the local livelihood, high levels of conflicts can occur due to predation. In Virunga Heartland, habitat destruction has resulted in the increase in conflicts between people and animals such as mountain gorilla, elephant and buffallo. Impacts on the subsistence of local farmers are economic devastation through destruction of crops, living in a state of fear, inconvenience and constant threat to life (MacFie 2003). Interactions and conflicts with mountain gorillas are a source of stress, which can result in the transmission of human diseases, and can lead to direct physical attacks, disabilities such as loss of limbs from snares, and even death. During the period 1996–2003, 18 mountain gorillas were killed in Virunga and Bwindi (Woodford et al 2002). In the Samburu Heartland, it was documented that lion (35%), leopard (35%), hyena (18%), baboon (4%), elephant (3%), buffalo (2%), wild dog (2%) and cheetah (1%) were responsible for killing livestock, leading to HWCs in the area.

6.5. MEASURES TO MITIGATE HWCs

There was a time when elimination of either people or wildlife was sought as a means to prevent HWC by totally removing either the animals involved in such conflicts by translocation of problem animals, scaring and repelling from the area or by evicting people from large tracts of land (e.g., when new National Parks or other conservation areas were to establish). For example, eradication of animals such as lions, leopards, elephants, buffalo, rhino and larger species of antelopes from large areas of Africa was made in the past. However, this is not an accepted practice in the recent years, unless such problem species are exotics. Killing of problem animals can reduce the density of such species in the locality for some time, but as an ecological consequence, if the habitat is suitable, there can be increased reproductive activities of the remaining individuals of the species leading to increased natality and decreased mortality to naturally compensate the population levels. It is also common to expect an increased immigration of such species from the surrounding areas into the areas of less density created artificially under management programmes. Hence, long-term practical use of eradication and killing of wildlife from natural habitats are of no use, and cannot be justified. Further, such human activities can also result in upsetting characteristics of natural ecosystems of the area including dramatic changes in the populations of other species linked with the food web. Such a phenomenon is called ‘mesopredator release’. For example, small to medium-sized predator populations can increase following the eradication of large carnivores from the area (Crook 2002). Regulated harvest is another option, but its impact is not properly assessed. The idea is to control populations of problem animals through sustainable harvest, after assessing all possible biological and physical factors of the area responsible for population increase and regulations. Individuals of the problem species that are to be harvested, timing of harvest and methods to be followed are to be assessed properly by a scientific team in advance of the process of harvest. It is also suggested that fertility of the problem animals can be controlled to limit the growth of their population. Fertility control of wild animals can be achieved through mechanical, surgical, endocrine disruptive or immunocontraceptive methods. Even though, theoretically it is a reasonably good method to suggest, in practice it is a difficult process to administer among wild animals.

Livestock depredation and crop raid are major issues provoking the local communities against wildlife conservation. As farmers have no other source of income, the only way to get their support for conservation is to give adequate compensation for the crop loss and for the livestock depredation. In general, government compensation is much lower than the full market value of the livestock killed, and the crop lost as a result of wild animal raid. The lengthy administrative procedure in the developing countries to release funds is another problem. To combat these problems, WWF India initiated a cattle compensation scheme in 1997 around various tiger reserves and wildlife sanctuaries, where the threat to the livestock was high, by providing funds to the livestock owner equivalent to the difference between a near market rate and the compensation provided by the government. In addition, WWF also made provisions for payments to the kin of humans killed in tiger attacks to help with the costs of the funeral and other rituals. The WWF and its partners have also established a compensation fund in Bhutan for livestock kills by snow leopards and tigers. However, a livestock compensation scheme in Hemis National Park in Ladakh (India), sponsored by the Ladakh Wildlife Department failed to be effective, primarily due the high costs involved. The rate of predation by snow leopard on cattle was high. Herders in remote areas had to walk up to three days to report their losses to the wildlife officials. The site inspection by the official was also not easy because of the remoteness of the area where depredation occurs.

Livestock insurance is another way to compensate depredation of livestock by wild predators. 'Project Snow Leopard' has set up an insurance scheme combined with eco-tourism for snow leopards in Pakistan. Under this scheme, farmers have to pay insurance premium based on the number of livestock into a fund managed and administered by the local community. The objective of this scheme is to compensate the losses to livestock depredation, so as to avoid retaliation among the herders. Also, additional income is generated through eco-tourism as an incentive to conserve the local snow leopard populations.

In the late 1990s, efforts were made to reduce the conflicts between human and large predators in the areas surrounding the Royal Chitwan National Park in Nepal by the projects aimed at reducing the dependence of humans on forest resources. In collaboration with the Park, King Mahendra Trust for Nature Conservation and the Buffer-Zone User Groups, WWF developed plantations in a degraded buffer zone area. When the plantations were six years old, they were thinned to yield firewood for the local community. In addition to the use by the local communities for firewood, successive thinning activities also provided income from the programme. Eco-tourism was also developed in the area and the income generated out of this was shared for communal development activities. Part of this fund was used to build schools and to develop health clinics, which were of direct benefits to local communities.

More than providing compensation for livestock losses and human sufferings, in the context of HWCs, it is important to establish effective barriers to avoid the entering of wildlife into farmlands and human settlements, and the entering of livestock into forests. This can be made either by providing strong electric fencing, preferably of solar or by establishing changes in livestock management practices to reduce the chances of contacts between predators and livestock. In Hemis National Park, Ladakh, the Snow Leopard Conservancy (SLC) is collaborating with herders to reduce depredation by snow leopards and wolves. Wire-mesh fencing was erected with stone pillars with the labour contributed by the villagers. This project is viewed positively by local communities. Similar wildlife conservation-oriented schemes are also in practice in other south-eastern nations to overcome HWCs.

Relocation of problem animals (such as tiger and leopard) is also being undertaken to reduce HWCs. In 2003, in collaboration with the Sumatran Tiger Conservation Programme, the Indonesian Government initiated a programme to translocate problem tigers to suitable forest habitats far away from human habitats. Translocated tigers were fixed with radio collars and monitored. Camera traps were also used to study them.

Land-use planning at the level of government is also important as a measure to reduce HWCs. The rural population should be aware that the areas near forest boundaries should not be used for planting edibles, which can attract wildlife into their farmlands. If they plant such crops in villages near forests, it will attract wildlife populations. Then, it is not the wildlife to be blamed, but the farmers who are responsible for developing such farms for crop raids by wildlife in such areas. Rao et al (2002) suggested that in such locations, the land-use pattern should replace those crops infested by wildlife with cultivation of medicinal plants, which are not raided by wildlife. Many of the medicinal plant species have the conservation status such as rare, threatened and endangered. As marketing of the products of medicinal plants may be a problem for rural farmers, welfare societies can be entrusted with this role to support the farming community. There is need for strict legal regulations for farming practices to reduce HWCs. A case study around Bardia National Park, Nepal has revealed that local people can be influenced to shift from traditional farming practices to those crops, which elephants do not like. This can be achieved only if the farmers are made aware of the benefit of the new farming system and the possibility of marketing the products, including the economic benefits that they can have if they shift to the new farming practice. Earlier, farmers around Bardia were farming maze to which elephants were attracted. When they switched from maze to mentha, not only the crop raid issue was controlled, but it also helped local farmers to earn a good income from the sale of mentha oil.

6.6. PHYSICAL BARRIERS

Wild animals can be effectively excluded from human settlements and farms if strong and properly planned physical barriers such as trench and fences are constructed. Both for trench and fence, the animal to be managed should form the basis for deciding the depth and width of the trench, and the type and height of the fence. Fences are being used in African Heartlands to exclude wildlife from human settlements and from cultivated or livestock areas. For example, fencing around the cultivated areas of Kimana and Namelok in Kilimanjaro Heartland has significantly reduced the crop damage by wildlife. Stone-walls have been used to exclude buffalos from invading cultivated areas in Virunga Heartland. Trenches and moats have been used to keep elephants from cultivated areas. Fencing has significantly reduced livestock depredation in Samburu Heartland. Routine maintenance of the trench and fence is also essential for this method to be effective on a long-term basis. Electric fences are mostly recommended, but they may not be feasible for local poor farmers due to the burden of additional expenses. Co-operatives of the local communities can be developed with government support, and solar fencing may be erected as a feasible solution to control HWCs. Fences of locally-available thorny or spiny hedges can control crossing of some of the wild animals into farm lands. Removing nearby cover including bushes around farmlands can also serve to protect crops from the attack of some of the wildlife. This is one of the methods suggested for rodent pest control in farmlands (Workeneh et al 2004).

6.7. VISUAL, AUDITORY AND OLFACTORY STIMULI TO SCARE WILD ANIMALS

Visual signals such as scarecrows; auditory signals such as chasing, lighting fires, crackers, exploders, bangers (beating drums), throwing objects at animals, disturbance shooting and distress calls of the same species; olfactory signals such as the smell of predators (e.g., urine of dogs and larger cats) are being used to deter problem animals from human settlements and farmlands. To some extent, these methods are effective at least in the beginning, but soon animals can become habituated to such stimuli. They learn that the signal is not a threat and hence, ignore them. Disturbance shooting is firing over the heads of crop-raiding wild animals, which is a method used by wildlife departments. Playing the recording of roaring lion and tiger in loud noise while walking along elephants is also being practiced by wildlife authorities, but it requires constant attention. It is used only when the animal is in villages and farms. It also requires people to move along the elephant tracks, and hence, may be dangerous in case of problem elephants, particularly during night raids. Even though these are useful methods to prevent problem animals from the field in the beginning, eventually they learn to overcome the effects. Some of these methods can keep inexperienced crop raiders away, but experienced individuals and family groups cannot be fooled by scare tactics (Sukumar 2003). For example, use of fog lights and horn sounds are effective against elephants, but only in the beginning. Behavioural adaptations of crop-raiding animals such as elephants, primates and wild boar, which are able to modify their behaviour in response to measures adopted by farmers, make them to overcome the effectiveness of the method.

6.7.1. Hot Chilly Smoke

Hot chilly smoke is found as an effective deterrent to crop-raiding elephants. Hot chilly, ground and mixed with grease, can be applied on a string and hung on fences. When the elephant touches the string, it causes irritation. Dry elephant-dung or cow-dung mixed with ground hot chilly produces noxious smoke when burned, which also acts as a deterrent against wild animals including elephants. Hot chilly can be burned with husk of grains in problem areas. The smoke of chilly can irritate wild animals and can keep them away from farm areas.

6.7.2. Beehive Fence

In Kenya, elephants are not confined to National Parks and Wildlife Reserves, but they frequently range in marginal areas, and crop raiding by them is common. Hence, it poses serious social, economic, political and conservation problems. To control such adverse effects of elephants on farming communities, efforts are being focused on effective farmer-managed deterrents that are suitable and affordable for the local farmers to practice. In northern Kenya, people believe that elephants are afraid of honey bees. Vollrath and Douglass-Hamilton (2002) experimentally proved that elephants avoid *Acacia* trees hosting beehives. Studies using play-back experiments have also demonstrated that elephants can be frightened by disturbed bee sounds (King et al 2007). King et al (2009) studied the effects of beehive fence in an Ex-Eroke community in the southern region of Laikipia (Laikipia plateau), Kenya. The land-use pattern in this region is a mosaic of large private and government cattle ranges, pastoral grazing lands and small holder farms. Wild animals including elephants compete for resources in farms in this region, and it is identified as a high-risk area for crop raiding and HWCs. Two small farms were selected in this region for the study, one as experimental and the other as control. Nine traditional log beehives were used along a 90 m long fence of the northern boundary of the experimental farm, cutting off all the entrance routes of raiding elephants. Kerosene-treated wooden poles were used to erect the fence. Over a six-week study period, the experimental farm experienced seven elephant raids involving 38 elephants and the control farm with no beehive experienced 13 raids involving 95 elephants. At the end of the season, the experimental farm had a relatively successful harvest of a variety of sorghum, beans, potatoes and maize, whereas the control farm lost almost 90% of the crop due to elephant raids. This study suggests that beehive fence was at least partially successful in controlling crop raids by elephants. As the hive can yield honey and wax, the expenditure for erecting and maintaining beehive fence can be covered by the sale of honey and wax. Beehive fencing is also tried in some of the elephant-problem areas in Wayanad, Kerala, where it is found effective. This method can also be deployed against other wild animals including primates. Beehives can be hanged on trees in primate-raiding crop areas, and when monkeys pass through such trees, the disturbed bees will sting animals nearby. Thus, they can be easily scared and ultimately such areas may be avoided by monkeys and other wild animals.

6.7.3. Guarding Crops

Guarding crops and livestock by the presence of people in and around farmlands is perhaps the most effective method to protect crops and livestock from depredation by wild animals. This is the method traditionally being followed by rural people. Simple watchtowers or sheds can be erected using locally-available resources such as bamboo and thatching grass, and people can stay throughout the night watching for the presence of crop-raiding wild animals around the field. They can alarm wild animals by beating empty hanging tins, strings and cowbells. Also, they can keep donkeys and dogs (to protect flocks of sheep and goats from predation by cheetah) to alert the presence of wildlife around as well as to threaten them. However, in areas where both diurnal and nocturnal animals are responsible for crop raiding and livestock depredation, effective guarding requires both night and daytime watching, which means farmers have to be in and around the farm on vigil the whole day. In this process, they can have sleepless nights as a result of which they may not be able to work in the daytime, other than resting in the sheds in the field. Farmers are also helped by their close relatives, including school-going children. If so, the children will miss their classes as they cannot attend schools during the crop-protection season. A study on Budongo Forest Reserve in Uganda showed that children and women were responsible for 30% and 34% of the guarding fields, respectively (MacFie 2003). In some areas, farmers employ guards to protect their fields, while others use dogs to frighten away and chase crop raiders. Guarding is costly through time that could be spent for other useful activities (e.g., domestic work of women, children attending school and adults who can attend to other work in their farm or elsewhere).

6.7.4. Creating Water Points

If crop damage by wildlife is limited only to the dry season, animals such as elephants may be attracted to crop fields during their search for water points. In such cases, it is possible to reduce conflicts if there are water points in forest areas, so that animals can be diverted to such locations instead of farm and human habitats. Similarly, if sufficient water points are available in and around human settlements, movement of people and livestock into wildlife habitats and waterholes in forests can be controlled. Thus, human and livestock interactions with wildlife can be eliminated or controlled and HWC can be reduced. For example, in Kilimanjaro Heartland, AWF rehabilitated the water supply at Imbaringoi in 2004 to serve livestock and the people in Kitirua Concession Area and to prevent livestock from going to Amboseli National Park in search of water. As an immediate effect of this measure, the encounters between livestock and wildlife in the Park were reduced, and thus, the level of HWC in the area was reduced. In 2004, with AWF co-financing, a water project was rehabilitated in Archer's Post, Kalama area of Samburu Heartland to supply water to the community areas, to create separate drinking points for wildlife and livestock, and to help boost tourism potential of the community areas. This project was aimed to reduce competition between livestock and the endangered Grevy's zebra (*Equus grevyi* Oustalet, 1882) (Muruthi 2005).

6.7.5. Landscape Modification and Management

Wild animals enter into human settlements in search of resources, and they are easily attracted to food and water resources, especially during the period when there is shortage of such resources in forest areas. In any area or region, the crops to which wild animals are attracted can be easily revealed and if farmers can avoid planting such crops, HWC can be controlled. In such cases, farmers have to opt for alternative crops, especially such crops for which there is an aversion for wildlife. This can be achievable by modifying the resource itself, the way it is managed, the resource's habitat, or the surrounding landscape (Muruthi 2005). Farmers in such regions should go for crops that are less palatable to wildlife, such as substituting chillies for maize, or even by changing the timing when a crop is planted and harvested. Damage by wildlife can be reduced by making changes near the resources so that the problem wildlife is more vulnerable to predation, easier to spot by people and dogs, and generally, less at ease in the area. For example, a livestock keeper can remove thick cover from near animal-holding areas. Similarly, the surroundings of farmlands should be cleared and should be free from bushes, not only to deter larger wildlife, but also to keep the surroundings free from rodent pests. Rodent pests are responsible for major crop loss all around the world, and also are a menace for farmers. If the surroundings are maintained free from bushes and shrubs, rodents will not find suitable nesting sites nearby farmlands, and thus, can be controlled from attacking crops during the reproductive and harvesting seasons (Workneh et al 2004, Chekol et al 2012). Small islands of crops scattered across a wildlife-inhabited landscape are more vulnerable to destruction than those that are clustered together. The long-term solution to HWC often lies in better planning of land-use in problem areas.

6.7.6. Translocation of Problem Animals

If there are only few problem animals in an area, one of the options to reduce HWC is to trap and translocate such individuals to other suitable areas where they can sustain. This method is advisable in case of predators such as leopard and tiger, which are mostly solitary. However, this is not a feasible method in case of large number of individuals of any species of animals such as primates that regularly attack crops. When one or few problem animals are translocated, it is possible for other individuals to use the area for foraging. Even if a troupe of monkeys are trapped and translocated into another area, another troupe may occupy the area in the absence of competition. Hence, trasnslocation is feasible only in case of individuals of species such as larger cats, to eliminate the problem of an identified individual of the species. There are also instances of returning the translocated individual to the site from where it was captured and removed. Translocation is also a risky process as there were casualties of the translocated individuals as a result of stress and effects of trapping. They may also develop problems for resident populations in the translocated range through introduction of disease or may develop competition for resources (Conover 2002).

6.8. COMPENSATION AND BENEFIT SHARING

Compensation for crop-raid, livestock-lifting, human casualities and other related damages is practiced as a conservation effort, and to compromise such issues with the local people. However, government compensation is often marginal, and hence, most of the farmers are unhappy with the amounts paid to them as compensation. As the government procedures are slow, it takes a very long time to sanction compensation, and on most of the occasions the attitude of officials is not in favour of the poor farmers. This programme is usually confined to specific types of losses only. For example, livestock killed by larger predators and crops raided by elephants are considered for compensation. However, larger extents of crops raided by a variety of species of birds and primates, and by wild boar are not often considered for compensation. In the recent years, there is also a rethinking in allocating funds for compensation against HWC, where the role played by farmers in protecting their crops and livestock is also taken into account before considering for compensation. The current view is that compensation should be considered only to those farmers, who have erected effective fences around their farm, and taken sufficient protective measures to protect their livestock from predators. It is known that efforts for crop protection and complaints about pests are often disproportional to the actual damage. Sharing benefits from wildlife conservation with the affected communities is an alternative to develop positive attitudes among the local communities towards wildlife conservation. Part of the revenues gained from eco-tourism, safari hunting and wildlife culling, if directly channelled through the local administrative bodies for local development projects such as developing primary schools, health centres, roads and communication facilities, local people view wildlife conservation as an activity bringing developments in their areas (Balakrishnan and Ndhlovu 1992). Further, in case there are employment opportunities for the local people as a result of eco-tourism and other related programmes, local people will offer additional support to such wildlife conservation activities. Appropriate government policies are required for such programmes with the involvement of local communities.

Well-designed HWC management plans that are integrating different techniques and are adjusted based on the nature of the problem can boost co-existence of wildlife and people. Potential mitigative measures relevant to the context and specific to the concerned area should be selected based on their effectiveness, cost and acceptability of the locals, particularly the affected communities. Reducing conflicts between wildlife and people is likely to reduce the negative attitude that many communities have towards wildlife and conservation. Improving food security by reducing wildlife-related impacts on crops and livestock will also reduce the need to seek alternative sources of food such as hunting of wildlife. Case-specific and region-specific methods should be recommended to mitigate the impacts of HWC and to reduce such problems for the welfare of rural communities as well as a variety of wildlife involved in conflicts with human beings all over the world.

6.9. HIGHWAY TRAFFIC THROUGH FORESTS

Highway-related fragmentation of forest habitats, traffic-related blockade of animal crossways and vehicle–wildlife collision seriously affect a variety of wildlife all over the world. These human development-related conflicts with wildlife are different from the issues already discussed earlier, as decisions and activities of governments rather than individuals are responsible in this context. Highway authorities and governments do not take appropriate measures to prevent adverse impacts of such activities on wildlife at the time of planning and developing roads through forest areas. Highways and fast-track roads are essential parts of development processes, but conservationists are concerned about sustainability of natural habitats with diversity of wildlife in the affected areas. Roadkills, particularly during night traffic, are common on highways passing through forests as several species of nocturnal animals cross the road during night. Small animals such as frogs, reptiles and birds, and a variety of small (e.g. hedgehogs, shrews, rodents) to large-sized mammals (e.g., elephants) are affected by such accidents. Common, rare, endangered and endemic animals are also seen killed on such type of roads.

There are few study reports on the intensity of roadkills of wildlife in some of the developed nations. However, such information is lacking in most of the developing parts of the world. In Netherlands, around two million birds are killed per annum on roads, whereas in Belgium, around four million larger vertebrates are killed per year. In Denmark, 1.5 million mammals, 3.7 million birds and more than 3.1 million amphibians are killed every year (Forman et al 2003). It was estimated that around 8.5 million birds were killed per annum on roads in Sweden alone (Svensson 1998). A four-year study by Rosen and Lowe (1994) in Sonoran Desert of southern Arizona revealed that on an average, about 22.5 snakes were killed per kilometre per year. Hedgehogs (*Erinaceus europaeus*), badgers (*Meles meles*), fox (*Vulpes vulpes*), wolves (*Canis lupus*), black bears (*Ursus americanus*), grizzly bears (*U. arctos*), polecats (*Mustela putorius*), rabbits (*Oryctolagus cuniculus*) and various species of deer are subjected to roadkills in large numbers in Europe and in the United State of America. Railway traffic is also responsible for killing a variety of wild animals in their natural habitats. Trains passing through forest habitats hit ungulates and even elephants, while crossing the railway tracks during their foraging movements.

In addition to the direct kill, wildlife is also affected by atmospheric pollution as a result of heavy traffic through forest habitats. There are both chemical and physical pollutions that spread in the habitats during road construction, maintenance and use. Vehicle emissions continue to spread the pollutants on either side of the road, whenever the road is in use. Major chemicals from vehicle emissions contain carbon monoxide, lead, hydrocarbons, nitrogen oxide and sulphur dioxide. The intensity of traffic, distance from road, direction of wind and roadside vegetation act as factors contributing and controlling levels of pollution on roadsides.

Various mitigative measures are adopted to control roadkill of wild animals. Road closure during nights and during certain seasons (depending upon the nature and behaviour of wild animals around) is being practiced in some forest areas. Speed limits and sign-posts on regular wildlife-crossing locations help unfamiliar drivers to drive carefully. However, close speed barriers such as road hubs for longer stretches of roads are not an advisable measure for speed control as it can affect human health (passengers and drives can suffer from slip-disc and back-bone problems), and responsible for more vehicle emissions on highways. The intensity of such problems depends upon the traffic volume and the number of passengers uses the route every day. Since past over three years, Mysore – Kozhikode National Highway (in Kerala and Karnataka States, India) is closed for traffic during night to allow free movement of wildlife in a stretch of around 10 km of the road in Wayanad Wildlife Sanctuary (Kerala) and Bandipur Wildlife Sanctuary (Karnataka), which along with Mudumalai Wildlife Sanctuary (Tamil Nadu) is developed as one of the Tiger Reserves of India. In addition to the ban on traffic during night, road bumps are made close to each other for a long distance to control speed of traffic during day time. Even though, in principle, it is a good beginning for wildlife conservation, its long-term environmental impacts on the forests can be severe. Further, it will also affect the health of large number of regular passengers and drivers. As Bangalore has become one of the most important IT cities of the world, large number of professionals and business people are frequently using this highway. This is also developed as a heavy goods transport route. Majority of the trucks carrying goods from the northern States of India to Kerala use this route from Bangalore to Kozhikode and back. Long-distance drivers prefer to drive during nights, but they are stranded before the points of entry into forest in the States of Karnataka and Kerala. As a result, the business loss and man-hours wasted are heavy.

What are better and corrective mitigative measures that highway authorities should consider to overcome such problems? Establishment of wildlife underpasses and overpasses at wildlife crossing points on highways passing through forests is the most effective step taken in many of the nations to overcome roadkill of animals, and to re-establish wildlife movement routes including migratory routes. Underpasses include culverts, tunnels and bridges, which are built for drainage at the time of road construction itself. If properly planned and effectively built with the basic understanding of the size and behaviour of the wild animals around, these structures can also be used by a variety of small and medium-sized animals for crossing the road. Amphibians, reptiles, insectivores, rodents, hares, rabbits, armadillos, river otters, opossums, raccoons, ground squirrels, skunks, coyotes, bobcats, black bears and mountain lions have been observed to use such underpasses in different countries in the world. The level of human activity and complexity of roadside vegetation also influences the level of usage of underpasses by wildlife (Clevenger and Waltho 2000).

However, it is better to plan overpasses for large-sized animals such as ungulates and elephants. Overpasses are usually made as green bridges of around 30–100 m width for better passage of larger animals, where natural habitat can be retained or re-established. Tunneled roads of around 100 m length on roads passing through hilly terrains can help to retain animal crossing tracks unaffected. In suitable locations, roads can also be made as flyovers (Fig 62) and tunnels (Fig 63), retaining natural vegetation cover of the landscape unaffected under the flyovers, and the traffic through such flyovers will not affect natural wildlife movements. In addition to under and overpasses and flyovers, suitable fencing on either sides of the road can also be made to make sure that wildlife would not use other locations to cross the road. Eventually, animals will learn to follow the under and overpasses alone to cross the road. To begin with, it may be more expensive, but in the long run, it will be more economical and public-oriented, rather than implementing ban on night traffic and other measures that affect human health and waste thousands of man-hours every day on major highways crossing through forests. Governments and highway authorities should consider such measures as a part of road construction processes, and additional budget should be earmarked for roads passing through forest areas. Cost-effective non-structural methods such as species-specific olfactory repellents, which can be sprayed on roadside vegetation, ultrasound repellents, road lighting and habitat modifications can also be tried to reduce roadkills of wildlife.

Well-designed HWC management plans, which integrate different techniques and are adjusted based on the nature of the problem, can boost co-existence of wildlife and people. Potential mitigative measures, relevant to the context and specific to the concerned area, should be selected based on their effectiveness, cost and acceptability of the locals, particularly the affected communities. Reducing conflicts between wildlife and people is likely to reduce negative attitudes that many communities have towards wildlife and their conservation. Improving food security by reducing wildlife-related impacts on crops and livestock will also reduce the need to seek alternative sources of food such as hunting of wildlife. Case-specific and region-specific methods should be recommended to mitigate the impacts of HWC and to reduce such problems for the welfare of rural communities as well as a variety of wildlife involved in conflicts with human beings, all over the world.

Fig 62. A view of a flyover highway through a forest area in Lausanne (Switzerland), allowing all natural animal movements in the area unaffected. It also protects the lake (bottom – right) from impacts of soil erosion, which would have caused, had the road constructed traditionally (Photo: M Balakrishnan)

Fig. 63. A tunnelled section of the Addis Ababa – Dessi road, Ethiopia, allowing free movement of wildlife over the tunnel in a hilly terrain (Photo: M Balakrishnan)

Questions for Review/Exercise

1. Have you understood meaning of the following?

   a. Human–wildlife conflict | b. Cattle lifting
   c. Crop raid | d. Hot chilly smoke
   e. Beehive fence

2. What have you learned about human–wildlife conflicts and their impacts on wildlife? Have you heard of such conflicts earlier? Can you cite few examples? Are there any such conflicts in your locality? If so, list all such conflicts.
3. What are your perceptions on the mitigative measures to be taken to overcome Human–wildlife conflicts?
4. Have you experienced any problem from wild animals in your locality? Can you suggest suitable mitigative measures to overcome problems from wildlife and help villagers?
5. What is your understanding about translocation of problem animals? (Where to translocate? How to translocate?)
6. What did you understand by 'compensation' in the context of human–wildlife conflicts?
7. What are your feelings on Figs 57 & 58? Are all human beings that cruel?
8. See Fig 61. How many species of non-human primates you have seen? Are they all pests of agriculture? Try to make a note on the feeding ecology of those non-human primates seen in your locality.
9. Is a highway passing through forest habitat in your locality? If so, have you seen any wild animal killed on the road as a result of vehicle hit? Using the mitigative measures you have learned, can you suggest suitable measures to control such problems in the area?

Wildlife Farming and Hunting 7

Farming wild animals, crocodile farming, ostrich farming, deer farming, deer species suitable for farming in India, rangelands, behaviour of farm animals and farm management, game hunting

---

7.1. FARMING WILD ANIMALS

Wild animals have been persecuted throughout the world for meat. Other than fish, it is the only protein-rich source among the food items for many of the rural and ethnic groups of people in the developing parts of the world. In addition to reptiles and amphibians, mammals and birds form the most important group of animals for this purpose. In the African continent, the practice of domesticating wild animals for the production of meat and byproducts such as skin and horn are in record since the 19th century. Later, wildlife farming was also given emphasis in the context of wildlife conservation. For example, crocodile farms were developed not to substitute protein-rich meat as human food, but mainly to extract its costly skin, which has high market value. Otherwise, crocodiles from natural populations will continue to be extracted for this resource. When they grow in excess in the farm, they can also be released back into their natural habitats. Thus, crocodile farms provide better opportunity for the extraction of the resource, and at the same time, provide better opportunities for the natural populations to grow and sustain. In southern Africa, ostrich is farmed for skin, which is a raw material for luxury leather goods, in addition to its flesh and eggs. Butterfly farming has no food value, but it is an economically-viable practice. Various species of butterflies can be sold at high price like aquarium fish as there are people who have interest to maintain them for pleasure of observation. They can also attract visitors. The concept of wildlife farming also aims to provide wildlife meat and other resources to the local communities at a cheaper cost than the market price for the sustainance of the poor local communities. Such practices can have the reward by way of co-operation and support of the local people in conservation activities.

Wildlife farming can yield economic benefits to rural people. Rural people can also get protein-rich meat from such farms. Wildlife farms can also be used as a tool for awareness creation among the public. For example, butterfly farms can regularly attract tourists and students as a centre of recreation and also as a study centre (Box 7.1 – Figs 64 to 66).

Wildlife farms also play a crucial role in rural land-use planning as these farms can be established as economically-viable units on lands that are presently under marginal to conventional livestock-farming operations or other types of land-uses. The potentiality of game farming for rural development in India was evaluated, and found that in many rural ethnic populations, wild animal meat was the main source of animal food (Naidu 1981). This includes the meat of wild boars, various deer species, hare, rodents, birds and reptiles, along with various hill-stream fish. In countries where diversity of habitats is high, potentiality of wildlife farming is also high, and it differs from area to area or from region to region. The most suitable and well-adapted form of wildlife can be used for farming.

Farming can be sustainable only if the reproductive potentiality of the selected species is high and they can be maintained at lower expenses. Otherwise, such farming will not be economically sustainable. For example, crocodiles can reproduce at a higher rate, and if properly managed and maintained, their populations will increase faster. Rabbit is being domesticated all over the world for meat and fur. They also reproduce at a faster rate and can sustain on a variety of food items including locally-available green fodder. In the Indian subcontinent, sambar deer and wild boar are perhaps the best to farm for meat and skin. Both these species are seen in large numbers in diverse natural habitats, and have high reproductive potential. Wild boar is a major vertebrate pest in most of the areas where they are seen. They sustain in and around farmlands spending daytime under marshes and bushes, and in croplands during night.

In Africa, the feasibility of farming snails was demonstrated in the 1970s (Ajayi et al 1978, Hodasi 1979) as a backyard activity to supplement rural household income, and as a commercial activity to supplement protein-rich food. Edible giant snails in Africa belong to the following two genera: *Achatina* and *Archachatina.* Both of them are common in south of Sahara. Snail populations are highest during the rainy season when they are collected in large numbers by rural communities.

They are marketed fresh or smoke-dried. Some people are very much fond of snail meat, and hence, snail farming is targeted for them. Snail-farming industry is growing rapidly in West Africa. The advantage of snail farming is that they are easy to store and transport alive. As they can also be produced in small units in households, they can have fresh harvest at any time for household consumption. Snail farming is also practiced in some parts of India, where people consider it as a delicacy.

Box 7.1

Butterfly Farming for Conservation and Awareness Creation

*G Mathew and E George**

Division of Forest Health, Kerala Forest Research Institute,
Peechi - 680 653, Kerala, India
(E-mail: gmathewkfri@gmail.com)
*Department of Horticulture, GKVK Campus, University of Agricultural Sciences,
Bangalore, Karnataka, India

Butterflies are highly fragile organisms with diversified habits, and require specific ecological conditions for survival. Natural forests, grasslands, canopies of trees as well as wet areas along the banks of rivers and streams are the typical habitats of butterflies. Habitat loss, commercial collection, and incidence of parasites and predators are the major threats for the survival of butterflies. It has been stated that extinction of a single species would eventually lead to extinction of about a dozen or more species that are linked with it, which in turn, can affect the functioning of natural ecosystems. Management and conservation of butterfly fauna are practiced by *in-situ* and *ex-situ* methods. In the former, the natural habitats are enriched through introduction of suitable butterfly host plants, while in the latter, stocks of butterflies are maintained through captive breeding and are reintroduced into suitable habitats.

Habitat enrichment can boost local butterfly fauna for *in-situ* conservation. The role of butterfly farm (garden) for *ex-situ* conservation and for promoting environmental education is well recognized (Newman

1967). A prototype of the garden was set up in a 0.5 ha patch of degraded deciduous forest in the Kerala Forest Research Institute campus at Peechi, mainly to standardize methodologies for establishing such gardens elsewhere in Kerala (Mathew 2001). The vegetation of this area included a few trees of *Bombax malabaricum, Ailanthus triphysa, Tectona grandis, Grewia tiliaefolia* and *Terminalia bellerica* along the slopes. The terrain was undulating and hilly in nature comprising of steep slopes and with a more or less plain ridge, which was more or less barren and devoid of trees. There was good ground vegetation mostly comprising of *Lantana camara* and *Chromolaena odoratum*. Foraging and breeding are the important activities of butterflies, and hence, due consideration was given to the habitat preferences of various groups of butterflies while designing the components of the garden. Openings with bright sunshine, shade, lianas, damp areas and hedges were set up. Plants introduced in the garden included nectar plants such as *Ixora* species, *Cassia* species, *Allamanda cathartica, Hibiscus* species, *Cuphea miniata, Zinnia haageana*, Marigold, *Mussaenda luteola, M. laxa* and *Clerodendrum capitatum* as well as larval host plants such as *Michelia champaca, Cinnamom zeylanicum, Zanthoxylum rhetsa, Aegle marmelos*, *Citrus aurantia, Wattakaka volubilis, Thottea siliquosa, Tylophora indica, T. camosa, Asclepias* species, *Calotropis giganteum, Carissa carandus, Ruta graveolens, Aegle marmelos, Albizia lebbeck, Cassia* species, *Murraya koenigii, Kalanchoe blossfieldiana* and *K. pinnata*.

Regular monitoring of butterfly populations was carried out by making transect counts during June 1998 to November 2000 to evaluate the effect of habitat enrichment. The number of species and individuals of butterflies sighted in the garden along with data on weather parameters such as temperature, humidity and rainfall were noted. The monthly seasonal index of butterfly citation was used as an index to compare the population trends of butterflies.

As a result of introduction of appropriate larval and adult host plants, there was sharp increase in the number of butterflies visiting the area. During the first half of this study (after 15 months), there were 4,509 sightings of individuals belonging to 43 species. During observations after 30 months, 5,993 sightings of individuals belonging to 50 species of butterflies were made. At the end of 36 months, 23 additional species of butterflies were recorded. Thus, the total number of species of butterflies spotted in the garden was 73, compared to a dozen species recorded prior to the introduction of host plants. The butterflies recorded in this study included eight species endemic to Western Ghats and 10 species having protected status under the Wildlife (Protection) Act of Government of India (GOI 1972) and subsequent amendments.

Fig 64. A view of the butterfly garden developed by the Kerala Forest Research Institute, Peechi

Several butterflies that are fond of bright sunshine such as *Troides minos, Papilio demoleus, P. polytes, Pachliopta hector* and *Pachliopta pandiyana* (Papilionidae); *Catopsilia* sp. and *Eurema* sp. (Pieridae); *Jamides* sp., *Talicada nyseus, Castalius rosimon* and *Everes* sp. (Lycaenidae); *Euploea core, Tirumala limniace, Danaus genutia, T. septentrionis* and *Parantica aglea* (Danaidae) as well as *Junonia hierta*, *J. altites, J. iphita, J. precis* and *Hypolimnas misippus* (Nymphalidae) were found in open habitats. Among them, danaids formed large aggregations with populations ranging from 8,000 to 15,000, on *Crotalaria retusa* and *Heliotropium keralense.* These plants are known to contain pyrrolizidine alkaloids, essential for the development of pheromone-producing glands (Ackery and Vane-Wright 1984).

The area having tall native trees providing cool shade mixed with sunlit patches offered favourable habitats for several papilionids (*Papilio buddha, P. polymnestor, P. helenus, P. paris, Troides minos, Pachliopta pandiyana* and *Chilasa clytia*), which were observed to fly rapidly amid bushes and treetops frequently feeding at the flower-heads of

Clerodendrum paniculatum, Mussaenda and *Ixora* sp. The cool, shaded areas amid dry leaves were occupied by the satyrids (*Orsotrioena medus* and *Melanitis* sp.).

Fig 65. Aggregation of blue tiger on *Crotalaria retusa* in the butterfly garden, Peechi

Butterfly families such as Papilionidae and Pieridae showed a more or less continuous population trend for most part of the year with resident population in the study area, while others such as Danaidae, Lycaenidae, Satyridae, Nymphalidae and Hesperiidae were seasonal. Some of the butterflies colonized the area following the habitat enrichment and host plant introduction were as follows: *Chilasa clytia* (Papilionidae); *Acraea violae* (Acraeidae); *Elymnias caudata* (Satyridae); *Talicada nyseus, Rathinda amor* (Lycaenidae); *Parantica aglea, Tirumala septentrionis* (Danaidae) and *Gangara thyrsis* (Hesperiidae).

Rearing of butterflies and moths has been in practice since long for commercial and conservation-oriented programmes. During the 19th century, collection of butterflies became a serious vocation for many people in the West, which led to the collection, identification and cataloguing of butterflies. Subsequently, butterflies were maintained in glass-houses and opened to the public as butterfly houses. Although butterflies have high fecundity, this potential is hardly realized under natural conditions. Because of the natural mortality factors due to unfavourable climatic conditions (drought, wind, temperature, etc.) and incidence of parasites and predators, survival rate of the butterflies in the wild is estimated to be only around 2%. However, by shielding butterflies from such adverse conditions, a successful butterfly farmer will be able to enhance the survival rate as high as 90%.

As compared to vertebrates, butterflies require only little space for farming, and they reproduce rapidly. Most commercial butterfly-houses are of walk-through heated glass-house type. It is expensive to construct and maintain. Such 'free-flight' butterfly-houses are very common in Europe, North America, Japan, Australia, Singapore, Malaysia, Hong Kong and Kenya. However, such technologically-sophisticated structures and capital requirements are not necessary in tropical conditions. The technological simplicity of butterfly farming, therefore, minimizes the strain on under-developed or developing countries to establish butterfly-breeding and farming programmes. Butterfly-breeding techniques are also relatively simple, which can be readily conveyed to the farmer with simple demonstrations.

Tropical butterflies require daytime temperature of 25°C and 85%–95% relative humidity. Some tropical species require almost 100% humidity to mate. Night temperature can drop to around 15°C. Butterflies should have access to moist sand, for which spraying the ground with a mist of water is the most suitable way. Many of the butterflies need to bask under radiant heat sources to bring their body temperatures up to 32–35°C for oviposition. Appropriate larval host plants should be provided in the cage. Eggs laid should be surface-sterilized by treating with 0.1–0.2% sodium hypochlorite (up to 10 min) or 10% formaldehyde (up to 30 min). Eggs can be stored either in small plastic tubes, loosely plugged with cotton wool, or in small plastic bottles having perforated lids. Reagent-grade sodium chloride as saturated solution may be used to maintain 75–77% relative humidity.

Continued supply of nectar plants for adults and food plants for larvae are essential for butterfly-farming. As caterpillars are highly voracious, they may finish off the host plants in a short period, and therefore, sufficient quantity of host plants should be available in the feeding chamber. Larvae can also be reared on artificial diet. The larvae, when fully mature, undergo pupation by attaching to a substratum with the help of fine silken filaments. After hardening, the pupae may be transferred to an emergence cage or to a pupal rack. They may be surface-sterilized using 2–5% sodium hypochlorite solution to avoid microbial infestation. Adults that emerge can flutter within the hatchery until wings become hardened, after which they may be maintained in the butterfly-release area provided with nectar and oviposition plants for mating.

Important factors to be considered while choosing nectar plants are the feeding preferences of the butterfly species and the length and timing of flowering. Lycaenids and pierids with short proboscis prefer smaller flowers (e.g., Compositae). Larger papilionid butterflies prefer flowers having a long pistil such as *Hibiscus*, *Ixora* and *Clerodendrum capitatum*. Species such as *Clerodendum capitatum, Cuphea*, *Marigold*, and *Ixora* spp., which flower almost throughout the year, are the commonly-used plants in butterfly farms.

Adequate number of nectar plants in bloom should be maintained in the butterfly-release area. When there is scarcity of flowers, artificial feeding may be attempted. Usually, sugar/honey solutions (10%) are presented as soaked pads of cotton wool or in a tube fitted with a coloured corolla. Butterflies often have strong species-specific colour preferences. As these colour preferences may change with age, it is important to provide a wide range of flowers. When given a choice of abundant floral and artificial nectars, butterflies show a preference for the natural substance (Calvert 1990). Moreover, for display purposes, it would be more appealing to observe butterflies exhibiting their natural feeding behaviour.

Some butterflies show preferences for substances such as fruits, berries, sap, dung and carrion. Rotting fruits, to which about 10% honey or toddy is mixed to allow fermentation, are ideal to get butterflies attracted to fruits. For many species, use of mud or sand mixed with salt or animal excreta are important for mud puddling.

The most common cause of failure of Lepidoptera captive-breeding programme is the inability to secure pairings and fertile eggs. Most of the butterflies require a suitable environment for the display of elaborate courtship behaviour. Suitable dimensions for a flight arena relate to the wingspan (x) of the species as: length 20–25x, height 10–15x, depth 10–15x. Large cages are to be avoided as some species tend to disperse to the roof and sides, and ignore potential mates. The presence of the larval host plants within the rearing cages stimulates the pairing in many butterfly species.

The principles emerging from scientific research into ecology have led to the view that it is impossible to conserve a species in its wild state without conserving the whole environment. For reasons mentioned above, current conservation strategy gives priority for protecting entire ecosystem rather than of a single species. Butterfly gardening is an approach that partially fulfils this aspect of conservation. Such farms also serve as centres of conservation and awareness creation among the younger generation. However, the approach is different for a single species, which is in danger of extinction. In such cases, direct and immediate actions are required to identify the causes of the threat, and to modify the conditions that endanger the species. Captive breeding and reintroduction of stock in the wild is a method adopted in such cases. Captive-breeding programmes have been successfully adopted for a number of species such as the heath fritillary (*Melicta athalia*), the large copper butterfly (*Lycaena dispar*) and the swallow tail (*Papilio machaon britanicus*).

Fig 66. School students as visitors in the butterfly garden at Peechi

References and Additional Readings

Ackery PR and Vane-Wright RI 1984 *Milkweed Butterflies, Their Cladistics and Biology*, London: British Museum of Natural History

Calvert J 1990 The horticultural aspects of butterfly houses, *International Zoo Year Book* **29** 18–22

GOl 1972 The Wildlife (Protection) Act, 1972 (Amendment, 1982), New Delhi: Government of India

Heal HG 1973 An experiment in conservation education: the Drum Manor butterfly garden, *International Journal of Environmental Studies* **4** 223–229

Mathew G 1998 *Mass Rearing of Selected Butterflies for Possible Reintroduction in Conservation Programmes*, Research Report No. 146, Peechi: Kerala Forest Research Institute

Mathew G 2001 *Conservation of Invertebrates through Captive Breeding: A Study with Reference to Butterflies*, Research Report No. 220, Peechi: Kerala Forest Research Institute

Morton AC 1991 Captive breeding of butterflies and moths: I. Advances in equipment and techniques, *International Zoo Year Book* **30** 80–89

Newman LH 1967 *Create a Butterfly Garden*, London: John Baker

7.2. CROCODILE FARMING

Until the 1970s, there was unregulated hunting of crocodiles all over the world for their skin in addition to the use of meat. This made them vulnerable to sustain in nature, and hence, efforts were suggested for their conservation. Control of international trade of crocodile skin was enforced under the Convention on International Trade in Endangered Species of Wild Flora and Fauna (CITES). Hence, crocodile farming was initiated in response to the demand for crocodile products. To begin with, the Food and Agricultural Organization (FAO) has provided assistance to develop crocodile farms in several developing countries including India. In Africa, crocodile farms were started first in Zimbabwe, but later spread to several other nations, particularly to the southern and eastern parts of the continent. Crocodile farming can be carried out in villages as a small-scale farming activity with minimum costs and capital. It can help in poverty eradication, and at the same time, it assures sustainable use of this wildlife resource for the welfare of poor and developing nations (Bolton 1978).

Crocodile farms are developed using eggs collected from the wild. These eggs are incubated and hatched in artificial conditions, and the hatchlings are raised in farms. As eggs are collected from the wild, it is emphasized that a proportion of the hatchlings should be returned to the wild to support natural populations. Thus, along with captive breeding and farming, wild populations of them are also improved significantly. The economic benefit out of this activity is very high and it can sustain the local communities involved in the process by enhancing their purchasing power and food security, in addition to the direct benefit in conservation of the farmed species in their natural habitats. It is a suitable way of income-generating mechanism for rural lowland communities in the tropics. The wide range of lands unsuitable for conventional agriculture can be used by the local people to earn their livelihood without major modifications in the habitat. Crocodile farms can also attract tourists, and the owners can get financial benefits by collecting entrance fees. It can also serve as a centre of education to create conservation-oriented awareness among children and youth.

Most of the crocodile ranching (Large-scale collection of eggs or young ones of crocodiles from the wild for captive rearing is known as crocodile ranching), and farming undertaken in different parts of the world have been successful in maintaining wild populations and protecting large extent of wetlands (Whitaker 1989). Mortality rate of the young and eggs of crocodiles in the wild is very high due to predatory effects in addition to hazards of flooding. Collection of eggs from natural habitats, incubating and hatching, and rearing in captivity can make up to 80% survival rate, and can provide seed stock for rehabilitation of wild populations (Webb et al 1987). Maintaining optimum temperature is essential to establish crocodile farms or ranches. Areas, where winter temperature falls below 10°C, are unsuitable for crocodile farming. Clean water is also essential for maintaining the health of the farm animals. Wetlands are more suitable for crocodiles.

During the development of hatchlings, periodical sorting and keeping individuals of similar size together would help the smaller and weaker ones to get sufficient food (Fig 67). Otherwise, the larger and dominant ones would depress the growth of others. They can survive even if crowded. For example, a 10 x 10 m^2 pen can accommodate about 100 individuals of about 2–3 years old. It is also reported that good food supply can enhance the growth levels of captive crocodiles and it can even be twice the level of natural populations (Blake and Loveridge 1975, IUCN 1983). Crocodiles eat a variety of food items such as fish and mammals. In captivity, they can be fed on chopped fish, meat, liver and heart of cattle and on poultry offal. Calcium by way of small fish or powdered cuttlefish shell sprinkled over the food can be a supplement for the hatchlings.

Poor sanitation, low water-temperature and poor-quality diet may affect crocodile farming as the animals can be affected by bacterial and fungal diseases. Parasitic infections of nematodes (lungworms and roundworms) are also common. Hence, proper veterinary care is also important in crocodile farming.

In India, gharial was on the verge of extinction during the second half of the 20th century. The Indian crocodile or the mugger was extremely rare and depleted in its natural habitats by the end of 1970s. The Government of India, with the support of FAO initiated the Crocodile Conservation Project through farming and ranching, with its headquarters in Hyderabad, Andhra Pradesh. This project helped to increase the populations of crocodiles in India (FAO 1983).

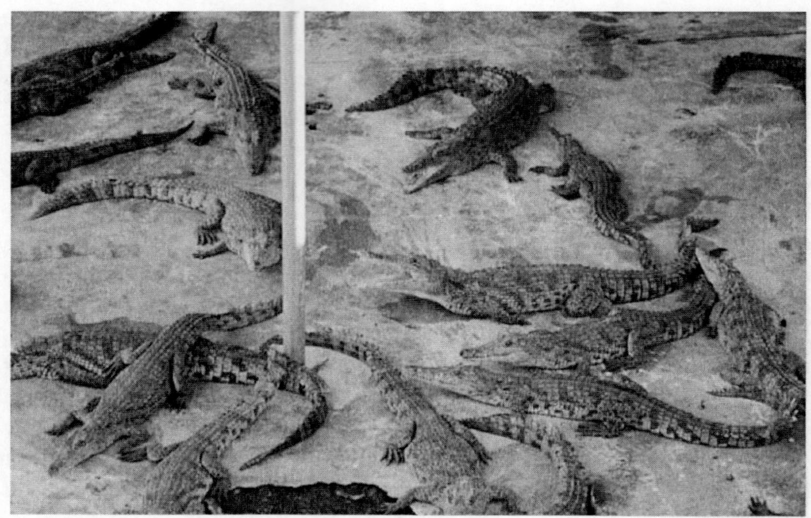

Fig 67. About six years old crocodiles in Arbaminch crocodile farm, Ethiopia (Photo: Sharon M Balakrishnan)

7.3. OSTRICH FARMING

Ostriches are farmed in many areas in southern Africa for their skin and meat (Fig 68). Ostrich skin is used in making luxurious and fashionable leather goods such as vanity bags. Ostrich meat is tasty, and low in fat and cholesterol contents. Zimbabwe and South Africa have ostrich farms running with high profits. The birds are slaughtered at the age of 14 months, when their skin has reached the desirable size; i.e., after the first picking of feathers. The pure wild breed hatched out in farms is also being used to restock wild habitats where their natural populations have become extinct or depleted.

7.4. DEER FARMING

Deer farming can be defined as husbandry of artiodactyls for the production of meat and byproducts such as hides, velvet, antlers and musk on a commercial basis (DeVos 1982), or even for releasing them back in the wild. Deer farming has been in practice since more than 100 years, especially in the European and Scandinavian countries. This has more relevance in developing countries as it can contribute towards the needs of rural people against protein deficiency, and at the same time helping in the process of conservation.

Fig 68. Ostriches in Langano ostrich farm in Ethiopia (Photo: M Balakrishnan)

The wide distribution of the members of the Family Cervidae shows the adaptability of various species of them to diverse geographical and climatic regions around the world. There are several species of them in the tropics and subtropics, which are dwindling fast due to human pressures of hunting and habitat destruction. The wide distribution of the members of the Order Artiodactyla provides feasibility to select the most suitable species for farming in any specified area. The red deer, maral, elk, fallow deer, sika, musk deer, rusa deer and reindeer are being farmed in different parts of the world. Among them, except reindeer, all are present in Asia. In addition to their adaptation for sustaining in different types of habitats, they can also be tamed easily, and hence, their husbandry and management are comparatively easy than farming other animals. However, during rut, males of the deer, especially of the red deer may become aggressive and even attack keepers. It is also interesting to note that the Finnish Laps follow the herds of their reindeer through their seasonal migratory routes throughout the year, permitting them to have a more or less natural way of life.

7.4.1. Deer Species Suitable for Farming in India

A number of deer species are distributed in the Indian subcontinent. Few of them have adapted to a wide range of habitats. For example, the sambar deer is seen in various forest habitats, and has a wide distribution in India. The spotted deer is also having a good range of distribution and can be maintained in captivity. Both these species breed well in captive conditions. The black buck is one of the endangered species adapted to dry weather conditions. Their populations have been dwindling in their natural habitats. Rao and Prasad (1982) have suggested the suitability of black buck for farming in selected areas in India. All these species are manageable in captive and semi-natural conditions of farming.

7.5. RANGELANDS

Rangelands are areas that are not suitable for conventional productive agriculture, but are suitable as a source of forage for free-ranging wild and domestic animals. These areas may experience low and unpredictable rainfall, poor drainage and extreme weather conditions (Stoddart et al 1975). Such areas may also be a source of firewood and other essential resources of the local rural communities. It is estimated that nearly half of the extents of terrestrial habitats on the planet Earth is of this category, where native grasses, shrubs and forbs predominate (Teer 1995). Natural habitats, where intensive grazing is practiced, may gradually change into rangelands as an impact of grazing. In grasslands, human interventions including deforestation would result in edaphic changes, ultimately leading to changes in the vegetative communities. Such areas are also visited by wildlife and in many such areas both wildlife and domestic animals may be seen competing for the available resource. As the extent of rangelands is large, such areas can be alternatively developed for wildlife with additional inputs of resources, where wildlife farming can be a resourceful activity. Such areas can be managed to support both livestock and wildlife without affecting other farming activities. Thus, rangelands can be developed and used as habitats of integrated farming activities.

Wildlife husbandry programmes should be well planned with four major qualities such as the economic viability, technical feasibility, social excitability and acceptance, and resource conservation. A technology suitable for the area and successfully tested on the farms can be implemented to increase production. It should make optimum use of soil, water, plant and animal resources. This should be within the capability of the local people and the forest tribes to operate. It should also maintain the ecological base. The supply-side approach to conservation, as recommended by economists, prescribes the provision of cheap substitutes for wildlife commodities in an effort to lower the price and harvesting pressure. Bulte and Damania (2005) developed a theoretical economic model to examine whether wildlife farming or ranching contributes to conservation. By applying the model to the case of rhinoceros poaching and ranging, they have demonstrated the potentially-ambiguous outcomes of rhinoceros-ranching initiatives, which may recover or suffer from additional depletion, depending on key parameters and the type of competition on output markets. They have also established that this type of ambiguity may be eliminated when policy makers restrict quantities of farmed output through a quota system, so that wildlife farming can unambiguously promote conservation.

7.6. BEHAVIOUR OF FARM ANIMALS AND FARM MANAGEMENT

The success of an animal farm depends on the manner in which the farm is managed to achieve higher reproductive potentiality of the farm animals. A proper management programme would require adequate knowledge on feeding, social and reproductive behaviour of the concerned species. Ecological requirements of the species in their natural habitat have to be studied and such ecological situations are to be provided for them in the farm. A semi-natural condition of the concerned species would be the most ideal situation for wildlife farming. The preferred natural food items and social situations are essential for most species of animals to maintain breeding potential. Information of daily food requirements of the species is essential to follow economically-viable husbandry techniques.

Feeding behaviour of an animal is directly related to the primary nutrient requirements of it for survival and all essential activities. Many of the activities of animals are directly oriented in procuring food. It is a fundamental component of the niche of the species. All animals regulate eating to maintain a constant intake of utilizable energy. Food consumption is an essential factor for growth and body weight, and the amount of consumption changes with the age of the animal. Such changes are relatively marginal among adults as growth is slow, and levelled off at older ages, and in relation to the lower metabolic processes, food consumption also decreases. This relationship can also be expressed in terms of calories of consumption per unit body weight.

Most of the animals feed on a variety of edible items available in their habitats. Usually they select only palatable portion of the food items. There are seasonal changes in the availability of food items in all wildlife habitats. The average daily requirement of the food is also important as the farm is to be managed economically. If extra food is supplied, it may form waste, and if less food is supplied, it may affect the growth rate of the farm animals. It is also essential to supply additional food for the young stock. Water should also be provided in excess as water scarcity may upset digestibility. Supplementary food with fat, protein, vitamins and necessary minerals can also be made available for farm animals at least when there is shortage of natural food items.

Information on social behaviour of the farming species is also highly relevant while dealing with animal farming. The patterns of social interactions vary from species to species. The level of social understanding in one species is different from that of the other. Some species are more sociable than others and many are gregarious, whereas a few are solitary. For the purpose of understanding farm animals, it would be advantageous to study various social groups such as the territorial groups, feeding groups, reproductive groups and parent–infant groups during different seasons of the year related to the physiological status of the concerned species (DeVos 1982).

Schaller (1967) studied social behaviour of some of the deer species in various forest habitats in India. The sambar deer is usually seen in small herds of 2–6 heads, even though occasionally larger herds of up to 10 or more are seen in the form of temporary aggregations. A herd is usually comprised of one hind, one or two yearlings and fawns. Herds of 2–3 adult hinds, and their yearlings and fawns are also seen. In some herds, bucks are also seen. The spotted deer lives in herds of 5–10 individuals. Larger herds of a number of family units living in the nearby areas are also occasionally seen in aggregations. All age groups are reported in the herd. Social interactions in the herds of black buck change with season. They have several types of herds such as mixed herds of both sexes with all age groups, breeding herds of one buck and several does, and all-male herds. The average size of the herd is 10–20 individuals. During the breeding season, the population is generally divided into breeding groups of one or more bucks and a few does and fawns, and they restrict their movements to a limited area. The leader, who is the most healthy and prominent buck, has the priority access to food and mate.

Social and reproductive behaviour of the black buck introduced into the Park of Cleres (Rouen, France) was studied by Dubost and Feer (1981) in a semi-natural enclosure of size 6.6 ha. The adult male associated with the harem becomes the leader. The herd also consists of subgroups that can most clearly be distinguished at certain times of the day, especially during the period of rest. Adult females form the core of the harem. They stay close to each other except during estrus, parturition and during the first bouts of nursing. Aggression between females of the harem is rare. Males of 1–2 years usually avoid adult males and keep themselves on the fringes of the herd. Bachelor herds are also seen consisting of males of all age groups – adult and young males, chased away from the harem. Information on reproductive behaviour is also important with respect to animal farming. Basic knowledge of the behaviour of farming animals can help farmers and managers to initiate measures to enhance and maintain better reproductive status and to take care of the new born.

Reproductive performance depends to a great extent on environmental and nutritional factors. Selection of breeding pairs can be based on the behaviour of both male and female; the superior ones form the breeding pair. This can help to maintain the farm with individuals of superior quality of genes for better growth rate and body size. Extraction of the animals for the products should be based on the optimum growth, so as to make the farming process economically more viable.

7.7. GAME HUNTING

Animal hunting is an act of seeking, following and killing animals for sport, consumption and for display. Historically, hunting of wild animal is known since the Palaeolithic period as a means of obtaining animal flesh for food. Hunting with weapons is the most popular, even though trapping and snaring are also practiced by rural communities.

Wildlife hunting is categorized into subsistence hunting, commercial hunting and sport hunting. The objective of subsistence hunting is to obtain food for the hunter and his family, whereas commercial hunting is to sell the resources of the hunted animal with commercial interest. In general, both these types of hunting are practiced by people of the lower-income group. On the other hand, sport hunting is a part of leisure activities by the wealthy people of the community including the rich urban dwellers. In case of sport hunting, cash is earned not by the hunter, but by the owner of the company who runs the business. In the developing countries, for example, in most of the African countries, this is mainly aimed at the effluent foreign tourists. Special terms such as 'game hunting' and 'trophy hunting' are also used along with this type of hunting, mainly to attract tourists. The word 'game' denotes the act as a part of sports. Hence, game animals are those larger birds and mammals considered under this practice. Trophy hunting is also game hunting, but the difference is that in this case what is hunted is considered more for its trophy, for example, the head of the animal with horns or tusks as the case may be. In this case, the whole animal is also allowed to be skinned, stuffed and preserved. The hunter has the right to take the stuffed body as reward. Such items are certified legally, and thus, they can be displayed even in private places including residences. Trophy hunting is an expensive sport, as the hunter has to pay the amount fixed by the hunting operator for each species of animal.

There are ecological and genetic impacts of hunting. In trophy hunting, the largest individual with the best quality product (e.g., horn, tusk and size) in the area is located to hunt (Milner et al 2007). On most occasions, the oldest individual with lots of experience in the area leads other individuals of the group as described by McComb et al (2001) in connection with the large-scale poaching of the African elephants. They have reported that the leader of the herd of elephants is not just one among the herd members, but is more than that. It is the repository of a variety of information on the home range of the herd. If this expertise of the herd is lost through the elimination of the leader through hunting, other members of the herd are not safe. The genetic characteristics of the best in the group will also be lost, instead of getting inherited into next generations.

Culling is also a type of wildlife extraction, but this is practiced when selected wildlife populations grow beyond the carrying capacity of the habitat, and the enhanced numbers of individuals are extracted for meat, skin and other products. Such controlled hunting practices are aimed to maintain the populations of larger mammals under control at optimum levels, and not to be responsible for habitat depletion. In such practices, only old males beyond the reproductively-potential age are culled. Hence, this practice does not affect the growth potential of the population. Further, it does not have any effect on the genetic quality of the future generations, unlike the case of trophy hunting.

Hunting operations also have positive effects in the modern perspectives. Income generation for conservation has been a hurdle for the managers of protected areas. Trophy hunting is considered as a good means of income generation, which can be subsequently used for conservation and management activities, in addition to support local communities (Lewis and Jackson 2005). Hunting operations are maintained based on the carrying capacity of the habitats. The operators are responsible for the development, conservation and management of the leased out hunting area and for maintaining better reproductive levels of the game animals so as to make them available for hunting whenever there are visitors with interest in hunting. Thus, it is considered as a development programme with financial gain, supporting the society, and as a part of environmental conservation.

Even though, legally-permitted safari hunting, game hunting, trophy hunting and culling are not being practiced in India, there are several African and European nations where these are part of wildlife management programmes. Some of the nations in Africa such as the South Africa and Zimbabwe have substantial foreign exchange earnings through wildlife hunting, where both government and private agencies run this business legally.

Questions for Review/Exercise

1. What are your perceptions on the following?

   a. Crocodile farming   b. Ostrich farming

   c. Game hunting

2. Why should we farm wild animals? Have you visited any such farms? If so, write a short account of the farm you have visited.
3. How can a wild animal farm help local people?
4. How beautiful are butterflies? How many species of butterflies you can identify?
5. Observe insects in your home garden. What are the common insects you have observed? Relate your observations of insects with the plants they visit. Observe them and understand the interactions and functions.
6. Have you heard of a crocodile farm in South India? Find out the information and make a short write up on it.
7. Have you ever seen an ostrich? What is the size of an ostrich egg? Is it the largest cell?
8. Have you seen a rangeland? How is it different from a conventional agricultural land? What are your suggestions to make use of rangelands more efficiently?
9. Have you visited any of the game hunting areas? (There are a number of legally-permitted areas for hunting specifically-listed wild animals in the respective areas with a fee for hunting). Do you have a game hunting area in your region? What are your perceptions on the ecological and genetic impacts of game hunting practice?

10. What have you learned from Boxed item 7.1? Have you visited any such farms maintained for conservation and awareness development among the public?

❑ ❑ ❑

Economic Benefits of Wildlife 8

Assessing economic benefits of wildlife, non-consumptive value of wildlife, consumptive value of wildlife, wildlife as pests, civet as economically-important wild animal, economics of wildlife conservation, hidden and unknown features of wildlife and wildlife habitats, carbon sequestration

---

8.1. INTRODUCTION

Natural biological wealth of the planet Earth is distributed all over the world in natural forests, plains, mountains, wetlands and other aquatic habitats. However, some of these are seen in higher concentrations in some of the geographical ranges such as the tropics, and especially in areas with more resources such as water, food and shelter essential for wildlife. Also, there are wild animals with specific abiotic requirements such as factors related to temperature and rainfall. This is evident from the locations of biological hotspots such as the Western Ghats and the Eastern Ghats of India, and the Ethiopian highlands where endemic fauna and flora are conspicuous. Wildlife is the wealth of nature, which is not uniformly distributed all over the world; it is distributed with higher concentration in the tropics. Thus, geographical range, mountainous terrain and climatic conditions are important factors in this respect. With species-specific adaptations, some of the species of wildlife are restricted in their distribution to some of the special ranges in such areas. For example, the Nilgiri tahr is limited in its distribution to certain ranges in the Western

Ghats only, while its counterpart, the Himalayan tahr is restricted in its distribution only to certain ranges in the Himalayan peaks. Effective conservation and management of such areas, in addition to the overall conservation and management practices, can help majority of the wildlife resources to survive and enhance their potentiality in quantity and quality. Constant regeneration of such resources is not only essential for natural ecological processes, but it also helps mankind to constantly extract economic resources on a sustainable basis. However, human activities have been negatively affecting the survival of many of them in forests, wetlands and seas, leading to non-renewable status of wildlife resources in the tropics.

8.2. ASSESSING ECONOMIC BENEFITS OF WILDLIFE

Wildlife has consumptive and non-consumptive values. The economic benefits of wildlife can be assessed using the following criteria (McNeely 1988):

- Assessing the value of natural products such as fire-wood, fodder, game meat, honey and minor forest produce collected by indigenous and other local people; these forest products have been used for their subsistence since time immemorial.
- Assessing the value of all the above and other products such as timber, tusk, horns, animal skin, medicinal plants and other useful resources that are illegally harvested for commercial purpose by local people as well as by those harvesting legally under government control.
- Assessing indirect values of natural and man-made ecosystem functions such as photosynthesis, watershed protection, soil protection and conservation, regulation of climate, plant–animal interactions, predator–prey interactions and such other properties of natural habitats.

The economic potential of wildlife is very high. For example, just in one year (1996), 35.2 million Americans of 16 years of age or older went for fishing, 14.0 million went for hunting and 62.9 million took part in at least one type of wildlife-related recreation (wildlife observation, picnic in forests, bird watching and photography), and the associated expenditure amounted to US $ 101 billion (transportation, equipment, lodging, etc). A similar study in Canada revealed that around 85% population of the nation participated in nature-related recreational activities to a tune of $11 billion. Out of this, $6 billion was for transportation and related activities, $ 3.1 billion for cameras, guns, fishing rods and such equipment, and $ 1.8

billion for other items used to pursue nature-related activities (McNeely 1988). These examples show the readiness of the financially-sound people who can spend money on wildlife recreation and related activities, and hence, once the resource is lost, a segment of the national economy in which large sums of money are exchanged will diminish.

8.3. NON-CONSUMPTIVE VALUE OF WILDLIFE

Being students of ecology, we are more interested in non-consumptive values of wildlife. Wildlife tourism is very popular all over the world. Some of the countries such as Kenya earn a lot out of this business, and it forms the second largest source of foreign exchange (next to coffee). The number of foreign tourists visiting Kenya annually is over 3,00,000 whereas it is over 2,50,000 in Zambia. The net profit of Kenya in this industry is over 100 million USD. It is estimated that the returns from Amboseli National Park are equivalent to \$ 40 $ha^{-1}a^{-1}$, almost 50 times of its potential agricultural value. In the same Park, based on tourist attraction, one lion is worth \$ 25,000 and an elephant herd is worth \$ 6,00,000 (McNeely 1988).

To attract rich tourists, there are some essential requirements to be satisfactorily met with. Of course, the resource in this case – wildlife itself is important. However, it is also essential to have excellent infrastructural facilities to attract European and American tourists, who are known as rich and ready-to-spend tourists. As far as Asian and African nations are concerned, the scenic beauty and topography are attractive, but in many of the developing nations, infrastructure facilities, particularly in eco-tourism centres, are poor to attract rich tourists. There is a need for further development of clean, well-built and well-maintained living facilities to attract rich tourists in eco-tourism centres.

Aarogyappacha (*Trichopus zeylanicus travancoricus* Burkill ex Narayanan) is popularly known among the local tribal people of Agasthyar hills of Kerala, South India as an energetizer. Fruits of this plant are used by the indigenous Kani people of the area for sudden flush of energy and strength after hard work. Scientists of the Jawaharlal Nehru Tropical Botanic Garden and Research Institute (JNTBGRI), Palode, Kerala gathered this information from the Kanni tribes during botanical expeditions in the area during 1987. Investigations using the active ingredients of this plant have revealed that it possesses anti-stress property. Scientists formulated a drug known as Jeevani using this plant and three other medicinal plants. The JNTBGRI rewarded the local Kanis with half of the benefits they got by selling the right to manufacture this drug to a private pharmaceutical company in South India.

Insects form a group of invertebrates having considerable economic importance. Some of them are noxious pests whereas a few are extremely useful to mankind. For example, several species of insects help in pollination. We use the potentialities of honeybees and silk worm moths as natural resources in honeybee farming and in silk industry, respectively. Butterflies are beautiful creatures that can be commercially reared. If properly developed, butterfly and other beautiful insect collections can be developed as a profitable hobby, which can also yield foreign exchange.

How can we convert these features of the once exterminated by poachers in terms of direct economic gains? Similarly, when a large tree in a tropical forest is cut, several organisms directly and indirectly associated with that tree and the interactions between them are lost. Even cutting a major branch of a tree may affect the interactions of organisms associated with that tree as well as in that area and the habitat.

8.4. CONSUMPTIVE VALUE OF WILDLIFE

Even today, wildlife meat is the only nutritive food of a large number of rural people. They depend upon wildlife around their settlements for their essential requirements. In addition to this practice, there is licensed hunting in some of the developed and developing countries, wherein the hunted wild animal forms the food of the hunter. Only in the case of licensed hunting, the hunting value is available. On the other hand, non-licensed hunting is not accounted for income generation by governments. A good sum of money is collected by culling practice in some of the African nations such as Zambia and Kenya.

Not only meat and animal byproducts, but also a variety of other products such as honey and medicinal plants are also heavily extracted from wildlife habitats all over the world. Such products are not necessarily channelled through local markets, and hence, the direct commercial values of such items are not assessed always. It also gives employment opportunities for millions of rural people, who are engaged in the collection process, both legally and illegally.

8.5. WILDLIFE AS PESTS

Although we consider wildlife as an excellent natural resource, some wild animals are noxious pests of agriculture, and hence, they create nuisance for the farmers in rural areas. A number of rodents are pests. More than 30% of the produced crop is lost or damaged as a result of rodent infestation. This is a heavy loss, particularly for the poor farmers. Adaptability of rodents made them to survive in almost all regions of the world, and hence, they can survive on a variety of food items. For example, the Norwegian rat (*Rattus norvegicus* Berkenhout, 1769), which is originally from northern China, has spread all over the world through the well-known Norwegian shipping activities. Several species of birds are also crop raiders during the cropping season. Wild boars, warthogs (Fig 69) monkeys, deers, elephants and hippopotamuses are known as serious crop raiders, at least during certain season in rural villages. In India, the wild boar and certain species of primates (e.g., bonnet macaque and common langur) are crop raiders in rural areas wherever they are present. Tiger and leopard are cattle lifters. Leopard is a common problem in certain villages, where they forage for domestic animals such as dogs, goats and sheep. Some individuals of such species may develop the habit of foraging in human settlements as they find it easy to get prey from human dwellings. Such individuals are being trapped and relocated by the foresters to help the rural villagers.

Fig 69. Warthog (*Phacochoerus africanus*), an African species is the counterpart of the wild boar of Asia. Even though they are considered as agricultural pests around forests, they have a major ecological niche in their habitats, responsible for rejuvenation of soil characteristics through their feeding patterns (Photo: Sharon M Balakrishnan)

8.6. CIVET AS AN ECONOMICALLY-IMPORTANT WILDLIFE

The use of odorous animal secretions dates back to more than 1000 years BC, out of which 'civet' and 'musk' have gained real commercial importance. Civets are extensively used for their perineal gland secretion (known as 'civet'), which is an important ingredient in perfume industry. 'Civet' increases durability of the perfume into which it is incorporated. It ranks among the twenty most-important raw materials in perfume industry, and is also one of the ingredients in a number of Ayurvedic preparations (a traditional Indian system of medicine). It is′ also used for flavouring tobacco.

In India, 'civet' comes from the small Indian civet (*Viverricula indica* É. Geoffroy Saint-Hilaire, 1803). Killing, capturing and rearing them are illegal. However, there are several civet holders in Kerala (India), maintaining them for the collection of 'civet' (Fig 70). In a survey of 200 households in different villages adjacent to forest areas, wildlife sanctuaries and plantations, 273 captive civets were located. Some of these households had experience in civet farming, but they had no animals at the time of the survey. Zoological gardens, Ayurvedic institutions, research establishments and a centre of worship had civets under captivity. Private civet owners are not interested in breeding them as the secretary output becomes less during the period of pregnancy and lactation. They depend on natural populations for restocking the captive colonies. If captive breeding is practiced and the captive civets are well maintained and managed, civet farms can form an excellent animal resource without adversely affecting the natural populations (Balakrishnan and Sreedevi 2007 a, b).

Captive breeding of civet cat and 'civet' production can yield considerable economic benefits. Ethiopia is known for the African civet population, captive civets (Fig 71) and for export of their anal gland secretion in large quantities. This forms an ingredient in perfume industry, and hence, is required in large quantities in countries such as France. There are several civet farms in Ethiopia (both legally and illegally established), but the civets are not properly treated by the farmers. Hence, there was a suggestion to ban the export of the gland secretion of African civets from Ethiopia.

No systematic investigation has been made to ascertain the economics of 'civet' production, either in Asia or in Africa. The loss of habitat and depletion of natural environment as well as inroads into wild populations for restocking are the major reasons for the less frequent sightings of them in their present range of habitats. Species become vulnerable to extinction due to direct human intervention and stochastic processes. Civet farming should be promoted by legalizing the existing farms and by improving their farming facilities. Captive breeding should be encouraged. Financial incentives at the initial stage can encourage the civet holders to breed them under captivity, which in turn, can reduce the pressure on natural populations. The major expenditure will be the capital investment on housing requirements. As this investment is insignificant as compared to other farming systems, civet farming, if made legal, is suggested to be a feasible project that can be initiated by marginal farmers.

Fig 70. Captive small Indian civets in Kerala (Photo: MB Sreedevi)

Fig 71. Captive African civets in Ethiopia (Photo: M Balakrishnan)

As there are legal restrictions for the trading of several wildlife species and their products, one should be careful in dealing with farming and related activities using wild animals. Any type of wildlife farming may be made only on getting legal permission from the concerned wildlife conservation agency.

8.7. ECONOMICS OF WILDLIFE CONSERVATION

The cost-benefit analysis seldom reflects the benefits of protecting natural areas. Social benefits of conserving biological resources are often intangible, widely spread and not fully reflected in market prices. Traditional assessment of national income does not find a place for ecosystem maintenance. What is reflected is only the price for the timber extracted from the ecosystem and sold out legally. However, this income denotes the price of destruction caused to the ecosystem and not for the resources retained intact in the habitat (McNeely 1988).

Politicians, policy makers and administrators have to depend upon the benefits that can foresee out of the reforms they set. However, as immediate and direct benefits of conservation are negligible, they cannot be properly convinced, especially in terms of the expenditure involved and the revenue that may be earned. Direct and immediate economic benefits are always weighed when development policies are planned. With the level of scientific information now available, it is difficult to assess the economics of conservation so as to convince administrators and politicians. This has been a major handicap in finding out resources for conservation all over the world. To overcome this problem, conservationists have initiated action programmes to demonstrate the economic value of biological diversity, which can be made available for the socio-economic development of the regions mandated for conservation.

8.8. HIDDEN AND UNKNOWN FEATURES OF WILDLIFE AND WILDLIFE HABITATS

All types of habitats need to be conserved for conserving global biodiversity. Usually, one may discriminate types of habitats and stress for conservation of wet evergreen and moist-deciduous types of forests. This is due to the fact that such habitats may have more species diversity, which may be habitat-specific and endemic ones. In contrast, arid and semi-arid habitats harbour only few species. However, due to the special adaptations of these few species to withstand adverse environmental conditions, they feature many potentially-valuable biochemical compounds such as the liquid wax of the jojoba shrub (goat nut or pig nut) and the natural rubber of the guayule bush. Many such organisms are threatened in their natural habitats due to various types of human interventions including livestock herding and conversion of land for farming activities.

Coral reefs, with an estimated half a million species in the 4,00,000 km^2, are being depleted at faster rates. Coral reefs represent the most-crowded habitat among the biological habitats, and the organisms of this habitat engage themselves in a biological warfare to ensure living space. During this process, the reef-dwelling organisms generate a variety of toxins, which are found to be of immense value in the field of pharmaceuticals. Even in tropical forest habitats, not only wet evergreen and moist-deciduous habitats are important, but all habitats are equally important for conservation. Studies on habitat association of animals have revealed that each habitat may contain species that are specific to that habitat (Balakrishnan and Easa 1986). Hence, large extent of land and aquatic habitats containing all types of natural organisms need to be conserved as integrated units for wildlife conservation.

The functions of ecosystems in one part of the world may also be beneficial to nations and regions in other parts of the world. For example, conservation of native gene pools in the tropics may also be beneficial for the temperate nations, as the resources will be more utilized by them in genetic engineering and in biotechnological advancements than the poverty-stricken nations that hold such rich resources. The depletion of habitats and the extinction reported during the past show that a major part of the benefits accrued out of them were shared by powerful global forces, primarily driven by markets in developed nations. Most of the costly biological resources such as animal skins, tusks, horns and several other resources gathered by commercial poachers, and the related costly byproducts are sold illegally in global markets. Thus, it is clear that people all over the world use the resources generated in any part of the world, especially those in nations that lack such resources. Hence, it is reasonable to argue that the heavy cost of conservation of natural resources in the tropics is also to be shared by the developed nations in the temperate world, or more uniformly by the global community.

What about the value of the genetic resources lost as a result of poaching or even by hunting on hunting licences? Poaching has always been aimed at the quality of the resources such as tusk, horn, flesh, skin and feathers. In case of Asian elephants, the tuskers have become less in many of the natural habitats, and hence, the chances of tuskers to be involved in reproductive processes are limited. As a result, tuskless males contribute for the reproductive potentiality of the population leading to increase in number of tuskless males. The trophy-harvested rams (*Ovis*

Canadensis Shaw, 1804) were of significantly higher genetic 'breeding value' for body weight and horn size than rams that were not harvested. Rams of high breeding value were shot at an early age before achieving high reproductive success. Decline in breeding values for weight and horn size occurred in response to hunting, which resulted in production of smaller-horned, lighter rams (Bolen and Robinson 2003).

8.9. CARBON SEQUESTRATION

Carbon sequestration refers to the process of capture and long-term storage of atmospheric carbon dioxide (CO_2) (one of the most important greenhouse gases) in suitable reservoirs, which occurs naturally through biogeochemical processes. Plants are excellent agents of carbon sequestration processes, which they perform through photosynthesis. During photosynthesis, plants assimilate CO_2 and ultimately store in the plant biomass. Thus, large trees act as excellent natural reservoir of CO_2. Even when they die and fall down on the surface of the Earth, they slowly release the stored carbon into the atmosphere during the process of putrification. On the other hand, if large trees are buried due to any of the processes by which large biomass is buried under soil, the soil will act as the reservoir for CO_2. Hence, natural forests, particularly tropical wet-evergreen forests have a major role in purifying atmosphere by reducing the effect of CO_2, and thus, controlling the effects of greenhouse gases to balance climate change. Subsurface saline aquifers, peats, lakes, oceanic habitats and oil fields are also natural carbon sinks. When carbon sequestration takes place through biological processes, it is referred to as biosequestration. As we have witnessed large-scale deforestation all over the world in the past couple of decades, human beings were responsible for release of naturally-stored large quantity of CO_2 into the atmosphere. As compensation for this mistake, and to overcome the problem of greenhouse gas effects, we have been involved in reforestation activities (Fig 72). The ultimate objective of such activities, in addition to awareness creation among the public, is to facilitate the enhancement of carbon-trapping capacity in human surroundings. In fact, it is the large quantity of CO_2 stored in the biomass and soils buried over and over for millions of years with additional pressure and temperature that coal and oil are formed and stored deep under the soil (Parr and Sullivan 2005). When we explore and mine coal or the petroleum resources including natural gas, the trapped CO_2 is released into the atmosphere, which leads to the accumulation of greenhouse gases. Even though water vapour is also one of the greenhouse gases, human beings have no major influence in the production and release of it into the atmosphere. The large extent of the oceanic water evaporates naturally and spreads all over. On the other hand, there is a major role for human activities in releasing increased levels of CO_2 into the atmosphere from time to time, and hence, Kyoto Protocol insists for control of such greenhouse gas emissions (Hohne et al 2007). Atmospheric carbon dioxide has increased from about 280 ppm in the year 1750 to 383 ppm in 2007, and is increasing at an average rate of 2 ppm per year (IPCC 2007). It is estimated that global deforestation has been responsible for about 20% of the overall greenhouse gases in the atmosphere. Hence, biosequestration is the best remedy to control the emissions of greenhouse gases and to balance climate change. The rate of biosequestration can be increased by increasing forest coverage and plant biomass all over the world (Renwick et al 2002, Azar et al 2006).

Fig 72. Students and teachers involved in a tree-planting programme in the Millennium Park in Addis Ababa. In the long-run, such afforestation programmes will have mitigative effects on various ecological parameters including a major role in carbon sequestration (Photo: M Balakrishnan)

The most important conservation challenge that we are facing is to protect a variety of wildlife of the world by conserving the ecosystems wherein they sustain. However, majority of the tropical nations are being forced to live on their biodiversity capital by exploiting and destroying at higher levels than they reproduce (leading to non-renewable status). Thus, they are destroying the natural wealth for the future. This trend is required to be changed by adopting appropriate conservation and management practices, wherein the resources can be sustainably utilized for national development, not only for the time being, but also for the future to support our future generations.

Questions for Review /Exercise

1. Have you understood the following?
   a. Non-consumptive value of wildlife
   b. Consumptive value of wildlife
   c. Carbon sequestration
2. How can we assess the economic benefits of wildlife? Evaluate economic benefits you expect from wildlife.
3. What are your perceptions on the value of a herd of elephants or a pride of lion, which attract tourists?
4. What is your understanding about the economic benefits of insects? Can you cite few insects that are pests and few others that are beneficial to us? Explain direct benefits we get from insects. Have you tasted honey? How are we getting honey? Have you seen natural honey beehives? What do you know about the social life of insects? Have you heard of honeybee dance (waggle dance)? Who invented honeybee dance, and for the related research shared the prestigious Nobel Prize?
5. Have you heard of silk? How is it produced? Have you seen mulberry plants? Have you seen silk worms? Have you seen silk-producing processes? Visit any of the villages where local people are involved in producing cocoons for silk production and witness the activities leading to the production of silk.
6. What is meant by non-consumptive value of wildlife? Can you cite examples? How does it differ from consumptive value?
7. Have you experienced any wild animal as a pest? Have you heard about this from your parents or from the villagers in your locality?
8. Try to locate some of the pest animals in your locality. What about rodents? Are they pests in your locality?
9. Which is the largest rodent you have seen? Do you know that porcupine is a rodent? Have you seen porcupine spine? What is it? Have you seen the Malabar giant squirrel? How does it look like? Did you notice its beautiful colour pattern?

10. Have you heard of civets? Which among the civets have you seen (small Indian civet or African civet)? Have you seen a civet in any of the zoos you have visited? What do you know about the useful properties of civets?
11. What are the purposes for which the perineal gland secretion of civets is used? Do you use perfume? Do you know that the perineal gland secretion of civets is used in perfume industry?
12. What are your perceptions about the direct and indirect benefits of forests and wildlife?
13. Can you try to explain what will happen to the organisms directly and indirectly associated with a large tree in a tropical forest when the tree is extracted?
14. Do you know that elephants are killed by poachers? What are the consequences of poaching wild animals? Explain with respect to the social life of elephants.
15. What are your perceptions about carbon sequestration? How does it contribute for environmental purification? What is the role of plants in this process?
16. What did you learn from Fig 72? Did you participate in any of the tree-planting programmes? How can such activities help in the process of carbon sequestration and in controlling global warming?

Case Study 8.1

Creating a Revolving Fund for Supporting Wildlife Management in Zambia*

The Luangwa Valley is one of the richest wildlife areas in Zambia, which contains national parks covering about 20%, and game management areas covering about 60% of the valley. The game management areas differ from parks in that they are zoned for wildlife utilization and allow human occupation. However, legally all wildlife is the property of the State, and to harvest the animals in game management areas requires licenses that are often prohibitively expensive to residents.

The wildlife resources of the Lupanda region of the Valley has supported safari hunting, which yields about $ 3,50,000 per year, but less than 1% of the safari hunting revenue was returned to support local village economies and a negligible amount went towards wildlife management costs. As a result, local support for conservation has been very low, and illegal hunting of wildlife, especially elephants and rhinos, has reached at such high levels that extirpation of some species is a real possibility. Despite strengthened law enforcement, villagers welcome poachers from distant parts of the country, as long as they share some of the harvested meat with the community.

Based on the premise, at least part of the revenues earned from wildlife should be returned to the National Parks and Wildlife Service (NPWS) to manage the wildlife resource, a Wildlife Conservation Revolving Fund was established in 1983. This Fund also enabled the NPWS to employ additional staff beyond the Government approved civil servants.

A Wildlife Sub-Committee was established in the chiefdom administered by a Unit Leader from the NPWS, and a village scout programme was initiated. The scouts were given a training course of six months, officially designated as wildlife officers, and employed throughout the year in their respective chiefdoms as the local 'custodians' of the wildlife resources of their villages. Additional manpower was also recruited from the local community on a seasonal basis to assist with other management needs, including building maintenance and construction.

Income for the Wildlife Revolving Fund came from harvesting of hippos, and from the auctions among safari hunting companies for the rights to hunt in the Lower Lupande Game Management Area, with terms of the auction including quotas of animals, which could be taken, and minimum levels of employment from the local communities. Forty percent of the proceeds from the auction was handed over to the local Chiefs for community projects of their choice and 60% was devoted to wildlife management costs.

Results were remarkable. Manpower increased from 11 to 26 from 1985 to 1987, and the number of field-days by staff increased from 176 to 717 man-days. Annual mortality of elephants and black rhinos, expressed as a number of poached carcasses found per year per square kilometre, decreased by 90% in the same period. In 1987, the total earnings for the Revolving Fund were $ 48,620, of which $ 14,840 was devoted to wildlife management, including $ 4,410 for the village scout programme. Overall, recurrent costs of wildlife management for the year were $ 9,870, considerably less than earned by the Revolving Fund. Villagers started supporting the NPWS wildlife management effort, and local headmen established security committees to prevent poachers from entering their areas.

Once economic benefits started to flow to the local villages, the reduced poaching of elephants has led to an increase in their populations to the level where sustainable harvest far exceeds the total costs of effective management programmes. In addition, about half the costs of supporting the village scouts was equivalent to the total sums derived from revenue earned by ivory collected by scouts from elephants that died naturally. While this source of revenue did not go back into the Revolving Fund, it did illustrate to the government the magnitude of funds, which could be recovered by this form of local involvement in wildlife management.

In summary, the Wildlife Fund acts as a legal mechanism for charging concession fees, selling wildlife products, and engaging in commercial ventures related to wildlife development. The Fund can then direct the income into appropriate channels to serve the interests of managing the biological resources of the area, and the interests of local communities co-existing with the wildlife. It therefore reduces the need to depend on Central Treasury for funds, which in recent years has been unable to meet the growing cost of conservation. A central factor in the success of the Fund was its establishment within NPWS. Using another agency or government body would have diluted the impact of the Fund, particularly its ability to employ local residents as legally authorized wildlife officers.

(**Reference:** Lewis et al 1987 as cited in McNeely 1988)

Essence of a case study is presented above as given by McNeely (1988). Read carefully and understand the economic incentive programme and the impact of it on the local villagers and the wildlife around, and review the following:

i. Are there similar wildlife areas in your locality/State/Country? Are there any such programmes to support local people from the earnings from wildlife resources in your area?

ii. If yes, have you realized any of the economic benefits of wildlife conservation through such programmes?

iii. In case there are no such programmes, can you think of planning a similar programme through which local involvement in wildlife management can be made available, and part of the earnings from wildlife resources can be transferred for local development?

iv. Check with the local forest and wildlife agencies and find out the existence of any programme involving local communities (participatory forest and wildlife management programme) in wildlife management and resource utilization. Would you be interested in joining any such programme if there is an opening?

v. Can you propagate the above case study? What have you learned from the above Zambian experience among the villagers and other communities with which you are in contact (e.g., school, college and other public forums)?

Protected Areas and Wildlife Conservation 9

Wildlife conservation and protected areas, history of PA system, world coverage of PAs, megadiversity countries and hotspots, protected areas and conservation initiatives, features of areas to be conserved, categories of PAs, IUCN revised categories of PAs – 1994, *in-situ* gene banks: a new category of PA, types of utilization, criteria for the selection of PAs, protected area management/park management, local people and PA management

---

9.1. WILDLIFE CONSERVATION AND PROTECTED AREAS

Wildlife conservation is aimed at maintaining diversity of all taxa in the area of conservation, whereas wildlife management is aimed at maintaining populations of selected species by manipulations so as to achieve better levels of populations for higher levels of extraction. Wildlife management and utilization have been in practice since long in the history of forest and wildlife administration. However, conservation, as compared to management, is a recently developed branch of biological sciences, with its application in biodiversity conservation. In case of management, manipulation of populations is done by the practitioners. Sustainable extraction and utilization are possible from well-conserved and scientifically-managed ecosystems. Wildlife management is mostly practical-oriented, whereas conservation is mostly theory-oriented. In case of management, usually larger species such as higher vertebrates are pinpointed whereas in conservation, all taxa are equally incorporated and considered.

The Global Biodiversity Strategy has defined Protected Area (PA) as a legally established land or water area under either public or private ownership that is regulated and managed to achieve specific conservation goals (WRI/IUCN/UNEP 1992). Protected areas are geographically defined land or aquatic areas that are designated or regulated and managed to achieve specific conservation objectives (UNEP 1992). The Fourth World Congress on National Parks and Protected Areas defined PAs as areas of land and/or sea, especially dedicated to the protection and maintenance of biological diversity, and of natural and associated cultural resources, and managed through legal or other effective means (IUCN 1994a). It is a clearly defined geographical area recognized, dedicated and managed through effective management techniques to achieve long-term conservation of nature with associated ecosystem services and cultural values (Dudley 2008, Dudley et al 2010). Protected areas provide key instruments for maintaining wider ecosystem resilience; i.e., the ability to absorb disturbances while retaining the same basic structure and functions (IPCC, 2007) at a landscape/seascape level. Protected areas can sustain a variety of habitats and landscapes of diverse plants and animals, and thus, they also form sources of wild varieties of domestic animals and plants. Thus, PAs act as storehouses of natural resources such as gene pool, water, minerals, forest produce and medicinal plants including microbes. Such areas contribute for ecosystem services such as absorption of natural disturbances, resilience, atmospheric purification and carbon sequestration. Special services that such areas provide are provision for scientific research, protection of wilderness, conservation of species and genetic diversity, and potential for ecotourism and recreation (Fig 73).

9.2. HISTORY OF PROTECTED AREA SYSTEM

In the year 252 BC, the Emperor Asoka of India passed an edict for the protection of animals, fish and forests. This may be the earliest-known documented instance of the deliberate establishment of what is today known as Protected Areas. Setting aside areas, as ‘sacred groves’ under religious and/or cultural grounds, is much older. It is widely being practiced and continued through centuries (Gadgil and Berkes 1991). Such PAs are present around religious establishments such as temples and are associated with deities. Also, they are present around churches and mosques, and are associated with burial grounds. Examples include sacred sites of Australian aboriginal peoples, tapu areas in the Pacific and sacred sites in Africa such as the Itshyanya Sanctuary in Rwanda, and thousands of sacred groves in India. In 1084 AD, King William I of England ordered preparation of the 'Doomsday Book', an inventory of all lands, forests, fishing areas, agricultural areas, hunting reserves and productive resources of his kingdom as the basis for making rational plans for the management and development of the country. The modern concept of PA systems, wherein wise maintenance and scientific management of ecosystems are emphasized, denotes nothing, but the following two ancient principles:

- Management of natural resources on the basis of scientific inventory.
- Conservation of resources to ensure their continued existence along with their sustainable use for the welfare of humanity.

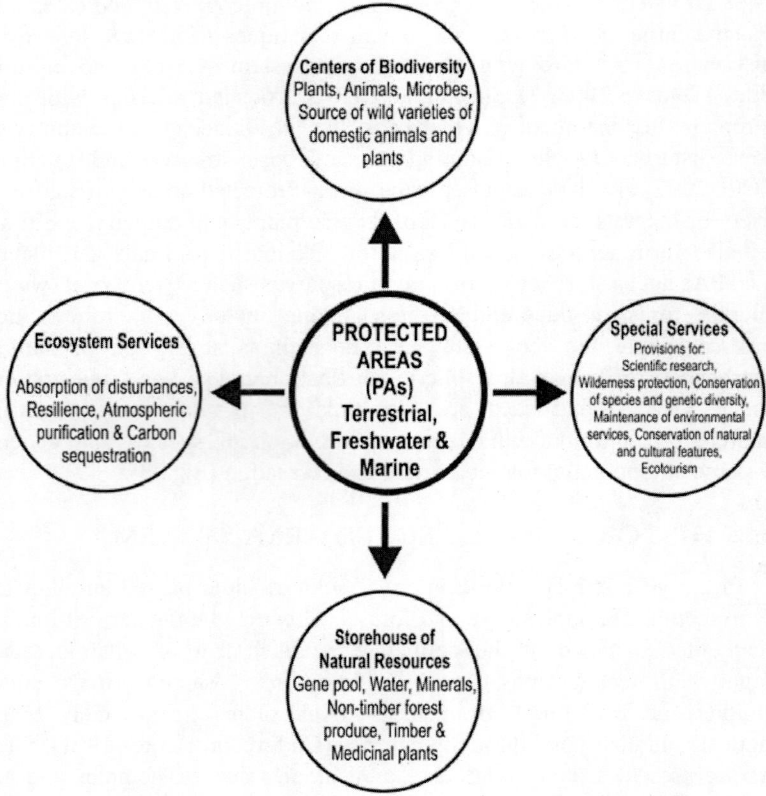

Fig 73. A schematic diagram showing the features, functions and sustainable uses of PAs

Conservation areas constitute an extent of natural habitats of diverse characteristics, where the governments have better legal control and direct involvement in management activities. Areas are brought under protective coverage based on the special characteristics of each of the areas such as geographical peculiarities, topographical features, historical importance, cultural features or biological constituents. These areas may not be able to sustain for long in the absence of protective cover. There were also many private reserves set aside by royal and wealthy land-owners for sport hunting. For example, the Gran Paradiso area in Italy was declared as a hunting zone in 1836 and then a Royal Hunting Reserve in 1856, much before it was declared as a National Park in 1922. The first modern Government-designated PAs were established in Australia and in the US. A reserve at Yosemite in Australia was declared in 1864 and the world's first National Park, the Yellowstone National Park in the US was declared in 1872. Then, Australia established its first National Park (Royal National Park) near Sydney in 1879 (Lockwood et al 2006). Subsequently, the concept of National Park gradually spread to other continents and countries. In general, PAs focus on conservation of biodiversity. However, objectives of species-specific conservation are laid down in the process of establishment of conservation areas such as the Eravikulam National Park in Kerala with the major aim to conserve the endemic and endangered Nilgiri tahr, or the Kazhiranga National Park in Assam (India) with the major aim to conserve rhinoceros.

Protected areas may be of large areas of several thousands of square kilometres (e.g., a biosphere reserve) or may be of smaller areas of less than a square kilometre (e.g., a sacred groove). In addition to the establishment of conservation areas, it is also necessary to document what are present in these areas. This was also realized long back as evidenced from the 'Doomsday Book' as ordered by King William I of England mentioned earlier in this chapter. It was possible for him to envisage the use of such a document as the basis for making rational plans for the management and development of his empire.

9.3. WORLD COVERAGE OF PAs

Even though the concepts of PAs and the PA-network have evolved long back, establishing PAs was not a thrust programme of governments until the late 1970s. It was only in the 1970s that concern over the need for existence of natural habitats with diverse organisms was emphasized in the context of the development of humankind. The Stockholm Conference on Human Settlements held in 1972 was responsible for a timely review of human activities and depletion of organisms and their habitats, which were in one way or the other responsible for the overall depletion of environmental qualities in human settlements and elsewhere. Not only the developing countries, but also the developed nations witnessed severe man-made environmental problems, which were detrimental in maintaining quality of human habitat. The per capita use of resources has been very high in the developed nations when compared to the developing nations. High levels of use of resources would affect the potential of the natural resources of even the developing tropical nations. Hence, it is established that protected conservation area network, irrespective of the region of occurrence would have global significance in improving the environmental quality.

The extent of PAs was less until the end of the first half of the 20th century. However, there has been a steady increase in the extent of PA network all over the world during the second half of the 20th century. It was during the 1970s and 1980s that a good proportion of the natural habitats were added to the PA system, globally. At present, there are over 10,000 PAs, recognized by the IUCN Commission on Parks and Protected Areas. Globally, there are over 1,20,000 designated PAs covering around 20 million km^2 of the land area, which is around 14% of the total land area of the planet Earth. Further, Marine PAs cover 5.9% of the territorial seas and 0.5% of the high seas (Coad et al 2009). There were 10 different categories of PAs recognized earlier. The IUCN (1994a) has revised the list into six categories. The extent of these areas is almost 147 million square kilometres. Region-wise, PAs contribute from 2.8% of the land area in North Africa and the Middle East to 12.6% in North America. The extent of PAs also varies widely even within regions. For example, in Europe, the nation-wise extent varies from 2% of the land area in Greece to about 25% in Austria. At present, it is in the PAs, where the world's biological diversity is concentrated and conserved.

Most of the biological diversity is concentrated in the tropical realms, where the human population is also dense, exerting more pressure on the finite land resources, and hence, the threat to establish and protect PAs is severe. Hence, management of PA is more challenging and necessitates region-wise approaches, rather than a global approach. As the world's biological diversity is concentrated in these regions, the people in the tropical realms have an added responsibility in ensuring maintenance of this rich natural heritage. Conservation problems in these realms are different from that of the other regions of the world. On the whole, the tropical nations and the people in the developing part of the world have major roles to play in conserving global biodiversity and the heritage that they have been maintaining through generations. Although conservation of natural resources including wildlife populations is stressed as a mandatory agenda of development, only few among the developing tropical nations have so far set aside at least 10% of the total geographical area under the protective cover. Botswana, Central African Republic, Benin, Malaysia, Tanzania, Zimbabwe, Bhutan, Senegal and Rwanda are the tropical nations of over 20,000 sq km in which over 10% of land area has been brought under the protective covers.

9.4. MEGADIVERSITY COUNTRIES AND HOTSPOTS

Mittermeier (1988), and Mittermeier and Werner (1990) have recognized that a very small number of countries, situated mainly in the tropics, possess a large fraction of the world's species diversity, and hence, they introduced the concept of 'megadiversity countries'. McNeely et al (1990) used country-species lists of vertebrates, swallotailed butterflies and higher plants and identified the following 12 megadiversity countries: Mexico, Colombia, Ecuador, Peru, Brazil, Zaire, Madagascar, China, India, Malaysia, Indonesia and Australia. These countries together hold up to 70% of the global diversity of vertebrates, swallotailed butterflies and higher plants.

Wildlife is not uniformly distributed all over the world or even in the tropics. There are certain areas having rich diversity than other parts of the world. Endemism is also not common with all regions. Some areas with specific vegetative and climatic conditions harbour more endemic species when compared to other areas. Some such areas with large number of endemics are more threatened due to more intensive human activities. Areas characterized with species richness, endemism and threat are identified as hotspots for conservation priorities (Myers et al 2000). There are 35 such hotspots so far identified (Table 4). Although 'hotspots' occupy a tiny fragment of the land, they are the exclusive showroom of one fifth of the world's plant species. During the past few decades of developmental era, we have altered even such hotspots in different parts of the world.

9.5. PROTECTED AREAS AND CONSERVATION INITIATIVES

Protected areas are essential for conservation of living resources. When an area is selected for conservation, it encompasses characteristics involved in that area, and thus, forms a representative sample of the region retained under conservation efforts. This process can help in the long-term

Table 4. A region-wise list of global biological hotspots

Region	Name of the hotspot area
North and Central America	California Floristic Province
	Caribbean Islands
	Madrean pine-oak woodlands
	Mesoamerica
South America	Atlantic Forest
	Cerrado
	Chilean Winter Rainfall-Valdivian Forests
	Tumbes-Chocó-Magdalena
	Tropical Andes
Europe and Central Asia	Caucasus
	Irano-Anatolian
	Mediterranean Basin
	Mountains of Central Asia
Africa	Cape Floristic Region
	Coastal Forests of eastern Africa
	Eastern Afromontane (Afromontane archipelago)
	Guinean Forests of West Africa
	Horn of Africa
	Madagascar and the Indian Ocean Islands
	Maputaland-Pondoland-Albany
	Succulent Karoo
South Asia	Eastern Himalaya, Nepal
	Indo-Burma, India and Myanmar
	Western Ghats, India
	Sri Lanka
South-east Asia and Asia-Pacific	East Melanesian Islands
	New Caledonia
	New Zealand
	Philippines

Region	Name of the hotspot area
	Polynesia-Micronesia
	South-west Australia
	Sundaland
	Wallacea
East Asia	Japan
	Mountains of South-west China

maintenance of biological and physical characteristics. It also ensures conservation of genetic diversities in the region. In general, conservation efforts can bring in positive effects in the maintenance of areas with stable physical and climatic factors, in protecting soils from erosion and in controlling floods and droughts. When an area is under effective conservation programme, human interactions should be limited and natural regeneration capacity of the area should be maintained. This would result in maintaining productivity of the area at a constant level so as to support diverse living organisms for their growth and development. Biomass in the area would improve in such situations. If conserved and managed properly, such areas would form excellent sites for productive research activities so as to learn more about nature, its wonderful biotic composition, adaptive features and natural evolutionary processes. Opportunities will also be available in using PAs for conservation and environmental education programmes. There would also be opportunities for sustainable use of natural resources for rural development and for involvement of local people in PA management, which can ensure the development of a self-supporting system of rural communities using natural resources. Protected areas can also provide opportunities for recreation and eco-tourism in the long run.

As conservation may not yield direct and immediate benefits, it is difficult to convince administrators and politicians about the need for conservation. Long-term policies of governments are essential for conservation. Majority of the developing nations are politically unstable, where government policies cannot sustain due to frequent changes of political parties in power. Those who are in power would have more interest in activities that may benefit by immediate visible returns. One can also see diversion of funds set aside for conservation into areas of defence and other non-conservation activities with intensive political support. Ultimately, instead of getting more support, conservation often finds its way with severe limitations of funds, resources, public support and expertise.

Conservation-oriented activities are to be more intensified in the developing tropical nations because of the rich biodiversity present in these nations. However, due to the limitations of resources, and due to the poor and comparatively illiterate human populations, who are not aware of the need for conservation, conservation measures are not properly taken into account in many of these nations. It is in this part of the world that modification of natural habitats is taking place at a faster rate, mainly due to explosive human population growth and resulted additional requirements of resources. On the basis of these negative trends in the tropical world, which are in contrast to the similar conditions in the temperate region, majority of the management options approved in the temperate regions may not be suitable for the tropical nations. Hence, it is essential to establish more regional-oriented conservation programmes for the tropics on a long-term basis.

9.6. FEATURES OF AREAS TO BE CONSERVED

Major questions to be answered while planning conservation activities are what and where are to be conserved? Unique features of ecosystems, for example, lowland rain forest, endemic island fauna, tropical alpine ecosystem and high-altitude dense forest-grassland composition are special characters that deserve attention of conservationists. Areas where a species of special interest is present, and areas of those species, which are rare or under threat, are required to be conserved (Figs 74, 75). Hence, areas where such species are located need to be brought under conservation activities. The major criteria for declaring Silent Valley of South India as a National Park was the presence of second viable populations of one of the most endangered species of primates, *M. silenus* in that area (Balakrishnan, 1984). They are also endemic, seen only in limited ranges of the wet evergreen forests in southern Western Ghats in South India. Similarly, the habitat of snow leopard and the habitat of musk deer are required to be conserved because of the small population size of these animals in their narrow range of the Himalayan habitats in Asia. Areas with unusual species diversity need to be considered for conservation. If species diversity is rich in an area, it would represent special characteristics that are responsible for the development of its rich diversity. In general, species diversity is known to be very high in Amazonian forests as compared to any of the other regions. Such a special diversity is a cumulative result of several factors, both physical and biological. Comparative studies in different regions would reveal the areas of rich biodiversity in each region,

Fig 74. A view of shola forests in Eravikulam National Park, Kerala. The hill-tops are grassy and look like carpeted from a distance. This is the best-known habitat of the Nilgiri tahr. Shola forest is a climax vegetation community (Photo: E Kunhikrishnan)

and such areas should be considered for bringing under conservation initiatives. Landscape and geophysical features of aesthetic or scientific values (for example, glaciers, hot-water springs and waterfalls) may be so special to an area. A waterfall may be a unique feature, which depends on the special feature of the landscape. If such features are affected by man-made activities such as dam construction, the natural beauty is lost forever. Entertainment and recreation are also important for a comfortable life and for the welfare of humankind. Sites suitable for recreation and tourism (e.g., lakes, beaches, mountain views, wildlife spectacles, beautiful landscapes and special natural blooming seasons) (Fig 76) are to be identified and conserved. It is also important to limit visitors in such areas based on the carrying capacity of the area. Otherwise, the beauty of the location will be lost, and the area will be depleted with the enhanced number of visitors and their activities.

Fig 75. A giant lobelia habitat in the Sanetti Platue in the Bale Mountains National Park (Ethiopia), which is also a typical habitat of the endangered Ethiopian wolf (Photo: M Balakrishnan)

Fig 76. *Strobilanthes kunthianus* (Nees) T. Anderson ex Benth flowers only once in 12 years in the Eravikulam National Park, in the southern Western Ghats, which is an area of the hottest among hotspots. A large extent of this Park was in blooming during the flowering season in August–September 2006 (Photo: M Balakrishnan)

Conservation is also to support research activities. Areas of long-standing research and important findings need to be conserved for future research as such areas would be ideal for detailed studies on various aspects of ecology and biodiversity. For example, long-standing studies on the Ethiopian wolf in Bale Mountains National Park for over 25 years have been instrumental in conserving the Bale Mountains region and the once critically-endangered Ethiopian wolf. It is currently under the 'Endangered' category of IUCN. Cultural sites such as temples and archaeological excavations are also to be conserved. Such areas are to be identified and the special features are to be notified for initiating conservation programmes. The relative value of each of such features and areas are to be analyzed and a priority list is to be prepared for the consideration of governments.

9.7. CATEGORIES OF PROTECTED AREAS

Conservation areas are established with specific objectives. Each of the categories is developed to serve the purpose for which it is established, and for conservation of its finite resources and for maintaining it sustainably as per the internationally-accepted norms. Different categories of management areas facilitate different aspects of management, including PA legislation, planning management strategies and decisions, controlling both type and intensity of utilization and justifying the benefits claimed for PAs and policies. The following are 10 categories of PAs established earlier:

9.7.1. Strict Nature Reserve/Scientific Reserve

Nature Reserves were established to protect nature and maintain natural processes in an undisturbed state in order to have ecologically-representative samples of the natural environment. Mainly, the objectives of establishing such areas were scientific study, environmental monitoring, environmental education and maintaining genetic resources in the dynamic and evolutionary state of nature. As the establishment of a Nature Reserve is comparatively easy, more natural areas are brought under this type of conservation category. This enables the nation to have representative samples of all types of habitats and species composition of all major habitats under protective cover.

9.7.2. National Park

National Parks are established to protect outstanding natural and scenic areas of national or international significance for scientific, educational and recreational uses. National Parks are relatively large natural areas maintained with their own natural features. These are the areas that have not been altered much by human activities during the development era. In this category of PA, using any of the extractive resource is not allowed. National Parks are established with specific functional intentions such as conservation of endemic species and specific landscape. For example, the Eravikulam National Park was established with the intention of conservation of the Nilgiri tahr and the Kazhiranga National Park for the conservation of the Indian rhinoceros. The best viable populations of these endangered Indian mammals are in these respective areas. Similarly, the Bale Mountains National Park was established for conservation of the best-known population of the Ethiopian wolf.

9.7.3. Natural Monument/Natural Landmark

Natural monuments are established to conserve nationally-important natural features because of their special interests or unique characteristics. Natural monuments are in relatively smaller land areas, where the monuments of special interest are conserved.

9.7.4. Managed Nature Reserve/Wildlife Sanctuary

Wildlife Sanctuaries are established to conserve various types of habitats with their biotic communities in the natural set-up. The idea is to conserve representative physical and environmental features of each region or State. Limited human activities may be permissible in Wildlife Sanctuaries, particularly in connection with the use of resources. As compared to other types of conservation areas, more Wildlife Sanctuaries are present in different parts of the world.

9.7.5. Protected Landscape and Seascape

Protected landscapes are established to maintain nationally-significant natural landscapes, which are characteristic of the harmonious interaction of man and land. These protected landscapes also provide opportunities for public enjoyment through recreation and tourism within the normal life style and economic activity of these areas. These are mixed cultural or natural landscapes of high scenic value, where traditional land-uses are maintained.

9.7.6. Resource Reserve

Resource reserves are established to conserve natural resources of the area for future use and to prevent or control development activities that would affect the resources pending the establishment of objectives based on appropriate knowledge and planning. This was a 'temporary category' until a permanent classification was assigned.

9.7.7. Anthropological Reserve/Natural Biotic Area

Anthropological reserves are established to allow the human communities to live in harmony with the environment without creating disturbance by modern development and advanced technologies. There are several natural areas where utilization of traditional resources by indigenous people is common. This category of conservation area is more appropriate to such areas.

9.7.8. Multiple-use Management Area/Managed Resource Area

Multiple-use management area was established to provide constant availability of water, timber, wildlife, pasture and tourism with conservation of nature. It was primarily oriented to support economic activities. It is possible to designate different zones within these areas to achieve specific conservation objectives.

9.7.9. Biosphere Reserve

Biosphere reserves are established to conserve large areas of various types of habitats with a variety of diversity of life and culture in large ecosystems. The objective is to conserve the rich genetic diversity available among the biotic communities on which evolutionary processes depend upon. Such areas are considered as areas for research, education and training, where cultures meet and diversity of traditions stand.

Biosphere reserves incorporate diverse habitats such as forests, grasslands, marshes, lakes, ponds and human settlements. In this type of conservation areas, indigenous human beings are also given appropriate place with their traditional activities including agricultural practices, maintaining and managing their indigenous domestic animals and pets. Thus, biosphere reserves are established in large areas where not only natural habitats, but human settlements and farmlands are also present. However, only indigenous people with their traditional farming and practices of maintaining traditional pets and livestock are included for conservation under this programme. During the past several years of development era, we have lost several varieties of traditional agricultural crops, which are replaced by hybrid varieties. The properties of traditional varieties are required to be conserved; otherwise the genes responsible for several natural characters will be lost during the years of human development.

9.7.10. World Heritage Site

World heritage sites are established to protect natural features for which the area is best known globally. Such areas are selected from among the world list of unique natural and cultural sites nominated by countries that are party to the World Heritage Convention (e.g., Simien Mountains National Park in Ethiopia and the Great Barrier Reef in Australia).

9.8. IUCN REVISED CATEGORIES OF PAs – 1994

Revised categories of PAs (IUCN 1994a) are as follows:

- ***Category I. Strict Nature Reserve/Wilderness Area:*** These areas are designated to protect nature and to maintain natural processes in an undisturbed state so as to have ecologically-representative samples of natural habitats for scientific study, environmental monitoring, education and maintenance of genetic resources in a dynamic and evolutionary state. There are two subcategories of such areas: Ia – PAs managed mainly for scientific research and monitoring; and Ib – PAs managed mainly for wilderness protection, subsistence and recreation.
- ***Category II. National Park:*** National Parks are designated to protect outstanding natural and scenic areas of national or international significance for scientific, educational and recreational use. These are relatively large areas that are not materially altered by human activities and where use of extractive resource is not allowed.
- ***Category III. National Monument/Natural Landmark:*** These are areas designated to protect and conserve nationally-significant natural features due to their special interest or unique characteristics. These are relatively small areas with special focus on protection of specific features.
- ***Category IV. Species Management Area:*** Species management areas are areas designated to ensure natural conditions necessary to protect nationally-significant species, groups of species, biotic communities or physical features of the environment, where

specific human manipulation may be required for their perpetuation. Controlled harvesting of some of the resources may be permitted in these areas.

- ***Category V. Protected Landscape and Seascape:*** These areas are designated to maintain nationally-significant natural landscapes that are characteristic of the harmonious interaction of man and land. These areas also provide opportunities for public enjoyment through recreation and tourism. Protected landscapes and seascapes are mixed cultural or natural landscapes of high scenic value, where traditional land-uses are maintained.
- ***Category VI. Managed Resource Protected Area:*** The managed resource PA is a category designed to include areas that ensure long-term protection and maintenance of biodiversity while providing a sustainable flow of natural products and services to meet the community needs. They are intended to be relatively large and predominantly unmodified natural systems, where traditional and sustainable use of resource is encouraged.

The current categories of IUCN PAs are the following:

- ***Ia. Strict Nature Reserves:*** These are PAs strictly set aside to protect biodiversity, geological and geomorphic features, where human visits and their impacts are strictly under control, which ensures protection of conservation values. However, scientific research is allowed in Strict Nature Reserves.
- ***Ib. Wilderness Areas:*** This category of PAs are large, unmodified or only slightly modified areas, retaining their natural characters in the absence of human interactions, and are protected to conserve natural conditions.
- ***II. National Parks:*** National Parks are large natural areas or areas near to natural areas, set aside to conserve large extent of habitats with their characteristic ecological processes, along with the balance of species and ecosystem characteristics of the area. These areas also provide a foundation for environmentally and culturally compatible, spiritual, scientific, educational, recreational and visitor opportunities.
- ***III. Natural Monuments:*** Natural Monuments are set aside to protect and conserve natural monuments such as landform, seamount and submarine cavern, and geological features such as a cave or an ancient grove.
- ***IV. Habitat/Species Management Areas:*** Habitat or Species Management areas are aimed to conserve particular habitats or

species. In this type of conservation areas, management interventions can be provided, if required, for the conservation of specific species or to maintain habitats.

- ***V. Protected Landscapes/Seascapes:*** This category of PAs includes areas with distinct characteristics of ecological, biological, cultural and scenic value that have developed as a result of interaction of people and nature over time. These PAs are established to safeguard and conserve such characteristic vital interactions.
- ***VI. Protected Areas with Sustainable Use of Natural Resources:*** These are PAs that conserve ecosystems and habitats together with associated cultural values and traditional natural resource management systems of large natural areas. Limited sustainable resource management activities are allowed in such areas to maintain low level of non-industrial use of resources.

9.9. *IN-SITU* GENE BANKS: A NEW CATEGORY OF PA

Conservation of genetic resources is the ultimate objective of conservation efforts. It is now realized that conservation cannot be effectively carried out if the resources are not conserved in the natural habitat itself rather than transferring and re-establishing in other areas. The relocation of organisms is advised only in essential situations. So, whatever is the ultimate objective of the type of conservation programme, conservation of organisms and their habitats in their natural set-up is to be achieved. The gene pool or gene bank is important in the efforts to use genetic resources for manipulation, particularly for agricultural developments (e.g., hybridization process, where useful properties are selectively used for development of new hybrid varieties). *In-situ* gene banks are the sites where wild gene pools of plants and animals are protected and maintained in their natural habitats, new domestication and biotechnological advancements (including genetic engineering).

'Gene pool' denotes the total number of genes responsible for different characters of an organism present in a group of interbreeding plants or animals. It is the sum total of the genes in a population. The gene pool in a population can be conserved only if the entire population is conserved in the natural habitat. Thus, several species present in such a community will be benefited, and the ultimate result will be not only conservation of the gene pool of one population, but the gene pools of several interdependent populations in that habitat.

Protection of gene pool is stressed in the context of practical application. Hence, species of economic importance are considered for gene pool conservation with the objective of future use in crop and livestock improvement, and in biotechnological and biomedical use. Thus, conservation of natural gene pool is envisaged for long-term maintenance of germplasms for future use.

9.10. TYPES OF UTILIZATION

Protected areas serve as excellent resources for various types of utilization (Balakrishnan 1993) (Table 5). When natural habitats are conserved and human activities are eliminated or at least minimized, such habitats can function in tune with the natural evolutionary processes. Such ecosystems can provide resources on a sustainable basis, and also opportunities for scientific investigations, environmental monitoring and educational programmes as in the case of Strict Nature Reserves and National Parks. Undoubtedly, these areas may also provide opportunities for human recreation. Conservation of special natural features is the objective of creating nature monuments. Wildlife Sanctuaries and Managed Nature Reserves are to provide opportunities for scientific investigations on ecosystems, and controlled and limited harvest of resources for human welfare. Protected Landscapes and Seascapes provide public opportunities for enjoyment through picnic and tourism activities. Traditional activities of human societies are maintained in harmony with nature in Anthropological Reserves and in Natural Biotic Areas, so as to continue such valuable traditions and information undisturbed by modern technologies and developments. Sustainable utilization of water, extraction of timber, wildlife and pasture are allowed in Multiple-use Management Areas and in Managed Resource Areas. Biosphere Reserves are established to maintain genetic diversity including those of domestic species in natural and man-made habitats, under the processes of natural and man-made selections, respectively. Protection of world's unique natural and cultural sites is envisaged under the World Heritage Sites.

9.11. CRITERIA FOR THE SELECTION OF PAs

There are several criteria that need to be considered while suggesting areas to be brought under conservation. When more than one area is suggested, a relative importance is judged by comparing the values of each area in terms of the approved criteria, and thus, a preference list is made available for conservation agencies for approval and implementation.

Table 5. Management objectives of different categories of PAs (Modified from: IUCN, 1994a)

Objectives	Categories of PAs						
	Ia	Ib	II	III	IV	V	VI
Scientific research	1	3	2	2	2	2	3
Wilderness protection	2	1	2	3	3	-	2
Conservation of species and genetic diversity	1	2	1	1	1	2	1
Maintenance of environmental services	2	1	1	-	1	2	1
Protection and conservation of natural and cultural features	-	-	2	1	3	1	3
Eco-tourism and recreation	-	2	1	1	3	1	3
Nature education	-	-	-	2	2	2	2
Sustainable use of natural resources	-	3	3	-	2	2	1
Maintenance of cultural and traditional attributes	-	-	-	-	-	1	2

(Ia = Strict Nature Reserve; IB = Wilderness Area; II = Ecosystem Conservation and Protection; III = Conservation of Natural Features; IV = Conservation Through Active Management; V = Landscape/Seascape Conservation and Recreation; VI = Sustainable Use of Natural Resources; 1 = Primary objective; 2 = Secondary objective; 3 = Potentially-applicable objective; = Not applicable)

9.11.1. Size and Shape

The size of the PA contributes considerably for the overall value of conservation. The area must be sufficiently large enough to support viable populations of organisms present. In case of larger organisms, conservation is possible only if sufficient area is available on the basis of their behavioural characteristics. For example, animals that require large areas for their regular movement and foraging activities (e.g., elephant, wild buffalo) cannot sustain in areas of smaller extents. Further, to meet the eco-requirements of such organisms, all essential ecological components are to be present in the area for their use. To incorporate these essential ecological requirements, a large area should be opted for conservation in place of a number of smaller areas, which may together constitute the extent of the large area (Fig 77).

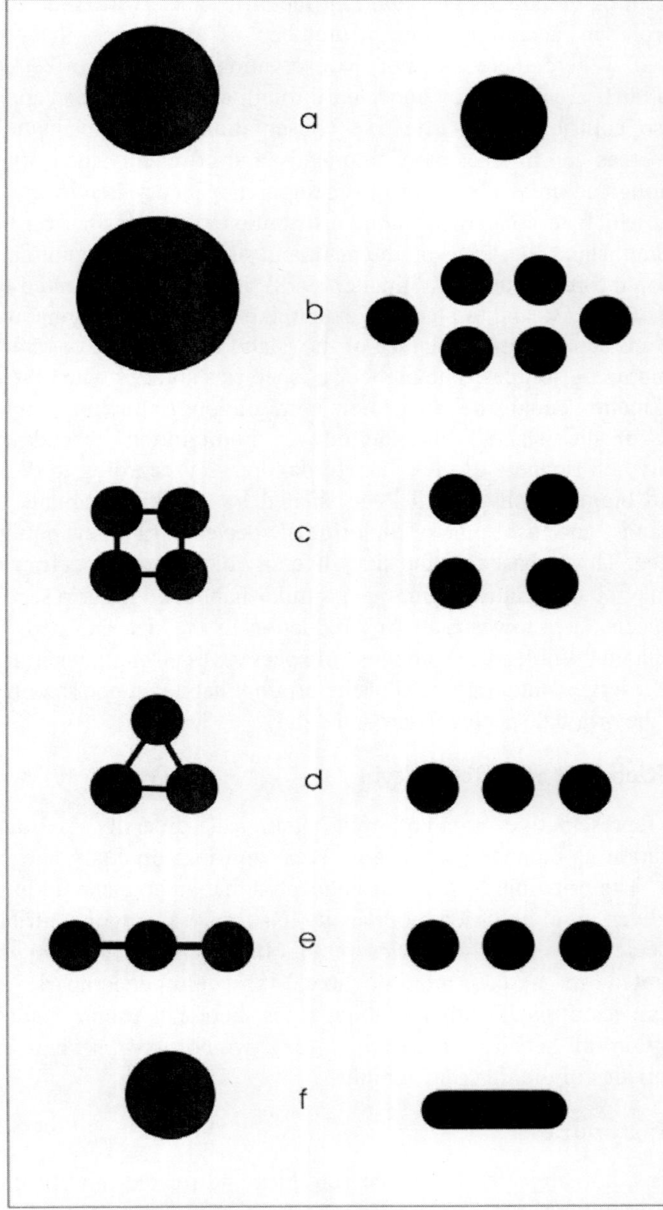

Fig 77. A schematic representation showing the shape and closeness of areas for conservation (Modified from Diamond 1975)

Extent of the area is an important factor in conservation as it plays a major role in accommodating a number of organisms that can be conserved. Where there are more conservation areas, the effect will be better if such areas can be connected through corridors. The shape of the area also contributes for effective conservation as organisms in more circular areas can interact more frequently and efficiently than organisms in an elongated area. The Island Biogeographic Theory is relevant in this context, which is concerned about distribution of organisms on oceanic islands and island-like areas on the mainland such as lakes, mountain peaks and isolated forest patches (Whittaker 1998). It has been illustrated that the number of species, and to a certain extent, the actual species inhabiting the islands, are dependent on the size of the island and its relative remoteness from colonizing sources. The number of species stabilizes when the rate of local extinction equals the rate of new immigration. Extinction is related to the size of the island while the rate of immigration depends on the proximity and richness of the colonizer landmass. According to the theory of island biogeography, small PAs isolated by modified habitats behave like 'islands' and lose some of the original species until a new equilibrium is reached. This is based on the size, diversity of the area, species richness and its degree of isolation from other similar habitats. Larger reserves lose fewer species at a lower rate. Any reduction in the size and depletion of natural habitat will lead to some loss of species. As a rough generalization, a single reserve containing 10% of the original habitat supports only about 50% of the original species (Diamond 1984).

9.11.2. Richness and Diversity

A representative area of rich biological diversity is ideal for conservation as compared to a large area with less diversity and species richness. The principle behind this concept is that an area should have rich habitat diversity to hold rich biodiversity. Naturally, this will contribute for rich species diversity and *vice-versa*. If diverse habitat types are incorporated in a conservation area, it would encompass species characteristics of each of these habitat types. Ecotones usually have larger composition of species diversity as they would have several species characteristics of the flanking habitats.

9.11.3. Naturalness

Areas, not subjected to human activities and related modifications so far, are important for conservation because such areas would contain characteristics of natural features. It may also be possible to restore natural characteristics, if areas of less human activities are brought under conservation.

9.11.4. Rarity

It is important to conserve organisms and habitats that are rare on a regional basis. In case there are several options in a region, an area with physical and biological characters that have depleted as a result of human activities should be given priority to be developed as a representative sample of conservation area. It is also important to conserve those species, populations of which have depleted due to human pressure. If an area with any such species is available, it needs protection so as to conserve the affected species and their habitats.

9.11.5. Uniqueness

Uniqueness of an area may be due to the special properties of habitats or its species composition. Specific features and properties may be due to the cumulative effect of physical and biological characteristics of that area. If the biome represented by an area is not adequately represented in other parts of the region, such areas and habitats are to be conserved, and hence, it is advantageous to bring such areas under PA system. For example, the Lake Malawi is considered as a unique habitat because of its large number of endemic species of cichlid fish. What is seen in Lake Malawi is the process of rapid speciation.

9.11.6. Typicalness

There may be large extents of specialized habitat types, but it may not be practically possible to bring all habitats and large extent of areas under protective cover. The alternative is to bring whatever extent of the typical habitat is possible under protective cover so as to act as a typical representative sample of the large extent of habitat available in the region. Such a protected habitat would form part of the typical bio-geographic unit.

9.11.7. Fragility

Some species may be fragile, which may not withstand the effects of human activities. Highly sensitive and habitat-specific species are also unable to withstand major changes in their habitats. High mountainous ecosystem, which may be seasonally subjected to soil erosion, may not withstand the changes for long, and would change the habitat associations.

Such habitats and species are to be supported by bringing them under protective cover.

9.11.8. Genetic Importance

Areas, which are gene-rich and which contain rare genetic composition (for example, occurrence of wild forms of domesticated plants and animals) are important for conservation, and hence, may be considered to be brought under PA systems. Thus, useful properties are made available whenever there is a need for hybridization or for any of the other purposes.

9.11.9. Historical Records

Areas that have been the subject for many years of investigations need to be conserved over areas that are less studied. Areas of long-standing studies are to be continuously monitored, and due to their importance as an area of outstanding scientific research, it is entitled to be conserved under protective cover. This would help the area for repeated monitoring and for comparative studies.

9.11.10. Ecological/Geographic Position

An area, which represents a geographical limit with high species diversity, should be considered for conservation. It makes a special area with representative species diversity of the region rather than having separate conservation area for each of the species in different areas in the region. The emphasis is to include within a single large area as many important and characteristic communities of diverse species composition as possible.

9.11.11. Indispensable

An area may have an indispensable feature (e.g., a vital watershed catchment, a major river valley or a winter migration ground of a bird species) that deserves attention for conservation. Such areas are to be identified and brought under protective area network.

9.11.12. Potentiality

Areas, which were of high potential, may have depleted as a result of human intervention in the past. In such cases, if protective measures and management can uplift the habitat quality, then such areas may be considered for bringing under protective area network. It may not be of much use if an area is already depleted considerably and may not improve to a satisfactory level even if brought under protective cover, and hence, the potentiality of the area is to be assessed before taking actions for improvement under protective management programme.

9.11.13. Intrinsic appeal

Although all organisms are of equal importance in nature, due to the high value of appreciation by human beings, certain forms (e.g., higher terrestrial vertebrates such as birds and mammals) are highly appealing to the general public. Such a consideration would be more appreciated for recreation and for the advancement of eco-tourism, which has economic potential for national development. Areas of such values of appreciation should be brought under conservation area network.

9.11.14. Modified Landscapes

If there is a special landscape modified for augmenting the biological values of the region in tune with a special land-use pattern, then that landscape is important for protection. Such traditional land-use pattern may be in connection with the national culture or tradition, which would have significant influence of the biogeography of the region.

9.11.15. Areas of Public Interest

Areas with special characteristics of faunal and floral composition may be beneficial for local people directly or indirectly. It may be advantageous because protection of such areas would be highly supported by local communities for conservation. Such areas would be more successful in conservation campaign with local support rather than bringing areas of no public interest for conservation. In general, it is argued that public support is essential for better protection and management of conservation areas. In the absence of local support, conservation efforts may not attain the expected success.

9.12. PROTECTED AREA MANAGEMENT/PARK MANAGEMENT

9.12.1. The Park Manager

Several people under the leadership of the Park Manager manage the PA. They comprise of managers, scientists, technical staff and protective staff of various levels. The Park Manager is responsible for all activities carried out in the Park, and hence, he should be able to watch all developments in different parts of the large extent of the area under his control. In case the extent of the area is very large, it may be difficult for the Park Management to control the interventions of poachers and other intruders as seen in some of the large African conservation areas. The Park Manager is also responsible for coordination of all activities in the Park. Only a well-trained manpower can regulate, control and manage conservation areas properly. It is also revealed that management authorities of conservation areas should have at least basic training in wildlife conservation and research activities so as to satisfactorily fulfil the objectives of conservation activities. There are several Park Managers, who have achieved such a position just by their seniority in the Department or through political affiliations and influence, but they are not able to effectively manage a conservation area established under scientifically-developed norms. People who have no real interest and basic knowledge in ecological and conservation principles should not be placed as managers of conservation areas. Unfortunately, several of our wildlife managers are of this category and this is one of the major problems in proper implementation of conservation activities in some of the developing nations, including India.

9.12.2. Protective Staff

Guarding the conservation areas comes directly under the activities of the protective staff. In most cases, officials concerned with protection is Forest Guards/Scouts, Foresters and Forest Range Officers. The Range Officer is responsible for executing protective measures and to report various issues in the Park to higher officials. They are to be well trained for the purpose, and to execute the orders received from higher authorities in the field. They are the people having direct responsibilities to gather information on human interactions, poaching and boundary encroachment from different sources. For this purpose, members of the protective staff should spend a great deal of time in the interior of the conservation area, where they control and monitor the resources and check the effectiveness of the regulations and policies of the Park. They are also responsible for controlling visitors and for eco-tourism management in the area. They are also responsible for giving protection to researchers in the area.

9.12.3. Park Ecologist

The Park Ecologist is responsible for research activities concerned with natural resources and ecology in the Park, and behaviour of organisms of special interest. It is advisable to have a good team of researchers under the Park Ecologist for regular research activities. This is important to gather information on various aspects of the Park including effects of implemented management activities. In case effects of such activities are negative, immediate remedial measures can be effectively taken if there is constant data on the processes. Otherwise, it would take long time to understand the problem and may be too late to take remedial measures against negative effects of the implemented programmes.

9.12.4. Park Veterinary Officer

The Park Veterinary Officer is to take care of any emergency situation in the Park such as the spread of an epidemic among wild animals or issues such as involvement of a wild animal in an accident. Immediate medical attention under an experienced veterinary doctor may be of some help to an animal in danger. The Park Veterinarian is preferred for treating infected wild animals even in captivity rather than an ordinary veterinarian, due to his additional experience in dealing with wild animals. The Park Veterinary Officer should also be consulted when it is essential to immobilize wild animals as required for radio-collaring, translocation and for research purposes.

9.12.5. Park Sociologist

The Park Sociologist is responsible for studies related to the Park users, particularly tourists and other visitors so as to suggest better amenities to the satisfaction of the visitors. His responsibility is to consult with all stakeholders of the conservation area including local people, tourism agents and scientists, and analyze their views to find out the best means to use the resources at the optimum levels by the Park users. He has to advise the Park management if some additional amenities are to be developed for the satisfactory use of visitors and for others involved in the conservation area.

9.12.6. Park Economist

The Park Economist is responsible for investigations on the allocation and utilization of the resources by various user agencies including visitors, indigenous people and other locals. Park Economist is responsible for proper allocation of the resources and for the optimum use. Park Economist has to spend time gathering and analyzing information on Park resources, user behaviour, ecological constraints and budgets, and advice the Park management for appropriate economic strategies to achieve the goals of the PA.

In addition to the positions such as Ecologist, Sociologist and Economist, positions such as Botanist, Zoologist, Geologist, Anthropologist and Marine Biologist may be required in special cases of conservation areas established with such specific objectives. For example, in those conservation areas instituted predominantly for conservation of historical or cultural resources, a Historian or an Archaeologist would be helpful for interpretation and related activities.

- ***Interpretation Officer/Education Officer:*** The Education Officer is responsible for preparation of education and information materials using data from the conservation area for the use of governments, students, visitors and tourists. Tourist Guides work under the Education Officer, pass on information to tourists and get feedback from tourists.
- ***Administrative Officer/Accounting Officer:*** The Administrative Officer is responsible for the day-to-day administrative affairs of the conservation area. Other members of the administrative staff work under the leadership of the Administrative Officer, and maintain regular records of receipts and expenses. The Administrative Officer is responsible for reporting all information concerned with the administration to the Park Manager and get his clearance.
- ***Park Infrastructure Maintenance Officer:*** This position is to be filled in by an Engineer, who is responsible for construction and maintenance of infrastructures such as buildings, roads, and tourist tracks in the Park, and the landscape of the area. The Maintenance Officer is responsible for discussions with Architects, Developers and Contractors for timely and satisfactory completion of various development projects in the conservation area.
- ***Law and Resource Policy Officer:*** The Park Law / Policy Officer is responsible for investigation and support of legal and policy issues of the Park management. The Law and Resource Policy Officer has to review the situation from time to time and appraise the Park Manager about the need for changes in the policies to be adopted for better resource use and subsequent strategies for development. All legal issues in relation to the conservation area should be dealt with the Law and Resource Policy Officer.

- ***Land Tenure and Acquisition Officer:*** The Land Tenure and Acquisition Officer is responsible for evaluation of land-use planning in and around the conservation area. When a new conservation area is designed, the Acquisition Officer is responsible for land acquisition and related activities, and to suggest the Park management on areas to be acquired and the methodology for an amicable land transaction procedure to be followed with the support of the local people.
- ***Public Relations and Extension Officer:*** The Public Relations Officer is responsible for the preparation of relevant information for the use of the public and the Park visitors. It is the duty of the Public Relations Officer to make essential publicity materials of the conservation area for the benefit of the public so as to attract visitors. He is also responsible for visiting such institutions and arranging classes for the benefit of students and the public so as to induce conservation ideas and to create general awareness among the youth and the public on the need for conservation. When a conservation area is established with specific motives, the ideas should be spread among the public in general and those who live in and around the area so that they become aware of the issue. Only when the public is aware of the issue and convinced, their support can be extended for the cause of conservation.
- ***Park Planner:*** The Park Planner is responsible for preparation of the Conservation Action Plan for the State and to prepare State policies with regard to conservation areas and conservation activities. The Planner is also responsible for the preparation of conservation plans and programmes, periodic evaluation, updating of management programmes and coordination of the planning of each Park in the State. Thus, the Park Planner acts as a coordinator for the PA network. The Park Planner is also responsible for the preparation of conservation strategy for the State and to streamline the process of its implementation.
- ***Park Artist or Exhibit Designer:*** The Park Artist is responsible for the preparation of exhibits, arrangements and layout of exhibits, and for maintaining museum specimens and other exhibits. He should work in collaboration with the Park Education Officer for proper setting of educational materials and other resources in the collection. He should also coordinate with the Public Relations Officer for advertisements of art and

museum exhibits, and exhibitions from time to time. New collections, whenever acquired, should be announced for the information of the public.

However, all these management positions are not available in most of the Parks. Further, in addition to the above positions, additional specialists may also be consulted for clarification in any of the specialized areas for proper establishment, development and management of the conservation area network on a state-wise or national network programme. The Park management should also incorporate scientists with expertise in the respective areas among the consultants for advises from time to time. The Park management is responsible for arranging such consultancy whenever required.

9.13. LOCAL PEOPLE AND PROTECTED AREA MANAGEMENT

In general, proper maintenance and management of a conservation area are results of cumulative effects of activities of a group of enthusiastic managers, scientists, engineers, historians, sociologists, anthropologists, economists, educationists, legal experts, security staff and technicians with the support and involvement of the local people. The support and involvement of local people, particularly the indigenous people and other stakeholders are considered to be of high value in conservation initiative in the wake of several recent examples. The well-known CAMPFIRE programme in Zimbabwe and the Lungwa Valley Integrated Rural Development Programme in Zambia are typical examples to show how far the involvement and active support of local people would benefit the Park management in conservation of forest and wildlife resources in the African continent (Balakrishnan 1994). The controversy over collection of thatching grass by local people in the Royal Chitwan National Park in Nepal is another example to show the inability of the Park management to function in the absence of public support.

Usually, the public has no place in forest administration and management. This is mainly because the administrators rarely give reasonable consideration to the socio-economic and cultural practices of the people traditionally living in and around forest habitats, and depending directly or indirectly upon forest resources for their subsistence. The present trend in the developing tropics, particularly in the African continent, is towards involvement of local people in forest management and conservation activities. This is because it is revealed that no conservation programme can be fully effective in the absence of the co-operation of local people. Incidents in Nepal and various African countries have revealed the difficulties faced by forest administration in implementing their policies when the public was excluded from forest-related activities. Traditional activities and views of local people have not been regarded by forest and wildlife officials in many of the resource management and utilization practices in most of the developing countries (McNeely and Pitt 1985).

Natives, who live in harmony with nature, are often in a commanding position to help conservation programmes. However, they may not be interested in modern conservation strategies. They must be made aware of the roles they can play in conservation activities. In India, the 'Hug-the-Tree-Movement' popularly known as the Chipco Andolan Movement, in the Central Himalayas inspired people, especially women, not to allow the contractors to fell trees. This movement produced miraculous results, far greater than a large sum of money and government infrastructure could have effectively achieved. This movement began many years ago in the desert conditions in Rajasthan. It may take some time to raise strong public opinion on conservation in areas of rich biodiversity; but once achieved, it can produce excellent results in the shortest period of time at a minimal cost. A point in favour of such an approach is that not all diversities can be conserved outside the centres of diversity. Even though diversities can be conserved elsewhere in the original habitat, artificially-created environments may lead to genetic drift.

Public support for conservation activities can be achieved if the public is informed of the situation in the right manner. Television and radio programmes, screening of educational films and display of effective posters under the banner of non-governmental agencies could achieve this objective. Students, youth and the public should be well informed of the need for conservation of natural habitats and the rich biodiversity, on a long-term perspective. It is established that whatever be the type of conservation programme, involvement of local people and their support are essential for its success anywhere in the world.

Questions for Review / Exercise

1. Have you understood the definitions and meanings of the following?

   (a) Protected Areas (PAs)

   (b) Biodiversity hotspots

   (c) Megadiversity countries

   (d) Local people and Park management
2. What do you know about the history of Protected Areas? Have you heard of Emperor Ashoka of India and the role he played in conservation?
3. See Fig 72 and elaborate your understanding on the features, functions and use of PAs.
4. What do you know about the current global extent of Protected Areas? What is your opinion on increasing the extent further?
5. Check Table 4 and find out whether your country is listed as one of the megadiversity countries? If so, check whether the whole country or only some special regions are listed? Try to understand special features of the country or region responsible for the consideration under biological hotspot. What are the criteria to be considered for biodiversity hotspots for conservation?
6. See Fig 73–75 and look for different types of natural habitats in your area and differentiate each other with its own special features.
7. Suppose if you have been involved in a Committee to select new areas to be brought under PAs, what will be the major factors you will suggest for selection? Give reasons for each of the criteria that you will suggest.
8. Check Fig 76 and find out the features such as shape and closeness of areas to be considered for bringing under a PA? What are the principles you have learned from this?
9. What are the current IUCN categories of PAs?
10. See Table 5 and find out the management objectives of each type of PA.
11. What have you learned about management of PAs? Discuss on the duties of each of the major Park officials.

12. What is known as participatory management of PAs? Can you give few examples from different parts of the world where participatory management activities have helped in PA management?
13. What are your perceptions on 'participatory management of conservation areas'? Who are the stakeholders under participatory management? What do you think about your role in the management of a PA in your locality?

❏❏❏

Conservation Biology 10

Conservation biology as a new discipline in biology, aims of conservation biology, an overview of extinction, a major extinction spasm, IUCN threatened species categories, keystone species, umbrella species and flagship species concepts, conservation perspectives for the future, conservation strategies, conservation agenda, application of conservation activities, the world conservation strategy 1980 and since then

---

10.1. CONSERVATION BIOLOGY AS A NEW DISCIPLINE IN BIOLOGY

Deterioration of habitats and rates of extinction of species were high since the second half of the 20th century as compared to the first half. The increase of human population and resulted pressures exerted on natural habitats were the major causes for majority of such extinctions. There were also opportunities to analyze the comparative levels of deterioration of habitats and rates of extinction of species in different continents, which provide evidences to the effect that the process of extinction was at higher rates in the tropical developing nations. Habitat depletion was at higher levels in the tropics, where the growth rate of human population was also higher in comparison to other regions of the world. There is a direct relationship between pressure of human population on land and the level of depletion of natural habitats. In the temperate world, growth rate of human population is comparatively less and there is no much additional stress on land, particularly in natural habitats. This is clear from the fact that some of the European countries still maintain over 50% of the land area under forest cover. During the 1950s and 1960s, encroachment into forestlands for agricultural extension and intensive land-use patterns were common in the developing countries, where intensive agricultural activities were carried out with the intention of production of more food to support the millions of people who were undernourished. It was only during the late 1960s that biologists realized the need for effective control of the use of natural habitats, in the absence of which many of the species of fauna and flora would have become extinct. It was in this context that ecologists have initiated conservation activities with the emphasis on protection of species and their habitats all over the world.

A prediction on expected changes in biodiversity is made on the basis of scenarios of changes in atmospheric carbon dioxide, climate, vegetation and land-use patterns, and the known sensitivity of biodiversity to such changes. It is expected that for terrestrial ecosystems, land-use changes will have the largest effect followed by climate change, nitrogen deposition, biotic exchange and elevated carbon dioxide concentration. In case of freshwater ecosystems, biotic exchange is expected to have more effect. It is also predicted that the Mediterranean climate and grassland ecosystems are likely to experience the greatest proportional change in biodiversity whereas the northern temperate ecosystems are estimated to experience the least (Cody 1986).

Conservation biology is an emerging field of biological sciences, which has its origin as an academic discipline only in the last decade of the

20th century. This discipline promises to help in deciding the facets of nature, tradition and culture that are to be conserved. It also envisages measures to be taken to conserve habitats and species diversity as far as possible in natural habitats and to reveal the benefits out of it for human welfare. During the past few decades, developmental processes were associated with high levels of extinction. Conservation biology emphasizes the need for halting such adverse biological impacts of development processes with a view to evolve action plans, which encompasses sustainable maintenance and management of natural resources. It also advocates the need for restoration of already damaged and depleted habitats and ecosystems. Ultimately, it is mandated for protection of vast variety of genetic diversity in various ecosystems and to make use of this immensely-rich biological treasure-house for the welfare of humankind.

Conservation biology is partly a response to a biological crisis, and partly a crystallization of older disciplines drawn apart by specialization and rapid advances in the biological sciences. This discipline has grown from several areas of activities such as zoos for setting up small captive stock of animals that are endangered or difficult to obtain; botanic gardens and agricultural departments worried about dwindling genetic resources; forestry, fisheries and wildlife services charged with deriving maximum sustained yields; conservation agencies mandated to set up and protect nature sanctuaries; and biologists awakened by the disappearance of their study species (Soulé 1985, 1986).

Conservation biology is multidisciplinary in its features and approach, and is integrated with human socio-economic and cultural dimensions. As it is a developing discipline, we are yet to develop appropriate action plans to streamline methodologies to be adopted. Its application principles are still rudimentary. Major conservation issues are integrated with other conservation values such as aesthetics, economics and recreation. However, as conservation is more concerned with saving threatened species and ecosystems, conservation biology has major roles in developing strategies for limiting human involvement in natural habitats, sustainable use of natural resources, and in conservation of all habitats all over the world.

Ecology and evolutionary biology are inseparable components of conservation. As compared to the limited and unstable interactions in a managed agro-ecosystem, there are more contrasting differences between the interactions among various organisms and communities in a natural habitat. In this respect, sacred groves represent a typical example of retaining portions of natural habitats in areas of human dwellings and in man-made ecosystems. Sacred groves contain representative samples of local communities in conserved units with added faith of human culture and religion. Natural ecosystems contribute by their own way to maintain equilibrium that human beings upset. In this respect, forests, wetlands and other natural habitats are vital in the process of decomposing and recycling wastes, and in recharging habitats and ecosystems. However, when these habitats are overloaded with wastes and pollution, they may turn out as carriers of pollutants leading to pollution and poison-emitting reservoirs, where life and biological activities would not sustain for long.

10.2. AIMS OF CONSERVATION BIOLOGY

As a young discipline, it is yet to be developed with strong basis, and many of the views expressed under conservation biology are yet to be proved conclusively. Ecologists, who are worried about their study areas and the maintenance of interactions of diverse species in the natural set up, have a tough time to answer the doubts raised from various corners. Such questions may be on the number of extinctions, the probable reasons of extinction processes, appropriate use of large variety of biodiversity, and the future and scope of conservation biology. Based on the overall assumptions, the aims of conservation biology are as follows (Soulé 1985, 1986, McNeely 1988):

- *To identify conservation problems, and areas and species to be conserved on a priority basis.* Conservation Biologists should analyze various biological issues in their region and find out the details of the problem for existence of habitats and rich species diversity. In case any special reason is identified with respect to any area or species, the issue is to be subjected for detailed investigation and data are to be analyzed properly before suggesting the need for conservation of the area or the species in question. They should make their views clear, and emphasize the areas and species to be brought under conservation measures.
- *To provide scientific conservation principles relevant to the area and species to be brought under conservation efforts.* Each area may be special with its own species composition relevant for conservation. However, the principles to be followed may be different from place to place. Conservation biology and the practitioners of this field of science should emphasize the principles relevant to the area and the ecosystem, which they aim as a part of their objectives.
- *To find out appropriate corrective measures to overcome the problems identified in the area or with respect to the species in*

question. Conservation is to overcome the problems faced by habitats and species for their long term survival, and hence, conservationists should be able to implement corrective procedures to make sure that the habitat and the species would sustain for long. Identification of the exact problem and corrective measures should be revealed by experienced ecologists or conservationists. This should not be judged by an inexperienced biologist, whose views may not be accurate with respect to the real problems in the area. Further, long-term experience of the conservationist in the area of conservation and the concerned species is important, and it should be positively weighed when decisions are made.

- *To bridge the gap between the science of conservation biology and management by making scientists responsive to conservation problems and managers responsive to biological issues.* It is difficult to convince politicians and administrators who are responsible for implementing and executing policy matters of the governments, and hence, conservation biologists have to present the problems and their views in a convincing manner in front of the politicians and administrators for favourable decisions. For example, it was with much effort that conservationists could convince Government of India on the need for conservation of Silent Valley forest area in Kerala during the 1970s and early 1980s, when the State Government of Kerala strongly felt that a big dam should be constructed in this area for the then proposed Silent Valley Hydroelectric Project. Finally, late Mrs. Indira Gandhi, the then Prime Minister of India, took a strong stand between the conservationists and the Government of Kerala to abandon the proposal of the project in 1982. Later, in 1984, late Prime Minister of India, Mr. Rajiv Gandhi, declared the area as Silent Valley National Park for conservation of the comparatively large area of virgin tropical forests of South India, wherein ecologists have located the second viable population of the most endangered and endemic lion-tailed macaque (Balakrishnan 1984). This was possible only through several discussions and debates held for almost a decade of effective propaganda for conservation.

To achieve these objectives and to conserve species diversity, a political will is essential along with executive decisions to design and formulate conservation strategies. In the absence of strong political and administration will, conservation initiatives cannot achieve what is expected out of them.

10.3. AN OVERVIEW OF EXTINCTION

Extinction can take place as a result of natural evolutionary processes. However, it takes tens of thousands of years for a new species to evolve, flourish and extinct naturally. Due to this long period of time, it may not be possible for one to witness the extinction of a species in large numbers during the comparatively short life span of a human being. However, the extinctions that we have witnessed in the recent past are different from the natural ones, as several species with fairly good populations have vanished in a short span of time. Those species, which have faced threat of extinction in the recent past, have been with useful and costly properties; for example, the beautiful skin and feather, costly horn and tusk, and tasty flesh. Populations of several species with any of such characters have depleted fast, if not extinct altogether as happened in the case of the passenger pigeon and the Indian cheetah during the first half of the 20th century.

An analysis of the probable reasons for such exterminations has revealed four major causes of extinction processes in which vested interests of human beings are directly involved. These are habitat destruction and fragmentation, overkilling, impacts of introduced species and secondary extinction, generally known as the evil quartet (Diamond 1984). As human beings are a part of nature, their traditional activities are also a part of nature. However, the modern man is far away from the traditional man, and his activities are different from those of the indigenous people. In case a sustainable level of human population is maintained and the natural resources are used on a sustainable basis, the nature maintains its ability to rejuvenate habitats and resources from time to time. The several-fold increase in human population and the corresponding enhanced level of extraction of natural resources in the recent past have been responsible for deterioration of the land capacity of the world, far from its natural rejuvenation processes (Gadgil and Guha 1992). If human activities were in tune with the land capacity for regeneration and rejuvenation, nature itself would have been able to correct the problems.

10.3.1. Habitat Destruction and Fragmentation

There was a time when human beings were living in natural habitats in harmony with wildlife, depending upon them for livelihood. However, since the advent of settled life and agriculture, there have been tremendous changes in the way of living, which have direct effect on the natural habitats and all forms of wildlife. Till the end of the 19th century, such effects were marginal because the human population was comparatively less. However, human population grew faster with the development of modern technologies, better living conditions and advancements of health sciences, exerting pressure on nature and natural resources. The level of such pressure was high in areas where human population was denser. This was more significant in the tropical areas as compared to other parts of the world. In the beginning, natural habitats were destroyed for agricultural activities, for collection of timber and for maintaining domestic livestock. During the era of agricultural developments, large extents of natural habitats were put under slash and burn practice. Also, wetlands were drained and the land-use patterns were changed. Introduced species occupied the place of natural fauna and flora, which contributed to large-scale manipulation and modification of natural habitats of various types. As a result of all these, areas of natural habitats depleted significantly in a short span of time. Such activities related to human development were mainly responsible for more than half of the recently recorded extinctions in the tropical world. Naturally, when the extent of habitat is depleted, wild animals find it difficult to forage in the reduced extents of their natural areas. The cumulative effect is the depletion of biodiversity in affected areas.

The habitat need not be completely destroyed for depletion of wildlife populations or biodiversity in general. When the extent of habitat is reduced, the species that were originally supported by the habitat would exterminate from the area. Fragmentation and isolation of habitats also have similar effects on biodiversity. Encroachment into natural habitats is common in all boundaries of forests, and this process continues year after year, progressively reducing the size of forest tracts. In such a situation, survival of several species depends upon their ability to move and disperse themselves into other suitable and marginal areas. However, those species with limited powers to disperse faster will be at the risk of extermination.

10.3.2. Overkill

Overkill is killing of organisms at rates faster than they could reproduce themselves. Human hunters targeted large mammals, birds and even other animals for flesh and byproducts. In the historical past, when man co-habitated with wild animals, technologies were not so advanced to kill large number of animals, and eventually, animals also learned mechanisms to escape from human hunters. During that time, human beings were only one of the predators in the natural habitats and animals were killed only for subsistence. As animals were exposed to a predator for several years, they evolved behavioural adaptations to avoid risk, and also, developed means to escape from predators. Modern man is more interested in commercial benefits coupled with the beneficial properties of species of economic interest. Modern technologies have developed fast and wild animals cannot adjust themselves with such fast developing technologies. Also, they cannot learn methods to escape, and thus, succumb to death in large numbers. For example, wild animals survived in Australia for thousands of years in close contact with native Australians. But, majority of Australia's large mammals and birds succumbed to the newly-arrived Europeans with specialized hunting technologies. Not only animals were affected by overkill, but also plants such as the wine palm and sandalwoods, extensively logged for their sap or wood, were also seriously affected with the vested interests of man in the past.

One of the well-known recorded extinctions in the history of human development is that of the passenger pigeon (*Ectopistes migratorius* Linnaeus, 1766). The last pigeon of this species in the wild was shot on March 24, 1900 in Ohio in the United States of America. However, there were few of this species in captivity in zoos. 'Martha', the last passenger pigeon in the world died in the year 1914 in Cincinnati Zoo, and along with that incident, we have witnessed the end of a species, which was seen in millions, if not in billions. We were concerned about the passenger pigeon as it was not just one among the avian species existed on the Earth, but was the most abundant bird ever existed. There was a time when flocks of them migrating in millions used to darken the sky for three full days as they flew past (Miller 2004). Since the middle of the 19th century, hunting of passenger pigeon for their meat, feathers (mainly for use in pillows) and bones (for fertilizer) was common. As they were flying in large flocks and nesting in long and narrow colonies, it was easy to shoot them down.

The Indian cheetah was badly affected by hunters aiming at its beautiful and costly skin. The elephant tusk is costly, and hence, poachers aim quality tuskers. Rhinoceroses have their costly horns. Several species of birds are affected due to their beautiful feathers. Snakes are killed for their costly skin. Whatever be the species exterminated in this way, it will have its repercussions in nature.

10.3.3. Impact of Introduced Species

Introduction of exotic species has adverse effects in the maintenance of native species. Several of the native species may affect just by the introduction of a single exotic species. In this case, there are two major reasons for the extermination of native species; the native species have no prior experience with the introduced species and that the introduced species do not have natural enemies in the new locality. For example, the introduction of *Acacia auriculiformis* in Kerala has affected several of the native species, which have been displaced from their original habitats (Thomas and Balakrishnan 1998). Similarly, the South American species, *Salvinia molesta* Mitchell is an introduced weed in certain wetlands of South India. The native species suffer because the introduced species have special mechanisms by which resources are better utilized, thus inducing shortage of resources for the former. In such cases, the introduced species grow faster and outnumber the native species. In this example, the *Acacia* multiplied in large numbers displacing the native species. Further, in the absence of natural predators, the introduced species grow faster, if other ecological requirements are satisfactorily met with. The best known example of impacts of introduced species on native fauna is that of cichlid fish in Lake Victoria in East Africa. The populations of cichlid fish constituted the most diverse in any one locality. The introduced predatory fish, the Nile perch (*Lates niloticus* Linnaeus, 1758) has exterminated a good number of the native cichlid population in Lake Victoria.

Until recently, we were concerned only about the introduction of natural organisms. However, in the past, in addition to introduction of natural organisms, scientists are involved in introduction of Genetically-Modified-Organisms (GMOs) under various agricultural and industrial development programmes. Even though, the Convention on Biological Diversity recognizes the development of biotechnology and its use as a tool in conservation biology, no one can predict the adverse impacts of GMOs in natural habitats in the future. Hence, it is important to be more cautious of the regulation, management and control of adverse impacts of the release of Living Modified Organisms (LMOs) in natural habitats. Such organisms may have far-reaching effects on no-target populations in the area of introduction. They may also disrupt the natural evolutionary processes as a result of the activities of the introduced transgenes. The argument in favour of such transgenes is that the introduction of Genetically Modified Microorganisms (GMMs) for bioremediation purposes would result in the addition of new genetic diversity (UNEP 1995), which can be considered under habitat restoration programmes.

10.3.4. Secondary Extinctions

All species seen in any area or habitat interact with each other directly or indirectly. This may be with the predatory behaviour, mutualistic activities, and competition for resource use, host–parasitic interactions or the influence of pathogens on healthy animals and plants. In a natural habitat, populations of all species are not equal; rather the populations of one species may outnumber the other based on the body size and adaptive behavioural characters. Usually, if living conditions are satisfactory, smaller organisms are more in number than larger organisms. Similarly, the population of prey animals is more as compared to the population of predators. Even among the predators, larger predators may require large foraging areas and their number may be considerably less as compared to smaller carnivores. However, due to the effects of changes in the population of one species, the populations of a closely ecologically-interrelated species would change. When one species is exterminated, it may lead to an increase in the abundance of its competitors (direct competitors and prey), and may deplete the populations of those species, which may directly depend on it for resources. For example, extinction of Hawaiian honeycreepers (Order: Passeriformis, Family: Fringillidae) adversely affected the endemic Hawaiian plants of the genus *Hibiscadelphus*. The honeycreepers were the pollination agents of the genus *Hibiscadelphus* species, and hence, their pollination was affected in the absence of honeycreepers. When one species is exterminated, it may negatively affect the populations of closely-dependent species. An investigation has revealed that the extermination of spider monkeys due to poaching resulted in the loss of heterogeneity of *Inga ingoides* (Family Fabaceae), a common tree in the lowland forests in certain areas of Bolivia (Pacheco and Simonetti 2000). It is known that some plants depend on special animal vectors for dispersal of their seeds. If the vector comes under threat, the plants are likely to have adverse consequences. Spider monkeys were responsible for seed dispersal of *I. ingoides*, which was essential for a thorough genetic mix in the population, and in the absence of spider monkeys, a series of genetically more uniform patches have developed.

Destruction of keystone species may result in cascade effect. For example, indiscriminate killing of the sea otter resulted in the over-population of sea urchin, which depleted kelp forests and all associated organisms. In this example, sea otters are the keystone species. In all ecosystems, all species do not exert equal influence to maintain the homeostasis (the term homeostasis has originated from *homeo* = same and *stasis*= standing, and hence, homeostasis refers to the ability of ecosystems for self-maintenance and self-regulation as are their component populations and organisms, thus maintaining the balance of nature). Certain species may have more influence to maintain homeostasis than others. Those species, which have a profound influence in the ecosystem, are known as keystone species. Destruction of keystone species results in the extinction of several other species as a chain of activities, which is known as the cascade effect (Narendran and Balakrishnan 2008).

10.4. A MAJOR EXTINCTION SPASM

At present, the global environment is not so safe for a prolonged optimum life of diverse organisms in their natural habitats. Human cultures tend to change the values of humanity as a result of vested interests. Until recently, we were of the opinion that environmental problems were severe only in the developing tropical nations, mainly due to the pressure of human population. But, now it is realized that even in the well-developed part of the world, the environment is not safe. If poverty is the main problem in the developing and poor tropical regions, advanced technological revolutions is the main culprit in the most-developed part of the world! Poverty-ridden nations and their people do not listen to conservation propaganda. Developed and industrialized nations, on the other hand, contribute for war by making and selling sophisticated defence equipment, which are instrumental in hurting the peaceful global environment. There is no scope for conservation initiatives and activities when the political atmosphere is not peaceful. There is no account on the damages of natural habitats and the rich organisms in the areas affected by intensive war. Did we know at least the identity of the habitat-specific organisms in the beautiful high-altitude Kargil area of Indo-Pak boarder, where India and Pakistan fought during a recent war? Only what we knew was the number of young soldiers killed and seriously wounded on either side, and nothing about the non-human organisms affected and vanished. Even though there were several discussions including international ones on war, no one was worried about the impacts of war on the unique types of habitats and their characteristic biodiversities. Similarly, the recent gulf wars in 1991 and in 2003 had wide-spread effects on both terrestrial and oceanic habitats. It is believed that such inhuman activities can trigger a 'nuclear winter', resulting in long periods of darkness, subfreezing temperatures, toxic smog and increased levels of radioactivity not only in the areas directly affected by the war, but all over the world. Environmentally-conscious nations should not be involved in war and they should not produce, propagate, sell or buy war-related tools. In the context of major 'business gains', environmental interests are blindly neglected by world leaders! No one can expect peace as a result of war. A war cannot be a pathfinder for democracy, human safety and global peace (Narendran and Balakrishnan 2008).

Due to various types of unscientific human activities in natural and human habitats, it is suspected that we are in the process of a mass extinction. A mass extinction may be a sudden and pronounced decline in abundance and diversity of organisms worldwide. Such a process would be detrimental to the global ecology and environmental conditions, leading to its repercussions on the existence of humanity itself. Such drastic changes already made by man on the planet Earth are far from repair and recovery. Extinction has been a fact since life originated on this planet, millions of years ago. In fact, the present few million species are the modern-day survivors of the estimated half a billion or more species that have ever existed. The only difference is that past extinction processes have occurred by natural processes, but man-made processes are the cause of recent extinctions.

Extinction rates based on birds, mammals and amphibians in regions such as Central and South America, Africa, south of the Sahara and tropical South and South-east Asia over the past 100 years show that the current rates are 50–500 times higher than those known in fossil records. It is also revealed that continental extinctions are becoming more common these days as in oceanic islands. Such threats are more common in the tropics than in temperate habitats (Baillie et al 2004). About 20% of the Earth's species is found in Latin American forests outside of Amazonia, and another 20% is in forests in Asia and Africa outside Zaire Basin. All these forests are threatened, and if they are to disappear, species loss in these areas could be to the tune of several thousands.

In addition to the case of the passenger pigeon mentioned earlier in this Chapter, there are few other cases of well-established extinctions in different parts of the world. In Madagascar, until about the middle of the 20th century, there were 12,000 plant species and around 1,90,000 animal species. At least about 60% among them is endemic to the eastern strip of Madagascar. Over 90% of the original primary forest habitats have been eliminated, and at least half of the original species have already been exterminated along with the habitat. Lake Malawi in Central Africa holds over 500 species of cichlid fish. Almost 99% of them are endemic. This lake is only about one-eighth the size of the Great Lakes in North America, which has only about 170 species of fish, of which less than 10% is endemic. There is no comparison between these lake systems and their fish diversity, but unfortunately, Lake Malawi is threatened. Western Ecuador had 8,000–10,000 plant species of which 40%–60% constituted endemic species. There may be around 10–30 animal species in association with any plant species, closely interacting with each other, and hence, the total species composition would be very high. Since the1960s, there has been large-scale manipulation of habitats in this area for plantation activities, and related species loss would be not less than 50,000. The Pantanal area of Brazil contains 1,10,000 km^2 of wetlands. This area is considered as the most extensive and rich among the wetland communities in the world, which supports the largest and most diversified populations of waterfowl in South America. Even though this area is denoted by UNESCO's 'International Area of Importance', it suffers from agricultural expansion, dam construction and other types of processes destructive to the habitat and its vast and area-specific diversity of life.

In the context of extinction and extermination of species, Soulé and Wilcox (1980) have remarked that death is different from termination of birth. Extinction has far-reaching consequences as one can foresee suspended or even terminated evolutionary processes due to selective pressures exerted on organisms. For example, hunting pressure on elephants is concentrated on the quality of tusks of the tusker. Ultimately, what is lost as a result of poaching is the genetic composition required for growth and development of quality tusk. Finally, the process of speciation will have to work with a greatly reduced gene-pool of species.

There is sufficient proof to establish that diverse types of human involvement in nature are responsible for several of the recent extinction processes. There are over 170 species and subspecies of birds that have gone extinct since 1600, of which 155 are island forms especially from Hawaii, New Zealand, the Mascarenes and West Indies. In fact, they had evolved in the absence of humans. There are also records of 10 more avian extinctions in North and South America. These species also had evolved in the absence of human beings. Regarding mammalian extinctions, there are records of 115 species and subspecies since 1600. Among them, 22 are from the West Indies, 19 from other islands and 44 from Australia and North America (Diamond 1984). Among the 29 recent mammalian extinctions in Africa, Asia and Europe, 26 were large mammals; many of them were victims of over-hunting. Thus, most mammalian and avian extinctions during the past few centuries have involved species that were not familiar to human beings, especially to Europeans and their advanced hunting technologies. However, Didham et al (2007) have cautioned against the recent tendency to focus on single-factor explanations for extinction threat without explicitly considering the full complexity of synergies between habitat conversion, invasion, climate change, disease, and a host of other extrinsic drivers of population decline.

The present phenomenon of extinction is called the 'sixth extinction' (Leakey and Lewin 1995). The expected magnitude of it is much greater than the previous mass extinctions. Even though, extinctions are also natural processes such as the mass extinction witnessed at the end of the Cretaceous period some 65 million years ago, which eliminated much of the previous flora and fauna including the dinosaurs, it gave way for the evolution of much of the recent biodiversity of flowering plants and higher organisms including mammals. However, the ongoing extinction process is expected to be different from the previous ones as species extinct at the rates faster than the origin of new species. In this context, the principles of ecological and conservation genetics propose measures to reduce the rate of human-induced extinctions and to conserve global biodiversity.

10.5. IUCN THREATENED SPECIES CATEGORIES

The IUCN has been responsible for collecting scientific information from all over the world on the populations and status of species and habitats. When the population of any of the species is alarmingly on a downward trend, the species is listed in the concerned Red Data Book of the IUCN. Red Data Books are frequently revised with the latest information about the listed species and their habitats. Preparation of the Red Data Book is done as a joint venture of a large number of scientists and conservationists by way of their contribution from the respective regions where they are working. Baillie et al (2004) have made an assessment on the status of global species based on the Red List Consortium (IUCN Species Survival Commission, Bird Life International, Conservation International and NatureServe) with a broader coverage of in-depth analyses. As per the latest Red List, among the major groups of species analyzed, the percentage of threatened species ranges from 12 to 52. It lists 12% of birds, 23% of mammals, 32% of amphibians, 42% of turtles and tortoises and 18% of fish as threatened. Among plants, only conifers and cycads have been completely assessed and the results show that 25% of conifers and 52% of cycads are threatened.

All such species listed in the Red Data Books are generally grouped under a common term 'threatened', and hence, is the threatened category of species. Even though IUCN has been revising the threatened category list from time to time; students may come across the earlier categories in various books and publications as it has been widely used in the past. Hence, it is appropriate to list the earlier categories and then go for the revised list so as to have a better understanding of these categories. Species identified as threatened are assigned under specific categories indicating the degree of threat as follows:

- ***Extinct* (Ex):** These are the species not definitely located in the wild during the past 50 years (criterion as used by the

Convention of International Trade in Endangered Species of Wild Fauna and Flora). Only a few occasions, the category 'Ex' has been assigned by confirming that the taxon is extinct in its natural habitats (e.g., the Indian cheetah).

- ***Endangered* (E):** These are the taxa in danger of extinction and whose survival is unlikely if the casual factors continue to operate. The taxa whose numbers have been reduced to a critical level or whose habitats are so drastically reduced in the recent past are also included in this category. This shows that such species are in immediate danger unless the causal factors are eliminated from their natural habitats. Further, the taxa which may have been extinct in the past, but have definitely been seen in the wild in the past 50 years, are included in this category.
- ***Rare* (R):** These are taxa with small world populations. At present, they are not endangered or vulnerable in their habitats. However, if they are not brought under special conservation measures, it is likely that their populations will be in danger in the immediate future, and hence, at risk. This necessitates them to be cared under protection. Usually, such taxa are localized within restricted geographic areas or habitats (endemic species), or are thinly scattered over a more extensive range of their distribution.
- ***Indeterminate* (I):** These are taxa known to be endangered, vulnerable or rare. However, information is insufficient to make sure which of the above categories is appropriate for them.
- ***Vulnerable* (V):** These are taxa likely to move into the endangered category in the near future, if the causal factors continue to operate in their natural habitats. These are taxa of which most or all the populations are decreasing because of over-exploitation, extensive destruction of habitat or other environmental disturbances. Also, the taxa with populations that have been seriously depleted and whose ultimate security has not yet been assured are included in this category. Further, those taxa with populations that are still abundant, but are under threat from severe adverse factors throughout their range are also included under this category. Endangered and vulnerable categories also include, temporarily, taxa whose populations are beginning to recover as a result of remedial actions, but whose recovery is insufficient to justify their transfer to any of the other categories.

- ***Commercially Threatened*** **(CT):** Commercially threatened are those taxa, which are not essentially threatened at present, but most or all of their populations are threatened as a sustainable commercial resource or may become so, unless their exploitation is regulated. This category includes only those taxa whose populations are assumed to be relatively large so as to be extracted as a commercial resource. This category has only been used for commercially-important marine species that are being overfished in several areas of their ranges.
- ***Insufficiently Known*** **(K):** Insufficiently known are those taxa that are suspected, but not definitely known to belong to any of the above categories, due to lack of information. Usually, such taxa are under review for inclusion under any of the specific threatened categories.
- ***Threatened*** **(T):** Threatened is a general term to denote species, which are categorized under endangered, vulnerable, rare, indeterminate, or insufficiently known.

10.5.1. IUCN Revised Categories of Threatened Species – 1994

The revised IUCN (1994b) categories of the threatened species are as follows:

- ***Extinct:*** There is no reasonable doubt that the last individual of the taxon has died.
- ***Extinct in the wild:*** The taxon is known only to survive in cultivation, in captivity or as naturalized population (or populations), outside its past natural range. A taxon is presumed extinct in the wild when exhaustive surveys in known and/or expected habitats, at appropriate times (diurnal, nocturnal, seasonal and annual) throughout its historic range have failed to record the presence of it. Such surveys should be over a time frame appropriate to the life cycle and life form of the taxon.
- ***Critically endangered:*** These are the taxa that may face extreme risk of extinction in the wild in the immediate future.
- ***Endangered:*** The taxon is not 'Critically endangered', but may face a very high risk of extinction in the wild in the near future.
- ***Vulnerable:*** The taxon is not 'Critically endangered' or 'Endangered', but may face a high risk of extinction in the wild in the medium-term future.

- ***Lower risk/Conservation dependent:*** The taxon is not 'Critically endangered', 'Endangered' or 'Vulnerable', but is the focus of a continuing taxon-specific or habitat-specific conservation programme, the cessation of which may result in the taxon qualifying for one of the threatened categories within a period of five years.
- ***Near threatened:*** The taxon is not 'Critically endangered', 'Endangered', 'Vulnerable' or 'Conservation dependent', but is close to be qualified as 'Vulnerable'.
- ***Least concern:*** The taxon is not 'Critically endangered', 'Endangered', or 'Vulnerable' and does not qualify for 'Conservation dependent' or 'Near threatened'.
- ***Data deficient:*** There is inadequate information to make a direct or indirect assessment of the risk of extinction based on the distribution and/or population status of the taxon. A taxon in this category may be well studied, and its biology may be well known, but appropriate data on abundance and/or distribution is lacking. Listing in this category indicates that more information is required and acknowledges the possibility that future studies would only reveal, which among the threatened category is appropriate.
- ***Not evaluated:*** The taxa that have not been assessed against the criteria are included in this category.

The category ***Lower risk/Conservation dependent*** has been subsequently dropped in later versions of the Red List Categories (IUCN 2001, Baillie et al 2004). As the total number of such species has been increasing from revision to revision, it is clear that conservation efforts are insufficient in most cases.

Even though conservationists have to concentrate on conservation of species, which are already threatened, it is also essential to safeguard all other species. Otherwise, they may also become threatened in the near future as human interests may vary from time to time. Some of the species that may not be over-exploited at present may turn out to be of high use in the decades to come.

10.6. CONCEPT OF KEYSTONE SPECIES, UMBRELLA SPECIES AND FLAGSHIP SPECIES

In addition to the IUCN threatened categories of species, certain species of animals or plants may be considered as more important for conservation efforts in any area. This is based on the relative importance of few such species, which may have an overall control or dependence on other species or the ecosystem in general.

10.6.1. Keystone Species

When we analyze the species composition in an area in relation to its function, we may come across with certain species that are disproportionate with regard to their abundance and function. In abundance, there may be only few, but in effect they may be supporting several other species in the ecosystem. For example, consider the case of *Ficus* in a tropical ecosystem. Even if they are not seen in large numbers, a remarkably large number of birds and insects in addition to mammals may directly depend upon them during the flowering and fruiting seasons of long duration. In other words, a species, which is not seen in large abundance in a natural habitat, may form a key element in hosting and feeding a variety of organisms in the habitat. The biomass of such organisms in the habitat need not be remarkably high as compared to its role in supporting variety of organisms in the area. For example, the species of palm and fig trees are responsible for maintaining all species of vertebrate fruigivores in a Peruvian rainforest, for a span of not less than three months of the year in the absence of any other fruiting tree (Soulé and Kohm 1989). No doubt, there is a major role for such species in the ecosystem maintenance. It may be this phenomenon of *Ficus,* which made it the temple tree in the Hindu religion. Wherever there is a temple, at least one *Ficus* species is planted and protected. The principle involved is conservation of specific species to support biodiversity in the area. It is not only their effects as prey and predator, they are also responsible for maintaining a physical environment wherein diversity is protected and organisms are dispersed appropriately. Keystone species can be cited from all types of habitats and from all ecosystems. The impacts of keystone species on its community or ecosystem are larger than expected from their abundance. Such impacts may be easily revealed when such species or the individuals of them are altogether eliminated from the ecosystem.

10.6.2. Umbrella Species

There are several organisms, which have a wider range of habitat use. For example, elephants or large cats such as tiger in general, require large foraging areas for their survival, being large herbivores and large carnivores, respectively. Hence, for elephant conservation, conservation of a large area is mandatory. Being a tertiary carnivore, the tiger also requires large area for foraging. There are two major programmes to conserve elephants, the African Elephant Conservation Fund and the Asian Elephant Conservation Fund. Similarly, Project Tiger is concerned with conservation of tiger populations in India. However, ecological principles do not agree with conservation of any single species, but the concept adopted here is to conserve large areas under the umbrella of a single species, which is in danger due to human interactions in its natural habitats all over the range of its distribution. In these examples, even though single species is emphasized, none of the species can be conserved in the absence of conservation of the ecosystem in which they inhabit. Based on this basic principle, the species that depend upon a variety of organisms directly or indirectly, and have influence over a wider extent of areas through which they range, are effectively brought under conservation measures. Under the cover of few such species, several of the species in the area can be protected. Such major species that are responsible for conservation of the area and indirectly responsible for conservation of several other species from the threat of extinction are known as umbrella species. The concept of conservation activities with reference to any such species is the umbrella species concept. The umbrella species concept is aimed not only to conserve a single species, but also to conserve a variety of species as a chain response of conservation initiatives under one of the major species. Umbrella species are also responsible for natural control of several organisms in their habitats on which they depend upon directly and indirectly. For example, tiger is responsible for the control of populations of larger herbivores, which in turn, is responsible for the growth, development and control of herbs, shrubs and grasses in the area. Elephants of both species require larger areas of forests for their survival. They feed upon the species on which they depend (most animals are selective feeders), imposing a regulation on their growth rate, development and dispersal. This may also regulate the populations of other organisms in the habitat, particularly small invertebrates including insects that are dependent on such vegetation communities.

10.6.3. Flagship Species

Even though a biologist cannot discriminate between organisms, there are several species of organisms that are more attractive than others to laymen as well as to the general public. The features that may be responsible for such special attraction of an organism can be borrowed for impressive conservation slogan or to develop a conservation motto, which can be passed on through millions of people all over the world. Such organisms can be a large animal or a smaller one; it can be a plant or an animal, but its special feature should be attractive so that the public appreciates it. For example, animals such as panda, cheetah, tiger, rhino, elephant and mandrill form excellent flagship species among animals. Beautiful butterflies can attract the minds of younger generations at anytime, anywhere in the world. Among plants, cacti, orchids, lotus, etc. serve as flagship species. Such species can also serve as indicator species because they can be observed carefully on a long-term basis and their population fluctuations can be assessed to reveal the environmental quality from time to time. In case there is need for conservation of any area, a prominent species of that area can be considered as a flagship species. For example, in connection with the conservation of the well-known Silent Valley forests (at present the Silent Valley National Park) of the southern Western Ghats, which is one of the hottest hotspots of South India, lion-tailed macaque was considered as the flagship species. Similarly, Ethiopian wolf forms one of the flagship species of Ethiopia, for the conservation of which the Bale Mountains National Park is established.

10.7. CONSERVATION PERSPECTIVES FOR THE FUTURE

If the current trends continue, human population is expected to be around 10 billion by 2050 AD, about 75% of which will concentrate in the developing part of the world, and hence, the stress on landmass in developing part of the world will increase several folds. Despite the improvements in connection with the ongoing global conservation strategies, no one can conclusively foresee the future of the rich biodiversity in tropical forests. The remaining tropical forest covers only an area of about 5 million km^2. The annual elimination of these habitats amounts to over 50,000 km^2. If such high levels of deforestation are not halted, the well-known centres of global biodiversity may vanish in a short span of around 50 years. Deforestation and several other human activities would also accelerate the process of global warming. In addition to the direct effect on species by deforestation, global warming has its own special threat on habitat and species diversity. Global warming projected to 2050 could increase extinction probabilities for over 1,100 species representing terrestrial regions from Mexico to Australia. It is predicted that if the global warming proceeds on an average level of projections, about 25% of the above species may disappear by the year 2050. This loss would exceed the expected loss from habitat destruction as each species can persist only under a particular set of climatic conditions (IPCC 2007). As warming alters climatic gradients responsible for distribution of species on the Earth and reduces the extent of climatically-suitable habitats, it is expected that many species may have to shift towards the polar region or towards higher elevations for their survival. However, if they make such a shift, it may take again hundreds of years if not thousands, for many of such species to get accustomed to the new surroundings in the higher elevations. If so, they have to pass through several generations following the principles of natural selection through the process of 'survival of the fittest'. During this struggle for existence, certain individuals may develop genetic traits, which may help them to survive in the newly-dispersed areas, and the progenies of them with such favoured genetic traits may contribute for survival of the species. However, there is no guarantee for such an optimism to be realized. On the other hand, there is considerable chance for the species to become extinct before the development of such favourable genetic traits in the populations for the survival of the species.

In addition to the illegal encroachment and over-exploitation of resources, policies of governments have also been equally responsible for the depletion of habitats. Politicians, policy makers and administrators have to depend upon immediate benefits that can foresee out of the reforms they set. However, as immediate and direct benefits of conservation are negligible, they cannot be properly convinced, especially in terms of the expenditure involved and the revenue that may be earned. Direct and immediate economic benefits are always weighed when development policies are planned. However, with the level of scientific information now available, it is difficult to assess the economics of conservation so as to convince administrators and politicians. This has been a major handicap in finding out resources for conservation all over the world, and in that respect, politicians alone cannot be blamed. To overcome this problem, conservationists have initiated action programmes to demonstrate the economic value of biological diversity, which can be made available for socio-economic development of the regions mandated for conservation (McNeely 1988).

All types of habitats need to be conserved for conserving global biodiversity. Usually, one may discriminate types of habitats and stress for conservation of wet evergreen and moist-deciduous types of habitats. This is due to the fact that these types of habitats may have more species diversity, which may be of habitat-specific and endemic ones. In contrast, arid and semi-arid lands harbour only few species. However, due to the special adaptations of these few species to sustain in adverse environmental conditions, they feature many potentially-valuable biochemical compounds. Even in tropical forest habitats, not only wet-evergreen and moist-deciduous habitats are important, but all habitats are equally important for conservation. Studies on habitat association of animals have revealed that each habitat may contain species specific to that habitat. Hence, large extent of land and aquatic habitats containing all types of natural organisms needs to be conserved as integrated units for biodiversity conservation (Balakrishnan and Easa 1986). One of the most important conservation challenges of the world is to conserve variety of natural habitats with its innumerable wildlife by conserving the ecosystems wherein they sustain. Unfortunately, majority of the tropical nations are being forced to live on the natural resources forgetting the natural regeneration potentials. This trend is to be changed by adopting appropriate conservation strategies and sustainable utilization of natural resources including wildlife resources (McComb et al 2001).

10.8. CONSERVATION STRATEGIES

There are two major conservation strategies being adapted for conservation of wildlife. These are *in-situ* and *ex-situ* conservation strategies.

10.8.1. *In-situ* Conservation

In-situ conservation is conservation of biological diversity as it is in the respective natural habitats without any modification by human beings. Natural habitats are *in-situ* conservation areas if human interactions are limited. *In-situ* gene banks are examples for *in-situ* conservation. It is in such areas that natural genetic diversity can be maintained for long-term periods. The dynamics in such areas are limited with natural changes, and hence, organisms in such conservation areas are strictly under the influence of natural evolution and related changes. The genetic components in such areas are subjected to natural changes. They develop in response to various natural interactions between diverse organisms in the surroundings and undergo evolutionary changes in tune with the changes from time to time. Hence, the organisms, which carry these genes, are the most fit to survive in that habitat for long. This is why conservationists argue to maintain natural habitats as far as possible for biodiversity conservation. This is true even with respect to desert and coral-reef ecosystems. Even natural desert communities are important for conservation as they also contain certain organisms that possess genetic characteristics for their sustenance in such adverse environmental conditions.

10.8.2. *Ex-situ* Conservation

Ex-situ conservation is conservation outside natural habitats. Zoos, Aquaria, Botanical Gardens and Gene-banks serve as *ex-situ* conservation centres. When a few individuals of a species of animals or plants are remained, it may be essential to translocate them into suitable habitats, where they can manage to develop and propagate at a faster rate than they do in their disturbed natural habitats. These species can be available for reintroduction in natural habitats in sufficient numbers at a later stage. Botanical gardens, Aquaria and Zoos are excellent *ex-situ* conservation centres, where breeding and propagation activities are also carried out. Such centres serve for intensive conservation and management of endangered organisms, research, re-introduction, education, awareness creation and public visits (Claxton 2011, Carrizo et al 2013, Reid et al 2013). When the natural populations of any of the animals are depleted considerably below their natural reproductive potentiality, the individuals available in the wild are translocated into appropriate captive conditions, where they can be monitored and managed for better growth and reproduction. When they grow in sufficient numbers, they can be re-introduced into suitable natural habitats. Care must be taken during such re-introduction processes because the captive born and reared ones may not be able to survive fast in natural conditions, where they are introduced. Similarly, any plant species, identified as in danger of existence in natural habitats as a result of depletion of habitats, can be propogated in botanic gardens under care and strict vigilance of its growth and development. Such plants can also be re-introduced into suitable natural habitats at a later stage. There are over 2,000 botanic gardens in the world, holding over 80,000 plant species, which are visited by millions of people annually (Brutting et al 2013). *Ex-situ* gene-banks are also *ex-situ* conservation sites. There are also *in-vitro* and *in-vivo* gene-banks that act as seed-banks, field gene-banks, pollen-banks, chromosome-banks and DNA-banks. In case of animals, sperms, eggs, embryos, tissues, chromosomes and DNA are cryo-preserved (freeze-dried) and conserved in gene-banks (Clarke 2009, Dulloo et al 2010).

10.9. CONSERVATION AGENDA

Conservation of natural resources is mandatory for national development. Today, conservation of biodiversity is a prime challenge of the tropical nations. About 4% of the Earth's land area is already brought under conservation. There is no question among any of the nations whether to go for conservation or not, but the question is only on how far each nation should mandate for conservation. Several of the nations have nature clubs, wildlife clubs and other organizations to educate the public, younger generation and students on the need for conservation activities, and the membership of such organizations have been constantly increasing. They are even exerting pressure on politicians and administrators for conservation activities. The 'green peace' movement is already on its way into politics carrying the banner of conservation of nature and natural resources.

Ecology is a major field of biological sciences, but it is yet to be advanced with respect to its application. Most of the institutions all over the world, even today consider this as a classroom teaching subject. Ecology should be considered as an outdoor-teaching subject, and each student should be exposed to diverse natural habitats and ecosystems to understand the diversity of nature and the various types of interactions taking place among the variety of organisms in diverse ecosystems in any of the geographical areas. At present, several ecological processes are impacted by human beings, especially in the developing regions of the world, where natural ecosystems as well as the human population are rich. Hence, it is of utmost importance for students of wildlife ecology and biodiversity conservation to have first-hand information on what is happening in the natural habitats. This can help our younger and future ecologists to understand nature in a better way, to judge upon the problems, to find out appropriate solutions and to achieve prosperity.

Conservation activities cannot be carried out effectively in the absence of scientific data concerned with the biodiversity of any area or region. Biology, systematics and ecology are basic and essential areas of research in this respect. Regarding each species, knowledge on population and trend, range and habitat of distribution and the status of specific habitat are essential. To advance further and to make a proper identification of species, molecular taxonomy is being developed as a useful tool. Unfortunately, countries with large number of threatened species, for example, Brazil, Cameroon, China, Colombia, Ecuador, India, Indonesia, Madagascar, Peru and Philippines are unable to invest effectively in conservation. The developed nations, which earn enormously using the biodiversity resources of the poor developing nations have moral responsibilities of unconditionally investing in conservation activities to safeguard the species and diversity they have been using legally and illegally since long.

10.10. APPLICATION OF CONSERVATION ACTIVITIES

The extent of damages made to diverse ecosystems and dwindling of populations of a variety of wildlife are already clear. However, we are yet to intensify the efforts to conserve such diverse types of habitats and the diverse types of organisms they contain. We have so far studied only less than 2% of the total organisms in the world. About 98% of the organisms are yet to be studied in detail, particularly for their useful properties that may be of some use in the future. Imagine a situation if cinchona had been extinct before humankind could make use of its properties in the form of quinine as a cure for malaria or if penicillium had been eliminated from the Earth before humankind could make use of it as an antibiotic.

The endod *(Phytolacca dodecandra* L'Herit) (Fig 78) is well known in Ethiopia as a detergent and as an antifungal agent. However, the recent invention of its property as a molluscicide has led to its extensive use in controlling moluscan vectors, and thus, in controlling schistosomiasis (Gemetchu *et al* 1994). It is necessary that the use of this species should be legally restricted; otherwise the species may be commercially exploited without any control and may get vanished in the near future.

Fig 78. Endod, an endemic plant species of Ethiopia, known for its molluscicidal property. Recent investigations have revealed its molluscicidal properties, and hence, its application to control snail-transmitted diseases (Photo: M Balakrishnan)

No one can guarantee the long-term existence of millions of species of animals and plants seen in the threatened tropical moist forests until they are studied for their properties to verify whether they may be of use in the future. Even if some of these may not be found useful at present, the same may come out as useful in the future when technologies advance. The types of our requirements may also change from time to time. It is already revealed that species and their genetic materials have promising roles to play during the coming decades as powerful economic tools, viable in the process of developments in agriculture, pharmaceutics and in industries of various kinds. This can be the strongest point to support conservation, as its direct applications have no limits in the present day scenario. There is considerable scope to use wild species and varieties for breeding to make hybrids of improved or specialized varieties of closely-related crop species. Several such varieties or strains that are resistant to pests and diseases have been established. Biotechnology and genetic engineering promise for better use of resources and management of natural resources, which in turn, can also support the nation to achieve economic progress in the decades to come.

10.11. THE WORLD CONSERVATION STRATEGY 1980 AND SINCE THEN

The World Conservation Strategy was developed jointly by IUCN, UNEP and WWF with the objective of conservation of living resources for sustainable development (Fig 79). Subsequently, several nations have evolved their own national conservation strategies with strong conservation agenda and action plans for implementation. Recent advances in biodiversity conservation strategies emphasize on developing innovative approaches for *in-situ* conservation of biodiversity resources. Incentive-based approaches have emerged as a possible option. Biodiversity conservation can be accomplished by means of a wide range of incentives through participatory management programmes such as the incorporation of the stakeholders into management activities and incentives for the local communities by way of sharing benefits, and local institutional and infrastructural developments such as schools, clinics, and roads.

In 2002, ten years after the Earth Summit, the World Summit on Sustainable Development, held in Johannesburg, assessed the progress of world-wide activities made for implementing biodiversity conservation, and noted unsatisfactory progress in this respect and reaffirmed the commitment of Parties to the Convention of Biological Diversity to reduce the rate of biodiversity loss by 2010. In 2004, the Parties to the Convention on Biological Diversity agreed to test and develop a specific set of indicators for assessing the progress towards 2010 biodiversity target. The change of categories in the status of threatened species is one of these indicators. However, the increasing number of species of threatened categories, year after year, project unsatisfactory results of conservation initiative, showing an increasing risk of extinction leading to biodiversity loss. Hence, Baillie et al (2004) was critic about meeting the target of reducing biodiversity loss, particularly considering data from birds, amphibians and cycads.

Fig 79. Emblem of the World Conservation Strategy 1980. The circle represents the biosphere of the planet Earth, and the three interlocking arrows represent biodiversity features for conservation. The large arrow represents the extent of conservation areas, the medium arrow represents the species diversity conserved and the small arrow represents the genetic resources conserved.

Questions for Review/Exercise

1. Have you understood the meaning of the following terms?
   a. Species extinction
   b. Overkill of organisms
   c. Secondary extinction
   d. Keystone species
   e. Umbrella species
   f. Flagship species
   g. World conservation strategy
   h. *In-situ* conservation
   i. *Ex-situ* conservation
   j. Fragmentation of habitats

2. What is your perception on the field of conservation biology as a new branch of biological sciences?
3. What are the aims of conservation biology?
4. What are your perceptions on the impacts of introduced align species? Have you seen any such introduced species in your locality? Do you have eucalypts in your locality? Is it a native species of your region? Check and understand the history of eucalypts in your country, if it is not a native species.
5. What are the IUCN threatened categories of species? Can you name some of the species of animals or plants, which have become extinct in the recent past?
6. Find out whether there are threatened species of plants or animals categorized under the IUCN threatened category in your locality or in your country? Prepare a list of such species.
7. What are your viewson the recent extinction processes of animals and plants? Is it a result of evolution? Or do you expect it as a result of human-related issues?
8. Can you briefly explain various adverse impacts of human activities in natural habitats? Can you cite examples from your locality or country?
9. See Fig 78. Are there such species of wild plants in your locality, which are known to have medicinal properties? List all such plants that are known to you and make short notes of the medicinal properties of each of them.
10. Have you heard of any animal having medicinal properties? If so, list such species? Is there any such species of animal in your locality?
11. See Fig 79, the emblem of the World Conservation Strategy 1980. What are your understandings about the three major objectives of conservation from this figure?

Abstract 10.1

Depletion of Biodiversity in an *Acacia auriculiformis*-Introduced Habitat

Rasmi Thomas and M Balakrishnan

Department of Zoology, University of Kerala, Kariyavattom, Thiruvananthapuram 695 581, Kerala, India

Abstract

A comparative study of the diversity of plants, butterflies, amphibians, reptiles, birds and mammals in an *Acacia auriculiformis*-introduced habitat and mixed vegetation was made. Vegetation, amphibians and reptiles were sampled using quadrat method, while butterflies, birds and mammals were sampled using line-transect method. A total of 181 species were identified. Sixty four species (35.4%) were common to both the mixed vegetation and *Acacia* monoculture, 114 species (62.9%) were seen only in the mixed vegetation and three species (1.66%) were exclusive to the monoculture. The mixed vegetation had 74 species of plants, 44 species of butterflies, four species of reptiles, 51 species of birds and five species of mammals, whereas *Acacia*-introduced habitat had 23 species of plants, 25 species of butterflies, two species of reptiles, 15 species of birds and two species of mammals. Insectivorous birds were the most abundant in both the study areas. Abundance values and species diversity indices substantiated the detrimental effects of exotic species on biodiversity of the area.

Keywords: Abundance values, *Acacia auriculiformis*, Biodiversity, Exotic species, Species diversity index

Reference: International Journal of Ecology and Environmental Sciences **25** 127–141, 1999

1. What is your understanding about the impacts of introduction of exotic species?
2. What are the impacts of introduction of *Acacia* in the study area presented in the above abstract?
3. Can you estimate abundance and diversity indices of wildlife in an area?
4. What are your perceptions on quadrat method and line-transect method to study wildlife?

5. Are you familiar with the keywords given at the bottom of the abstract?
6. Is *Acacia auriculiformis* a native species of India? If not, find out the country of its origin. Do you have any such introduced species of plants/animals in your State/region/area? If so, make a list of such species, and try to learn the impacts of them on the native fauna and flora.

11 International Conventions on Wildlife and Nature Conservation

---

Reasons for international collaboration, Organizations and institutions – International Union for Conservation of Nature and Natural Resources (IUCN) – The World Conservation Union, World Wide Fund for Nature (WWF), United Nations Environment Programme (UNEP), World Commission on Environment and Development (WCED) – conventions (Treaties and Agreements) – The African Convention on the Conservation of Nature and Natural Resources (The African Convention, Algers 1968), Convention on International Trade of Endangered Species of Wild Fauna and Flora (CITES, Washington 1973), Convention on Wetlands of International Importance Especially as Waterfowl Habitat (The Ramsar Convention 1971), The Convention for the Protection of the Ozone Layer (UNEP Vienna Convention 1985), Convention on Biological Diversity (CBD 1992), United Nations

Framework Convention on Climate Change (UNFCCC 1992), United Nations Convention to Combat Desertification (UNCCD 1994), Kyoto Protocol (2005)

11.1. REASONS FOR INTERNATIONAL COLLABORATION

International collaboration for environmental conservation is of relatively-recent origin as compared to political and commercial co-operations and agreements. With few exceptions, establishment of institutions and agreements for conserving nature, wildlife and natural resources, in general, has been initiated from the second half of the last century. Natural resources such as contiguous habitats are shared among countries by migratory birds and mammals, which cross international borders without passport, visa, and emigration and immigration regulations. The East-African plains, constituting one of the large ecosystems shared by many countries, witness vast migrations of larger herbivores. Several species of birds migrate from the northern hemisphere to the southern hemisphere in the beginning of winter season and return to their original breeding ground by the time winter is almost over. This process of migration gives them prolonged daytime for their activities, which would have been otherwise restricted in the northern hemisphere where daytime would be less affecting foraging activities. Some wildlife populations occur temporarily or permanently in areas beyond national jurisdiction – the global commons – the oceans, the atmosphere and Antarctica. Several of the marine organisms are of this category. A great deal of the most important commercial fishery operations take place in international waters, and depend on rules and regulations accepted by all the nations involved. In the absence of international agreements and understanding, such living resources might be exploited and exterminated for short-term profits.

Natural resources in one country may be harmfully affected by activities in another, be it physical encroachment or management policy. For example, rivers passing through more than one country and determining many natural ecosystems may be dammed or diverted in an upstream country. This may cause ecological problems in the downstream countries. In recent years, water and air pollution, especially in industrialized regions, have caused environmental disasters in areas far away from the 'producer' country, destroying both aquatic life and forest habitat. The acid rain problem in Europe is well known. Nitric oxide and sulphur dioxide from burning of fossil fuels in Britain, and in central and eastern Europe have killed large number of lakes in Norway in the last century. It is estimated that only 15% of the acidification originates from inland sources and the rest is from other regions. The Chernobyl nuclear plant accident, occurred in 1986, resulted in the spread of radioactive waves through a number of European countries. To combat such global environmental problems, international collaboration, agreements and understanding are essential between nations, regions and at a wider circle under the banner of the United Nations.

11.2. ORGANIZATIONS AND INSTITUTIONS

- ***International Union for Conservation of Nature and Natural Resources (IUCN) – The World Conservation Union:*** Founded in 1948 as an independent international organization, IUCN has a membership of voluntary organizations as well as 120 governments and government agencies. It is an advisory scientific body to other international organizations and national governments on issues of conservation and best use of natural resources. Its headquarters is located in Glands, Switzerland. The IUCN is responsible for the preparation of the 'Red Data Books' on animals and plants. Red Data Books deal with the timely status of threatened categories of plants and animals all over the world. The information published in Red Data Books is based on contributions by individuals, scientists and organizations dealing with natural organisms and their conservation.
- ***World Wide Fund for Nature (WWF):*** The World Wildlife Fund was established in 1961 and later renamed as the World Wide Fund for Nature. It is functioning in close collaboration with IUCN. The WWF is mainly responsible for raising funds to finance action plans for specific conservation programmes of endangered organisms. The headquarters of WWF is also located in Glands.
- ***United Nations Environment Programme (UNEP):*** The UNEP was established after the UN Conference on Human Settlements held in Stockholm in 1972. Its main purpose is to promote and coordinate activities in the field of environment and natural resources. The UNEP does not implement projects directly. Its headquarters is located in Nairobi, Kenya.
- ***World Commission on Environment and Development (WCED):*** Unlike other agencies mentioned above, the WCED (which is also known as the Brundtland Commission, after the

name of its Chairperson, Mrs. Gro Harlem Brundtland, former Prime Minister of Norway) was set up in 1984 with specific objectives. Its final report entitled 'Our Common Future' was submitted in April 1987. This report had major impacts on global environmental concerns. The members of the Commission were strong politicians, lawyers, planners and scientists representing all main geographical regions of the world. It was intended to put the environment on the political agenda. One of the major recommendations of the Commission was to expand the extent of protected areas several folds to incorporate representative samples of the Earth's ecosystems under protective coverage.

11.3. CONVENTIONS (TREATIES AND AGREEMENTS)

11.3.1. The African Convention on the Conservation of Nature and Natural Resources (The African Convention, Algers 1968)

During the colonial times, in the first half of the 20th century, many agreements were established for the protection of African wildlife. As most African nations gained independence after World War II, the need for a new treaty more relevant to modern Africa became evident. In 1968, member States of the Organization of African Unity (OAU) signed the African Convention of the Conservation of Nature and Natural Resources in Algiers. This Convention is primarily concerned with wildlife (plant and animal life), but it also embraces conservation of other natural resources such as soil and water. This Convention emphasizes the need for protected areas, and special conservation measures for the species listed in an annexure to the Convention, which is divided into two Classes. Class A includes the species that require total protection because they are facing extinction or may be threatened in the future. The species in Class B are also given full protection in the beginning, but hunting and capturing are permissible under special authorization. The Convention also covers subjects such as conservation education and research, and the need to integrate conservation into development plans. Unfortunately, similar to other international treaties, it also suffers from the deficiency in not establishing an administrative structure to enforce regulations. Its practical value for African wildlife might have been considerably greater as a central machinery responsible for overseeing and promoting enforcement. Further, as the signatories to the African Convention do not need to submit reports to the OAU on their implementation of the objectives of the Convention, it is difficult to assess the extent to which its provisions have been enforced by Member States. However, several new protected areas have been set aside in the Member States as insisted by this Convention. Majority of the African Nations are also party to a large number of other global Conventions.

11.3.2. Convention on International Trade of Endangered Species of Wild Fauna and Flora (CITES, Washington 1973)

In addition to the destruction of biotopes and habitats, international trade of rare and vulnerable species and the products out of them is the most serious threat for organisms to sustain. Most of the organisms and their products involved in international trade are considered as luxury items, trophies or valuable on the ground of medicinal effects. Such exports are mainly from the tropical regions to the developed countries in the North.

The CITES was concluded in Washington in 1973 and came into force in July 1975. The principle behind this Convention is that if there is no demand or market, there is no trade. This Convention regulates international trade of wild animals and plants, which are listed in three Appendices. With a few exceptions, it prohibits commercial trade of species listed in Appendix I (species threatened with extinction). It allows a controlled international trade of species listed in Appendix II (species not threatened at present, but may become so in the future). According to Appendix III of the Convention, a member State, which has domestic legislation regulating the export of those species, not listed in Appendix I or II, can seek the support of other member States to enforce its legislation. The member States are required to establish management and scientific authorities at the national level, and are responsible for ensuring that the conditions laid down in the Convention are satisfied before granting a permit for international trade of the species listed under Appendix I and II. Further, the CITES Secretariat in Switzerland oversees the permit system on an international level.

There are 104 countries, which are parties to CITES. It is one of the most successful global wildlife conservation treaties.

11.3.3. Convention on Wetlands of International Importance, Especially as Waterfowl Habitat (The Ramsar Convention 1971)

Wetlands are the most productive among the ecosystems. However, in the industrialized and intensively cultivated parts of the world, more than other habitat types, natural wetlands such as marshes, shallow lakes and estuaries, tidal flats, seashores and mangroves are polluted, modified, fragmented and destroyed. The threats are mainly from drainage, landfills and pollution. Aquatic birds, especially some of the waterfowls – ducks, geese and waders – are intercontinental migratory birds, depending on wetlands in the tropical and subtropical regions during the winter months. To protect the large populations of such migratory birds, it was essential to establish international understanding and regulations. In case their migratory habitat in the developing tropical part of the world is lost, their populations would vanish in the absence of alternative habitats. This was the idea behind the establishment of the Ramsar Convention. It was signed in Ramsar (Iran) in 1971 and came into force in December 1975.

The Ramsar Convention obliges governments to promote the wise use of wetlands, to stem encroachment and to designate suitable wetlands within their territories for inclusion in an international list. These wetlands shall be given special protection, normally as nature reserves. The designation of a wetland reserve as a 'Ramsar Site' will help to defend the protected status of the reserve.

At present, there are 2,186 Ramsar sites in 168 countries in the world. Upon joining the Ramsar Convention, each Contracting Party is obliged by Article 2.4 of the treaty to designate at least one wetland site for inclusion in the List of Wetlands of International Importance. The Convention's goal is to achieve Ramsar listing for as many wetlands throughout the world as possible to meet the criteria of international importance. It was expected to establish at least 2,500 sites in the list, covering at least 250 million hectares, by 2010. To bring this much of the world's wetland resource under the umbrella of Ramsar status, the Convention urges all States to join the Convention, if they have not already done so, and to make significant efforts to implement the systematic approach for the development of the list in each country as adopted under the strategic framework of the Convention. The current extent of Ramsar sites is only about 200 million hectares. The United Kingdom has the largest number of Ramsar sites (170), whereas Bolivia has the greatest extent (1,40,000 km^2).

11.3.4. The Convention for the Protection of the Ozone Layer (UNEP Vienna Convention 1985)

The objective of this Convention is to protect human health and the environment against adverse effects resulting or likely to result from human activities that modify or are likely to modify the ozone layer. For this purpose, it is essential to co-operate by means of systematic observations, research and information exchange in order to better understand and assess the effects of human activities on the ozone layer, and the effects on human health and the environment from modification of the ozone layer. The Convention urges to adopt appropriate legislative or administrative measures and to co-operate in harmonizing appropriate policies to control, limit, reduce or prevent human activities that have or are likely to have adverse effects on the ozone layer. This Convention also envisages that the Parties undertake to initiate and co-operate through competent international bodies, the conduct of research and scientific assessments on the physical and chemical processes that may affect the ozone layer, the human health and other biological effects deriving from any modifications of the ozone layer, particularly those resulting from changes in ultra-violet solar radiation having biological effects (UV-B), climatic effects deriving from any modifications of the ozone layer and any consequent change in UV-B radiation on natural and synthetic materials useful to mankind, substances, practices, processes and activities that may affect the ozone layer, and their cumulative effects, alternative substances and technologies and related socio-economic matters.

11.3.5. Convention on Biological Diversity (CBD 1992)

The CBD was formulated during the Earth Summit held in Rio de Janeiro, Brazil in 1992. The objectives of this Convention are conservation of biological diversity, sustainable use of its components, and the fair and equitable sharing of the benefits arising out of the utilization of genetic resources, including appropriate access to genetic resources and appropriate transfer of relevant technologies, taking into account all rights over those resources and technologies.

Biological diversity is the variability among living organisms from all sources including inter alia, terrestrial, marine and other aquatic ecosystems, and the ecological complexes of which they are part of. This includes diversity within species, between species and of ecosystems. Biological resources are all resources of biological origin. Under this Convention, each Contracting Party, in accordance with its particular conditions and capabilities, shall be required to develop national strategies, plans or programmes for conservation and sustainable use of biological diversity and integrate, as far as possible and as appropriate, conservation and sustainable use of biological diversity into relevant sectoral or cross-sectoral plans, programmes and policies. It is envisaged that each Party shall – identify components of biological diversity that are important for its conservation and sustainable use; monitor the identified components of biological diversity through sampling and other techniques by paying particular attention to those requiring urgent conservation measures and those which offer the greatest potential for sustainable use; identify processes and categories of human-related activities that have or likely to have significant adverse effects on the conservation and sustainable use of biological diversity; monitor their effects through sampling and other techniques; and organize and maintain database of the biological organisms identified. The major action plans of this Convention are: establishing protected areas to conserve biological diversity, rehabilitation and restoration of degraded ecosystems to promote recovery of threatened species, preventing introduction of alien species, and controlling and eradicating those alien species that threaten the ecosystems, habitats or species and regulation and appropriate management or control of the risks associated with the use and release of Living Modified Organisms (LMOs) resulting from biotechnology. It also expects due respect towards traditional knowledge and practices of indigenous communities concerned with the rural biodiversity.

The CBD offers to the world an opportunity to translate the rights and aspirations of farming, fishing and forest communities, and traditional healers into a political reality. However, in 2002, ten years after the Earth Summit, the World Summit on Sustainable Development held in Johannesburg assessed the progress of activities made in implementing biodiversity conservation world over, and noted unsatisfactory progress in this respect and reaffirmed the commitment of Parties to CBD to reduce the rate of biodiversity loss by 2010. In 2004, the Parties to CBD agreed to test and develop a specific set of indicators for assessing the progress towards 2010 biodiversity target. The change of categories in the status of threatened species is one of these indicators. However, the increasing number of species of threatened categories, year after year, project unsatisfactory results of conservation initiative, showing an increasing risk of extinction leading to biodiversity loss.

The United Nations has declared 2010 as the International Year of Biodiversity and the period from 2011 to 2020 as the UN-Decade on Biodiversity. The Nagoya Protocol was adopted in the 10th Conference of Parties to CBD held in Nagoya, Japan. The Nagoya Protocol on access to genetic resources and the fair and equitable sharing of benefits arising from their utilization is a supplementary agreement to CBD. It provides a transparent legal framework for the effective implementation of one of the three objectives of CBD; i.e., the fair and equitable sharing of the benefits arising out of the utilization of genetic resources.

National Biodiversity Strategies and Action Plans (NBSAPs) are the principal instruments for implementing the CBD at the national level. The Convention insists Parties to prepare a national biodiversity strategy, which is to be mainstreamed through strategy planning and activities of all those sectors whose activities may have impact on biodiversity. There are 173 Parties including India who have developed NBSAPs. The United Kingdom, New Zealand and Tanzania have advanced further in conserving individual species and their specific habitats. The United States of America, even though not so far ratified CBD, has been in the forefront in the implementation programmes through species recovery and conservation mechanisms. The Conference of the Parties (COP 11) on biodiversity held in Hyderabad, India in 2012 has initiated the preparations of a 'World Wide Views of Biodiversity'.

Currently 192 countries and the European Union are Parties to CBD, and all UN signatories other than the US, Andorra and South Sudan have ratified the treaty.

11.3.6. United Nations Framework Convention on Climate Change (UNFCCC 1992)

The UNFCCC makes a distinction between the climate change attributable to human activities altering the atmospheric composition and the climate variability attributable to natural causes. The objectives of UNFCCC are to stabilize the concentrations of atmospheric greenhouse gases (carbon dioxide, methane, nitrous oxide, hydroflurocarbons and sulphur hexafluoride) at a level that would prevent dangerous anthropogenic interference with the climate system within a timeframe sufficient to allow ecosystems to adapt naturally, to protect food production and to allow economic development in a sustainable manner. This Convention came into force in March 1994. Currently, over 186 nations are Parties to this Convention. It is mandatory for the Parties of the Convention to protect the climate system for the benefit of present and future generations of humankind. The Parties are required to take precautionary measures to anticipate, prevent or minimize the causes of climate change and mitigate its adverse effects. Specific needs and special circumstances of the developing countries to the adverse effects of climate change should be given full consideration. Policies and measures to protect the climate system against human-induced changes should be appropriate for specific conditions of each Party and should be integrated with national developmental programmes, taking into account that economic development is essential for adopting measures to address climate change. The Convention also envisages that the developed countries would recognize the need for a return of emission levels of greenhouse gases to the levels of 1990, to modify long-term global emission trends.

11.3.7. United Nations Convention to Combat Desertification (UNCCD 1994)

The 1992 United Nations Conference on Environment and Development (Earth Summit) recommended the elaboration of the United Nations Convention to Combat Desertification (UNCCD), which was adopted in June 1994 and entered into force in December 1996. This is the first and the only internationally and legally-binding framework set up to address the problem of desertification. There are over 180 countries that are Parties to this Convention.

The UNCCD has its objective to combat desertification and mitigate the effects of drought in countries experiencing serious droughts and/or desertification, particularly in Africa through effective action programmes, so as to achieve sustainable development in affected areas. The Convention envisages for actions involving international co-operation and a partnership approach for improving land productivity, rehabilitation of land, conservation and sustainable management of land and water resources. Such actions are expected to prevent long-term consequences of desertification such as mass migration, species loss and accumulation of greenhouse gases.

Desertification is a process of land degradation in arid, semi-arid and dry subhumid areas resulting from various factors including climatic variations and human activities. It is a gradual process of loss of soil productivity and thinning of the vegetative cover, mainly due to human-related activities along with the climatic variations such as drought and flood. It is a worldwide problem directly affecting over 250 million people and one-third of the Earth's land surface (over 4 billion ha). It affects the livelihood of over one billion people, who depend upon land for most of their needs. Further, the impacts are more severe in the poor developing countries in the tropics. Although desertification affects Africa the most, where two-third of the continent is desert or dry lands, it is not a problem confined to the dry lands of Africa alone. For example, over 30% of the land in the United States is affected by desertification. It is important to fight against desertification as it is one of the root causes of socio-economic problems, threatening the environmental equilibrium in the affected regions. When the top soil is mistreated, blown and washed away, it takes centuries to build up again.

11.3.8. Kyoto Protocol (2005)

Kyoto Protocol is a protocol to the International Framework Convention on Climate Change with the objectives of reducing greenhouse gases that cause climate change. It was adopted on December 11, 1997 and entered into force on February 16, 2005. Altogether, 182 parties have ratified this protocol. Among them, 36 developed nations are required to reduce greenhouse gas emission to the level specified for each of them in the treaty. Among the developing nations, 137 (including India and China) have ratified the protocol, but they have no obligation.

The treaty recognizes that developed countries have contributed the most to the anthropogenic build-up of CO_2 in the atmosphere (around 77% of emissions between 1750 and 2004), and hence, many of the developed nations have agreed to reduce their greenhouse gas emissions in two commitment periods. The first commitment period was during 2008–2012. The second is during 2013–2020. The protocol was amended in 2012 to accommodate the second commitment period, but this amendment is yet to be made effective. The 37 countries with binding targets in the second commitment period are Australia, all members of the European Union, Belarus, Croatia, Iceland, Kazakhstan, Norway, Switzerland, and Ukraine. On an average, CO_2 emission per person in developing countries is only 2.9 tones as compared to 10.4 tones of the developed nations. However, there are differences in the opinion on the target countries, particularly as China and India with higher national contributions of greenhouse gases, are not legally bound to reduce their emission levels. Due to this difference of opinions, the US has not so far ratified Kyoto Protocol, even though signed the Protocol long time back. Canada withdrew from the Protocol in 2012.

Questions for Review/Exercise

1. Have you understood the following agencies?

   a. IUCN b. WWF c. UNEP

2. Have you understood the following international conventions?

   a. Convention on Biological Diversity (CBD)

   b. The African Convention, Algers 1968

   c. CITES, Washington 1973

   d. Ramsar Convention 1971

   e. Kyoto Protocol

3. What are the purposes of international collaboration and conventions on environment and wildlife related issues?

4. What is expected by implementing the Convention on International Trade of Endangered Species of Wild Fauna and Flora?
5. What are the objectives of the Convention on Wetlands of International Importance?
6. What are your perceptions on ozone layer and the importance of it? What are the adverse effects of depletion of ozone layer?
7. What do you know about deserts? Are deserts expanding? If so, are there human roles in such expansion of deserts? Have you heard about the United Nations Convention to Combat Desertification (UNCCD 1994)? Discuss the importance of this Convention during a seminar presentation.

Traditional Methods in Wildlife Ecology 12

Ecological monitoring, what are assessable by general ecological monitoring, species–area curve, population studies, counting animals, pugmark tracing, facts from faeces, population estimates from faecal density, population estimation, population indices, estimating biodiversity, habitat and ecosystem diversity, species diversity, trapping wild animals

---

The ecosystem and the organisms it contains are to be constantly monitored from time to time and from season to season to reveal the status of the species, populations and abundance of various species present, dynamics and other changes and to understand both positive and negative effects of various man-made activities. This is of importance in the context of management of wildlife habitats and sustainable use of resources. Such investigations are more important as we do not know the dynamics of many of our wildlife populations and the impacts of human activities in their habitats.

12.1. ECOLOGICAL MONITORING

Ecological monitoring is aimed to reveal climatic parameters and to analyze habitat structure and status of any area. It would also reveal the type of habitat, various vegetation communities present in the area, and variations in response to seasons and various human activities. Such monitoring can also be aimed at revealing various aspects of animal populations such as the number, structure, fluctuation and distribution patterns. There are several migratory animals such as birds, wildebeests and white-eared kob (*Kobus kob* Endeben, 1777). There are also several local migrants among animal species. Temporal and altitudinal migrations are common in animals in response to even daily variations in temperature and even in relation to the temperature differences during different times of the day. Human beings with their superpower make several alterations in environments and the impacts of these activities are reflected on various organisms and ecosystems in different ways. Detailed and constant monitoring in ecosystems can reveal these factors including the levels of involvement of human element and the effects of such activities on habitats, particularly on ecosystems and species in the habitat. For example, constant observations in some of the coastal areas can reveal the effects of untreated effluents from industries discharged directly into the coastal waters. Similar problems are seen in many of the riverbeds and riverine systems in the tropical world. Even though pollution is widespread all over the world, the level is much higher in the developing part of the world.

Ecological monitoring can be undertaken by three major means such as ground monitoring, aerial monitoring and satellite monitoring. Ground monitoring is by observing ground truth by direct or indirect methods. In this means, the observer has to make close contacts with all areas in the specific ecosystem to have a close observation and to make in-depth analyses of the ecosystem and organisms of all types seen in the ecosystem. In such studies, macroorganisms may be easily visible, and hence, such organisms would get more attention of observers. On the other hand, microorganisms are often not studied mainly because of the smaller size, and as many of them are invisible to ordinary human eyes. Studies on microorganisms require special attention and facilities.

Aerial monitoring is involved with the use of aircraft for the purpose of monitoring. The advantage of this is extended coverage of larger areas in comparatively shorter time periods. In case a large area is to be surveyed for its large mammal populations, it may take several days, weeks or even months to complete only the preliminary observations. On the other hand, if an aircraft is systematically used, the survey of a comparatively large area can be satisfactorily completed in a few hours. However, the question is how far this means of monitoring is effective and how much would be the expenses involved in it. The effectiveness need not be comparable to the ground means as the survey is actually done from an aircraft at a height. In case there is any obstruction from the aircraft to the ground, a clear observation is impossible. Hence, these means can be more applicable in large extents of savannah forests rather than in thick wet evergreen or even moist-deciduous forests, where visibility from the aircraft is often blocked by trees and other vegetation communities. Even in highly elevated and in undulating mountain ranges, these means may not be applicable. Close observation is also impossible as the observer is always at a distance from the site to be observed.

Satellite monitoring is done by analyzing the satellite images of the region. Satellite images of all areas of the world are available in different scales, and these maps are the source of information in this means of analysis of ecosystems. The advantage of this method is to have general information on larger areas. Such maps are easily available for the researchers. However, what is revealed need not be correct to the minute levels of accuracy as compared to the ground truth that may be gathered by ground monitoring. Further, this method may give only approximate picture of the habitat and not to the levels of organisms in any type of habitat. Satellite imageries are analyzed using the colour patterns developed. The colour pattern is developed based on the altitude of the area and the density of vegetation cover. The dark brown colour combination denotes areas of high-altitude dense forests. It need not be 100% accurate as there are several other factors that may also influence the development of the specific colour pattern. For example, in an area of 1,000 m elevation, an evergreen forest and a rubber plantation may reveal more or less the same colour pattern, at least during certain seasons based on the leaf cover. Such problems encountered in satellite imageries can be eliminated only by double checking the ground truth using ground monitoring.

On the whole, even though it is more laborious and time consuming, ground monitoring is the one traditionally being used by ecologists. It is more accurate than the other means of monitoring. It may also be less expensive as compared to the other means. Further, it is necessary to re-check the data gathered by aerial and satellite means by ground assessment for confirmation and for fine details. For the purpose of interpretation of the remote data such as the one gathered by aerial and satellite data, ground monitoring is essential to reveal the ground truth.

12.1.1. What are Assessable by General Ecological Monitoring?

Permanent, semi-permanent and seasonal attributes can be studied and assessed by ecological monitoring. Among the permanent attributes, topography of the area can be easily analyzed during a general survey of the area. The soil types, drainage pattern, waterholes in the area, animal features such as termite mounts, nests and burrows are visible for observation. The soil type and drainage patterns can partly be responsible for the development of specific vegetation types. For example, a river belt will have riverine vegetation, which is different from the typical evergreen and moist-deciduous types of vegetation. The soil type and climatic factors would contribute for the development of a dry deciduous habitat. The elevation gradient and topography may contribute for the specific high-altitude dense forest and grassland compositions. General features of these different types of habitats, the characteristic vegetation formations and associated animal combinations can be studied in detail by intensive long-term ecological monitoring.

Semi-permanent attributes include plant physiognomy such as vegetation type, community composition, zoogenic features such as wallows and salt licks, distribution of non-migratory species of animals, human settlements such as villages, farms, ranches and roads. Even though majority of these looks like permanent attributes, these are not static. Even if the farm area is repeatedly used for farming, there occur various stages of farming activities and shift of farming practices at least once in 3–4 years. Villages also show changes in the patterns, facilities and the supporting population.

Seasonal attributes are also known as ephemeral attributes. These include rainfall, insulation, soil moisture, evapotranspiration, plant phenology such as greenness and defoliation, plant productivity (biomass, part composition, chemical composition and energy content), presence and distribution of migratory species, animal behaviour, population structure and productivity, fire and availability of surface water. Physical factors are also seasonal factors. Rainfall, temperature, humidity and moisture are strictly seasonal in relation to the geographical location of the area. Plant phenology is well pronounced in temperate habitats. Phenological changes are also well evident in deciduous forests in the tropics. Both plant and animal productivity depends upon various factors including climate and season. Seasonality in reproductive processes is well known in majority of the organisms, particularly in temperate habitats. In relation to seasonality, animals migrate from one part of the world to the other (e.g., bird migration). This is in response to the shorter day-length for their feeding activities in the northern hemisphere during winter months. They return to their original breeding sites by the end of the winter period. This behavioural adaptation is repeated every year. Most of the behavioural activities of animals are in relation to their environmental parameters. Population structure of animals, particularly of larger mammals, also changes significantly in response to season. During the dry season, when there is shortage of forage, larger herds of herbivores split into smaller family units and range in a variety of habitats and in all available ranging areas. They aggregate again when forage is available in plenty in the beginning of the rainy season, and form larger herds. Forest fire is strictly seasonal, seen only during the summer months. Availability of surface water in foraging ranges in natural habitats is also seasonal. There would be only limited areas in the range where surface water is available during summer months. Larger herbivores would depend upon limited perennial water sources in their habitats, for which they have to walk longer distances during the noon hours in the hot Sun.

All these features listed under various attributes are important characteristics of habitats, ecosystems and of organisms, which are to be studied in detail for understanding the nature of ecosystems and the encompassing biodiversity. Hence, ecological monitoring can help us to understand the components of biodiversity and their interactions from time to time. Long-term studies in natural habitats are essential to reveal the wonderful pattern of life in any type of habitat.

In case the study is to assess populations of any of the specific organisms, for example, butterflies, frogs or mammals with habitat specificity (Balakrishnan and Easa 1986, Sreekumar and Balakrishnan 2001, Wasonga et al 2006), the observers should concentrate their investigations in specific types of habitats, where they can expect their study species. Habitat-specific organisms may be seen only in those habitats with which they maintain close association. Such preferences are mostly based on the food preferences of the concerned organism.

12.2. SPECIES–AREA CURVE

When large extent of habitat is involved for ecological characteristics, it is essential to find out how much the minimum extent of the area to be analyzed is. The species-area curve gives us the basic information on the number of species of plants and animals seen in restricted extent of habitats. For example, when a 2 x 2 m^2 area is analyzed for species composition, we get information such as the presence of three species of animals and eight species of plants. When the area sampled is enlarged to 4 x 4 m^2, the respective figures may increase as five species of animals and 12 species of plants. A further increase in the extent of area to 6 x 6 m^2, the figures may again increase as 10 species of animals and 25 species of plants. Similar trends of increase in the number of species of organisms can be seen until a certain larger extent of the area is studied for wildlife diversity, and finally further increase in the extent of area need not result in the increase in species composition, but may continue to have almost the same number of animals and plants. This shows a levelling off with respect to the number of species in the area in relation to the extent of the habitat, which gives us an idea for the minimum extent of the area to be considered when diversity and population assessments are made in wildlife habitats (Fig 80) (Stromberg 1995).

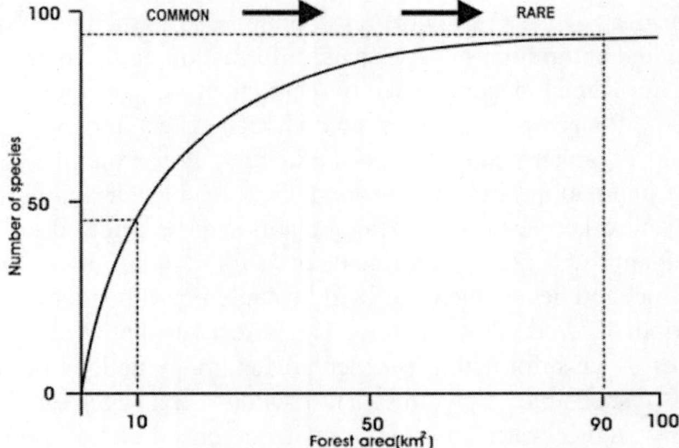

Fig 80. Species–Area curve showing the approximate extent of the area to be surveyed for wildlife studies; note the steady number of species arrived at around 90 x 90 km^2 area for large mammal counts

The species–area curve is made by plotting the number of species on the y-axis and the extent of area samples on the x-axis. In relation to the increase in the size of the quadrat or area sampled, the total number of species in the plot would increase, until it would level off in relation to a specific extent of the area. This is the extent of the area in that habitat that wildlife ecologists should sample to get better ideas on the species composition in their respective study areas.

12.3. POPULATION STUDIES

Ecological monitoring and related studies are undertaken with special emphasis on studies of populations, behavioural patterns and adaptations of organisms to the habitat. Studies of populations and individuals of species are essential components of biodiversity assessments. Larger organisms are easy to observe and assess as compared to smaller organisms and microorganisms.

Populations of larger organisms (e.g., larger mammals) are usually studied using line-transect method. Both direct and indirect evidences are used for population assessment of larger animals. Despite their large size, many of the larger animals are shy and may not be easily observable during field visits. Hence, indirect evidences of them can be identified and analyzed for the presence of such species and for the assessment of their approximate population size.

Population studies are carried out to find out approximate populations of organisms, density, growth rate, population structure, fluctuation, migration and emigration of organisms. Information on these parameters is essential for their management. In case there is an increase in the population of any species in an area, it can affect the populations of interacting organisms directly or indirectly. For example, when the herbivore populations increase beyond the carrying capacity in a forest area, the entire herbaceous species of plants will be affected by intensive feeding. Plants will not get sufficient time for growth, development and reproduction, and hence, the result will be depletion of habitat qualities in a short period of time. This, in turn, can affect the herbivore population themselves. Such information is essential for timely actions of managers, who are responsible for forest and wildlife management. Animals, particularly higher vertebrates migrate from one area to another when environmental conditions are not satisfactory. In this context, migration of birds is well studied in different continents. Perhaps birds are the animal class in which this behavioural adaptation is widespread, and they are the only class of animals that are capable of migrating to the areas thousands of kilometres away, including inter-continental, and return to the original area when the environmental conditions become favourable. Migration and emigration alter the population size of a species in any area, thereby affecting the density and thus, the available food resources. Frequent wildlife census alone would reveal major fluctuations in wildlife populations, and hence, wildlife managers have to undertake such studies and gather information on population status of each of the major groups of animals in the area under their jurisdiction. Hence, wildlife managers have to depend upon appropriate methodologies for this purpose.

Even in areas where wildlife density is fairly high, it may not be possible for an observer to make direct counts on animals. Wild animals are shy and are also afraid of human presence in their immediate surroundings. They try to escape from human beings, hide under forest cover and under shrubs or may move fast from the area to escape from human intruders. Hence, direct counts on many of the wild animals would be difficult to make. To overcome this difficulty, we have to depend upon indirect evidences, which would also give correct identity as far as majority of the wild animals is concerned.

One may be doubtful about the authenticity on indirect evidences of wild animals, especially when the species is not seen in the area. A tiger may not be visible in its habitat, even if the habitat has a good population of them. Even though tiger looks dangerous and ferocious, it is shy, particularly in the presence of human beings. Immediately on sensing the presence of human beings in its habitat, the tiger either flees away or hides under vegetation cover. It will never remain for long in an area after sensing the presence of human beings in its immediate surroundings. In such a circumstance, what are left for its identity for the human observer are its own volatile scent, pugmarks in loose soils, droppings, remains of kills and sites where the animal has scent-marked with its urine mixed with the whitish precipitate of the anal gland secretion. All these are species-specific, and an experienced wildlife observer cannot mistake it for the presence of any other animal, even for a leopard.

Human footprint is so specific, which cannot be mistaken for any other organism. The footprint of sloth bear is somewhat similar to that of a human child, but a human child cannot be seen alone in any forest area so as to leave footprints. The hoof-marks of artiodactyls look more or less similar, but are species-specific in size and shape, and hence, hoof-marks of one species can be easily distinguishable from that of another species. It is also to be remembered that in specific regions and habitats only one or few of the related species can be seen. Differences in the size of the animals reflect in differences in the size of pug, foot or hoof, and hence, it is easy to identify the pugmarks, foot-marks or hoof-marks of different species seen in the same area or habitat.

Pugmarks may not be evident in thick forest habitats. It cannot be seen even in grasslands. It may be clear only in loose soils, particularly on the banks of rivers and waterholes. Also in soils of different types and with varying levels of gravel and water content, the shape of the pugmark may vary to some extend. Hence, it is better to look for different indirect evidences of various animals such as burrows, pugmarks, droppings, kills and signs of eating in wildlife habitats to support observations with additional data.

Only where there is a tiger, one can expect to see tiger droppings. Tigers are solitary in their habits. Being a large carnivore, they require large foraging areas; at the core of which they defend against other individuals of the species. Opposite sexes are seen together only during the breeding season. Hence, it is easy to sense the presence of a tiger in a natural habitat if a tiger dropping is seen in the area. Artiodactyls such as sambar deer, barking deer, mouse deer, wild boar and gaur are seen in most of the peninsular tropical forest habitats of South India. However, the droppings of them are different in shape, size, structure and quantity. The droppings of the Nilgiri tahr and of the Indian hare (*Lepus nigricollis* F. Cuvier, 1823) are more or less similar, but can be distinguishable on close examination. Further, the distribution of Nilgiri tahr is limited only to some of the high altitude grassy and rocky habitats whereas the Indian hare has a wide distribution in diverse types of habitats in the Peninsular India.

Burrows and nests are also species-specific, and an experienced Wildlife Biologist can easily identify them from one another. Diggings of various species of rodents are common in grasslands and in deciduous forests. However, the size of the burrow mouth, the nature of soils drawn and deposited at the burrow mouth, and the open and closed nature of the burrow mouth are species-specific. The burrows of pangolin (*Manis crassicaudata* Geoffroy, 1827) cannot be suspected for those of any of the rodents. The occupied and abandoned burrows can also be identified easily by observing the signs of movement of the animal in the loose soil at the burrow mouth (Figs 81 to 82). The Malabar giant squirrel makes large nests on narrow forking branches of tall trees. One individual may make and maintain more than one nest at any time. They may also transfer the nestlings from one nest to the other occasionally to protect the nestlings from birds of prey.

The calls of birds are also well-studied. An experienced Ornithologist can identify several species of birds by listening to their calls. Also calls during various social contexts are context-specific. The distress calls of a bird are different from the calls related to their reproductive behaviour. Mammals also produce species-specific calls during various social interactions. For example, the alarm calls of a sambar deer when a predator is located in the immediate surroundings are different from the contact calls it usually produces. The alarm call of the Malabar giant squirrel is indicative of the presence of predators around; on which other species, particularly artiodactyls depend upon and try to escape from the vicinity. The calls of bonnet macaque, lion-tailed macaque, Nilgiri langur and common langur are different from one another. They can be easily distinguished by their calls even from a distance. Among them, the contact calls of lion-tailed macaque are very specific and closely resemble to that of human howling. The Nilgiri langur makes long whooping calls. Further, all these primate species may not be present in any one area or habitat, as each of them is habitat-specific. Bonnet macaque is more frequent in villages and in forests near village boundaries. Lion-tailed macaque is an evergreen-dweller. Nilgiri langur is comparatively more widespread and is present in deciduous and in evergreen habitats. On the other hand, common langur is usually restricted to dry-deciduous habitats. In case of doubts, when depending upon calls heard from a distance, the habitat association can also be taken into consideration to eliminate mistakes and to confirm the indirect evidence.

Fig 81. A non-occupied aardvark burrow; note dried leaves at the burrow mouth with no sign of animal activity around (Photo: M Balakrishnan)

Fig 82. An occupied burrow of aardvark; note loose and fresh soils around the burrow mouth and signs of animal movements (Photo: M Balakrishnan)

12.3.1. Counting Animals

If provided with sufficient time, experienced Field Ecologists can count all visible organisms in an area accurately. However, when the extent of the area to be counted is large, accuracy can be lost. Larger organisms can be more accurately counted, as their number will be less as compared to smaller organisms that can be encountered in an area. To overcome such practical difficulties, methods such as total counts (absolute count) and sample counts (partial count) are available. Total count envisages making an absolute count of all individuals of a species present in the study area. On the other hand, if total count is impossible due to large area to be surveyed or due to lack of visibility of organisms in the area, sample count method is to be followed, where sample areas are intensively surveyed for populations of the study species. In this regard, floral count is comparatively easy as terrestrial plants do not move from one locality to another, and hence, can be better analyzed in a given sample area. In case of animal counts, one may not be able to see some of the species of animals in the counting blocks at certain time, as they are not static, and hence, repeated counts are essential on a seasonal basis.

In case of both total count and partial count, line-transect method is followed to scan the given area for its populations of various organisms. In this general methodology, straight-line transects are laid in the area to be studied (Fig 83). These transects are to be marked with semi-permanent markers such as light-coloured paint or by tying coloured ribbon on twigs, shrubs or tree branches in the study area. Transects of prefixed length (e.g., 1 km, 2 km, 4 km or 5 km) are to be laid at fixed distances and are to be regularly observed for the organisms to be studied. All direct and indirect evidences of the species to be studied should be noted down in an observation chart (Table 6). It is better to use a map of the study area superimposed on a numbered grid system, so that the observer can be sure about the location of observation on the grid with respect to the time of observation. If the area is limited, and if there is sufficient manpower to complete the survey in the whole area within the time limit, total count is always advisable. On the other hand, if the area is large and the manpower is insufficient to complete the survey within the time limit, sample count is recommended.

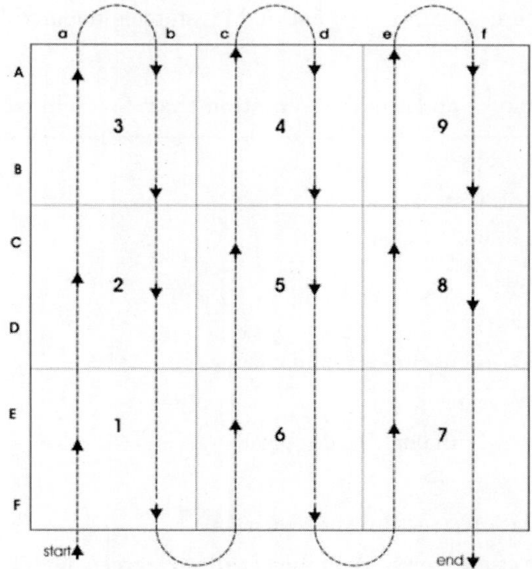

Fig 83. A schematic diagram showing a study area of nine grids with the directions of transects for absolute observation of wild animals

Even if it is total count, one cannot expect to achieve a cent percent accuracy in population counts of animals in wild habitats; even human population census is not cent percent accurate. What is revealed is only the approximate population of the species present at the time of observation.

Whether absolute or partial count method is followed, the estimated population figures are not much different, if the counts are made properly (Fig 84). To assess populations of wild animals, and to analyze observed data, the entire area is to be divided into smaller manageable blocks (grids) of equal size. Counts can be based on each block or grid and the population in each block can be separately recorded. For example, data for sambar deer population in grids 1–9 by total counts can be 14, 16, 14, 12, 14, 13, 11, 10 and 7. The total population of sambar deer in the study area is the sum of the counts in all grids i.e., 111 individuals. Alternatively, the entire area can also be assessed for the population of the species by sample counts. For the purpose of sample count, the grids 1, 4, 6 and 8 may be randomly counted. Then, the mean population in a grid as per the sample data is 11.25 animals. Hence, the total population in the nine grids is 110 animals. To get a better picture, sample counts can be repeated on a random basis at least five times and the mean population for sample units can be assessed. Average of these 20 sample estimates may be 110 animals, a figure close to the total of 111 animals counted by total count method.

Table 6. Format of an animal observation chart for field work*

1	Date of observation
2	Place of observation
3	Time of observation
4	Altitude
5	Habitat type
6	Vegetation parameters and changes
7	Species seen
8	Number in the herd/pack/group
9	Identity of sex
10	Population structure and age group
11	Indirect evidences (droppings/pellets/faeces/diggings/burrows/calls/any other)
12	Activity at the time of observation
13	Any other field aspect
14	Special notes, if any

* A good map of the study area, essential field tools, collection vials/bags, GPS and a camera are to be carried by any ecology field worker as far as possible.

Line-transect method is not that easy, particularly when the area to be assessed is large. Fixed-length transects are laid covering all types of habitats as far as possible. More the number of transects, better would be the results, and as far as possible, transect length should also be extended to a larger area to get a better coverage of all types of habitats in the area. The major problem in such transects is visibility of animals. Surveys are also planned to get information of all species of animals in the area. A general reconnaissance survey would face the problem of visibility of different species of animals. To overcome this problem of transect width coverage, the width of transect is to be fixed on the basis of the species concerned. For example, an elephant when active can be observed from a distance as the observer can easily visualize its presence by the large size, by hearing the calls made when active, by the movements of shrubs, and by the sounds of breaking branches of trees or bamboos during its feeding bouts. On the other hand, when the elephant stands still in a tall grassland habitat, without making any movement and sound, on detecting the presence of human beings in the surroundings, the observer may not be in a position to detect its presence even if it is close to him. Due to the difference in size, an observer may not detect the presence of a sambar deer feeding at a distance in a forest area, but he can see an elephant. Similarly, one may be able to see a deer at a distance, but he cannot see a hare at that distance. To overcome this problem during field work, the area to be covered on the transect for each animal is to be fixed by establishing a visibility index with respect to the concerned species and habitat attributes. Visibility index for each of the species in the same habitat is different. It may also vary when the same species is in an active state and when it rests. Based on the visibility index, the area coverage for each species is to be assessed to estimate the populations in the study area.

Majority of the wildlife observations is based on indirect evidences of animals in their natural habitats. Very rarely we get sufficient data by direct observations in the wild. Pugmarks, droppings/pellets/faeces, diggings/burrows, calls/sounds, etc. are the indirect evidences for wildlife in natural habitats.

12.3.2. Pugmark Tracing

Footprints, pugmarks or hoofmarks are species-specific. In case of felids, it is known as pugmarks, whereas in case of artiodactyls, it is known as hoofmarks. The presence of pugmarks can be considered as proof for the presence of the species in a natural habitat (Figs 85, 86).

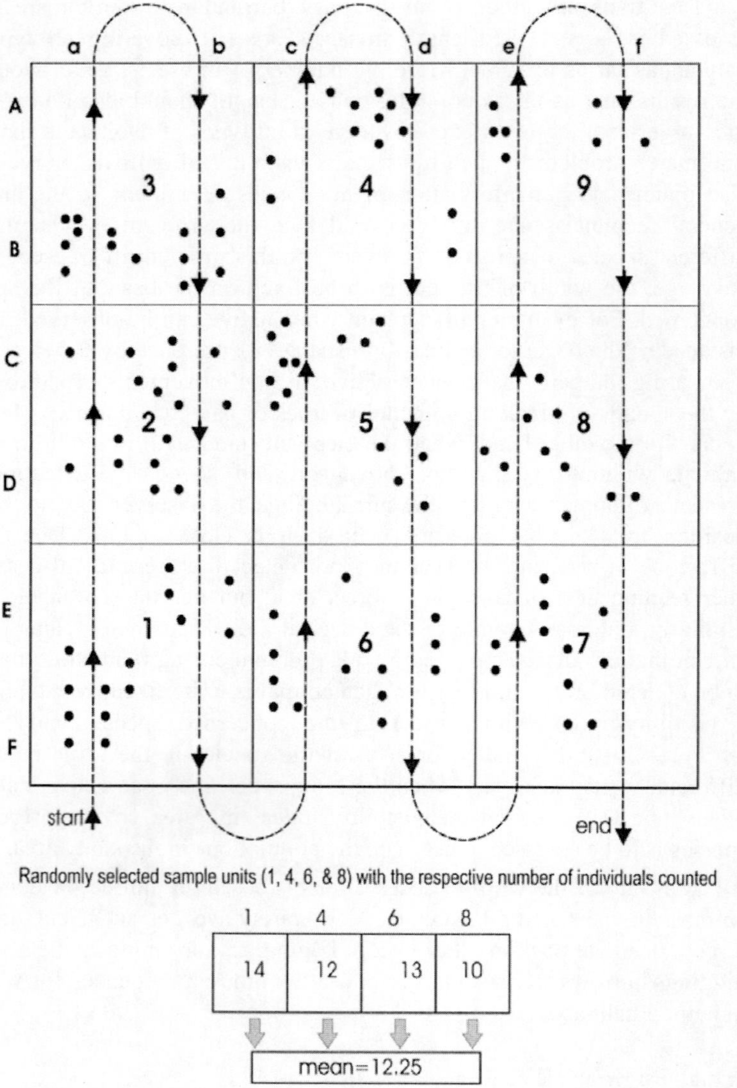

a
b
c
d
e
f
A
B
C
D
E
F
3
4
9
2
5
8
1
6
7
start
end
Randomly selected sample units (1, 4, 6, & 8) with the respective number of individuals counted
1 4 6 8
14 12 13 10
mean=12.25

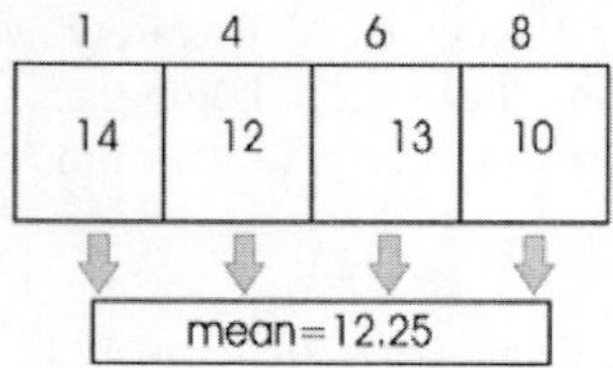

Fig 84. A schematic representation of transects and the absolute count of wild animals in the nine grids of the study area. Randomly-selected study grids with the mean count for a grid are shown at the bottom for estimation of the population of the study animal in the area

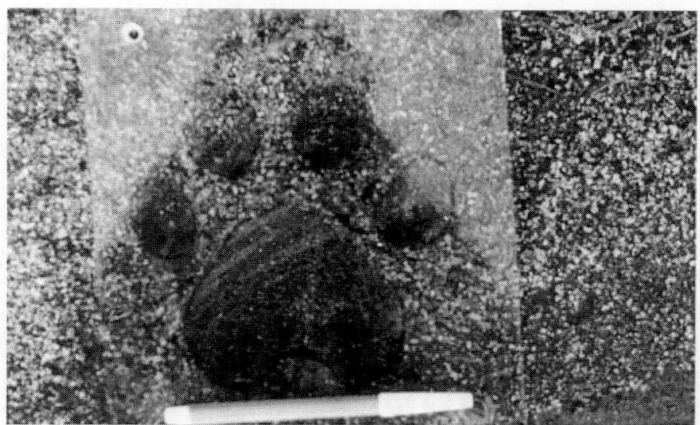

Fig 85. A tiger pugmark observed on the banks of Periyar Tiger Reserve, Kerala (Photo: M Balakrishnan)

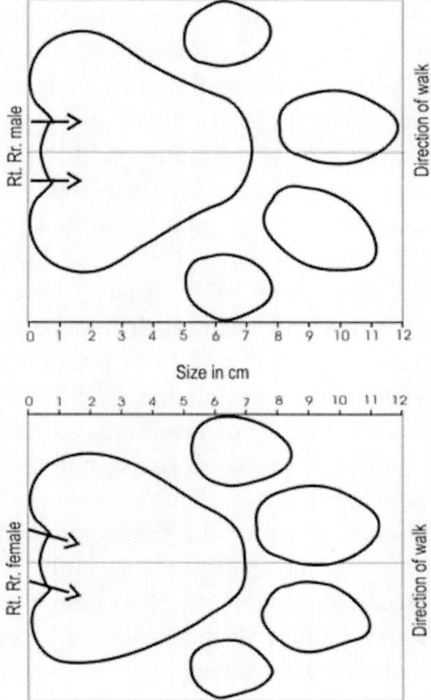

Fig 86. A schematic representation of pugmarks of a tiger and a tigress; note that the more squarish pad is of male (upper) and more rectangular-shaped pad is of female (bottom)

Pugmarks can be carefully studied to identify the individual tiger. Even though the broad layout of different pugmarks appears similar, differences can be inferred through detailed observations using the tracings and the moulds made out of the pugmarks laid in natural habitats. A pugmark seen in a forest habitat can be traced on to a glass plate using a simple equipment, which any one can fabricate (Panwar 1979). It comprises of a colourless rectangular glass plate of approximately 20 x 20 cm^2 size and of 3 mm thickness, with four holes, one on each corner, each of which carries a screw holding on to the plate by a nut and washer. Each screw has an additional fly nut. This glass plate can be placed over the pugmark in the natural habitat and its outer boarders can be traced using a free-flowing pen or a good glass-marking pen. This figure can be traced on to a tracing sheet and the size can be measured using a measuring tape.

To get a good tracing, a good pugmark laid on loose soil should be available in the field. Only an observer with good experience in tracing pugmarks can distinguish minute differences of individual pugmarks, which are to be looked into while making individual identity. To get a complete picture of the pugmark, the observer has to look for the marks of hind paws. In case of quadrupeds, in a normal pace of walk, the hind paw of one side would be placed at the same place where the front paw of the same side had earlier rested. Hence, there may be a partial or even complete superimposition of the fore-pugmarks by the hind-pugmarks. When the animal is moving faster, the rear foot overshoots the fore-pugmark and in such an event it will be possible to see both the pugmarks of each side, when they are placed in somewhat loose soil. As the rear pugmarks are seen clearly in any of the cases, these are considered for tracing. Further, there is also a structural difference in the forelimbs and hind limbs. Among carnivores, the forelimbs are more massively built so as to use them as a weapon during predation. The forepaws are bigger than the hind paws and when a tracing of the forepaws is enclosed in a linear frame, it would tend to be square-like in shape in both sexes. However, the tracing of the marks of hind paws of males would fit in a prominently square-like frame, whereas those of the females would fit in a relatively rectangular frame. Thus, when one depends upon the rear pugmarks, there is a possibility for sexual identity. Hence, it would be advantageous for the observer to depend upon the rear pugmark of tigers and such other carnivores during population assessments.

As there may be only marginal differences in the shape of the pugmark of different individuals, care should be taken while drawing pugmarks. The most perfect hind pugmark should be selected for this purpose. Even slight twisting due to the presence of small gravel or a small hole can cause the pugmark to twist or get splayed on one side. As the paw is a supple organ and adjusts itself to the ground available for resting, it may splay out and appear bigger on a slushy ground, and look smaller and compact on a hard subsurface with a thin cover of dust or sand.

A tiger pugmark would show the impressions of the pad and only four toes, as the toe in the rear (fifth) is placed well above the pad and does not touch the ground while walking. Cats have retractile claws, and hence, they do not leave imprints during a normal walk. While drawing, the glass plate is to be placed above the pugmark and the fly nuts should be adjusted so that on pressing the plate comes down close to the pugmark. The tracer should sit placing his/her knees on either side, or bend over the plate so as to see the pugmark from vertically above it. The outline of the pugmark can be traced on the glass plate carefully by moving eyes with a glass-marking pen so as to see vertically from above. Then, this tracing can be transferred on to a sheet of tracing paper, which will form a permanent record of the pugmark observed in the field. For its identity, the date of observation, place of observation and the name of the observer should be marked. Any inference on sex, age and other relevant information in relation to the observation should also be recorded on the tracing sheet for future reference.

It is difficult to distinguish a leopard pugmark from that of a tiger cub. The largest leopard pugmark may not be bigger than that of a six month old tiger cub. Further, the tiger cub may not be seen alone as its mother will always accompany it. Hence, comparatively smaller pugmarks unaccompanied by bigger pugmarks can be interpreted as that of cubs with the mother tigress.

Information on the date, place, size, length and inclination of the flat top edge of the pad, shape of the lobes, the placing of the two conical tips in the bottom edge and the direction in which they point are helpful to identify individuals.

Using pugmarks seen in the field, pugmark moulds can also be prepared for permanent record. A mould is made out of Plaster of Paris. On locating a good pugmark, the depressions can be filled with good quality Plaster of Paris soaked well with water. The date of observation, place from where recorded and the name of observer can be marked on the surface of the mould before drying. A well-dried mould can be taken out; adhered soil particles can be washed out; and kept as a permanent record of observation. All essential measurements of the pugmark can be made using such moulds. A comparison of several such moulds prepared from pugmarks observed in different parts of a conservation area can be compared using parameters mentioned above, and individual distinction can be made to assess the population size.

Such moulds can also be made using pugmarks and footprints of any other animal in the area for comparative purposes. However, if there are large numbers of individuals of the same species in an area, it may not be that easy to distinguish individuals as that of tiger or other large carnivores, whose number may be comparatively less in any area.

12.3.3. Facts from Faeces

Animal droppings of each species are very specific. Faeces of one species cannot be confused to that of any other. Hence, faeces are widely used as indirect evidence to identify and to assess wildlife populations. Putman (1984) published a good account on various inferences that can be made out of observations and analyses of faeces of wild animals. As far as a wildlife observer is concerned, animal dropping is one among the most important tools in wildlife research. For example, the droppings of carnivorous animals can be easily detectable by the presence of undigested hairs, pieces of bones and parts of hooves of its prey (Fig 87). Such undigested parts of the prey would help the investigators to identify the food habits of carnivores. Analyses of carnivore droppings would also reveal the percentage dependence of various prey animals in the area. The age structure of the herbivores forming prey can also be assessed from detailed analysis. A variety of undigested food parts can be seen in the droppings of herbivorous animals (Fig 88).

Elephant densities can be estimated from dung analysis even in dried out riverbeds (Rasmussen et al 2005). By closely observing dung of elephant herds or droppings of artiodactyles, the observer can distinguish the approximate population and age structure of the herd. Differential digestibility of different foodstuffs and proportional representation of fragments in the faeces would reveal a clear idea of various types of forage they consume. The day of visit to the area and continuous movement track of the herd in a large forest area can also be revealed by observing the dung sites, and the nature and level of moisture content in the elephant dung. The pattern of habitat use of the herd can also be revealed. Detailed microscopic studies of food-remains in the faecal matter would reveal the diet composition. Within any given species, weight of individual dung or pellet is linearly related to the weight of the individual defecator. This would help us to infer the age structure of the herd.

Analyses of faeces of lesser cats such as the viverrids would reveal their food habits. They feed on fruits as well as on smaller animals. The droppings of toddy cats can be seen on exposed roots and on exposed rocks along the animal tracks in the forest. The droppings contain parts of fruits with seeds. These seeds are not crushed and damaged while eating and passing through the intestine. Such animals act as seed-dispersing agents, and are responsible for carrying seeds of fruits they eat to distant places far away from the mother plant. This is essential for the seedlings to come up, as they may not grow to maturity when germinated under the shade of the mother plant (seed shadow). There are also examples of plants, the seeds of which should pass through the digestive system of mammalian frugivore for germination. In others, passing through the mammalian intestine may accelerate the process of seed germination (Sreekumar and Balakrishnan 2002).

Fig 87. Dropping of tiger showing hair of the prey and pieces of bones (Photo: M Balakrishnan)

Some of the wild animals do not defecate all over their habitats. They maintain specific sites where they defecate every day. The African civet is an excellent example for this behaviour by defecating at specific sites in their habitat, and such locations of defecation are called civetries (Fig 88). More than one civet may use one civetry for a period. Also several civetries may be seen in any one area. The approximate size of the population of the civet in the area can be estimated by counting the fresh droppings every day on all the civetries in the area. The age structure of the population can be revealed from the size of fresh droppings accumulated on the civetry.

Fig 88. Fresh droppings of African civets in a civetry, showing undigested remains of various food items (Photo: M Balakrishnan)

12.3.4. Population Estimates from Faecal Density

Estimates based on faeces are considered as potentially viable in wildlife population studies. The principle behind this technique is that the population size of the concerned species can be estimated based on the rate of defecation and the level of faeces accumulation over a period of time. A count on the faeces of a known wildlife population can be made and compared to the faecal accumulation in natural habitats to assess the number of individuals responsible for the quantity as well as bowls of faeces in random quadrates. Population estimates can be made from regular studies on faecal accumulation. The rate of accumulation of faeces in fixed sample plots would give a picture on how many individuals of any particular species have visited the quadrate over a known time period. The quadrate size should depend upon the faecal density. Quadrates of 16 m^2, 25 m^2, 49 m^2 or 100 m^2 may be opted based on the species to be studied. Even though different types of quadrates have also been tested, in general, square quadrates have been found to be better than rectangular and circular quadrates. In case of animals that maintain territory and latrine area (e.g., African civet, European wild rabbit), accumulation of faeces in such sites can be studied regularly to find out the population status and the population fluctuation of the species.

The major factor to be considered for this type of estimates is defecation rate of the species. The consumed food, type of food, quantity of intake, moisture content of the food and the level of water intake during the period can have influence on defecation rate. It may also depend upon various age group and health status of the species. There are two means by which the faecal accumulation can be observed. The best method among these is regular and constant observations, several times a day, to check for new droppings in the quadrate. But, there may be practical difficulties for more frequent observations, and in such cases, the frequency of observations in a day and number of days to be observed are to be fixed in advance. However, in the later case, another important problem of decaying of faeces is also to be analyzed for which the decay rates of the particular dropping in different environmental conditions are to be revealed. Dung of the same species decays at widely different rates during different season in different habitat types under different climatic conditions. Hence, it is important to analyze in detail the decay rate of the dung of the concerned species in the area of observation to make the findings more accurate.

The faecal accumulation over a period of time divided by the defecation rate of the species would yield an estimate of population size responsible for producing the recorded faecal accumulation. Alternatively, rates of faecal accumulation may be studied in areas of known population density, before application of the method in unknown areas. For this purpose, faecal accumulation plots can be established in experimental areas where the absolute population of the species is known. Regular counts of faecal deposition by known individuals can be made so as to develop a 'conversion factor' for the number of animal hours per plot represented by a 'single pellet or dung group'. This conversion factor derived can be used to estimate the unknown populations of the same species elsewhere, more or less depending upon the same type of food and related factors responsible for defecation.

Even though experienced wildlife managers and trackers claim that the sex of the individual responsible for the dung can be distinguished, it is not that easy to make sure. It is easier to distinguish the droppings of a young one from that of an adult of the same species. Body size and age are correlated to certain extent, but the proportional body size in relation to sex may not be clearly distinguishable by observing droppings.

12.4. POPULATION ESTIMATION

12.4.1. Line-transect Count

This is one of the most widely used methods in wildlife census. During this process, the investigators have to fix the study transects, consisting of fixed length (e.g., 1 km, 2 km, 4 km and 5 km) and width (e.g., 1 m, 5 m, 10 m, 50 m, 100 m on both sides of the transect), based on the habitat type and the animal to be observed.

If the study animal is small in size, transect width should be small, so as to make a close observation. On the other hand, if the study organism is large in size, transect width should be large enough to expect at least few of the organisms within the transect coverage. Several transects should be observed to make a better coverage of the area. Randomly laid transects are better to eliminate bias. For example, roadside or a lakeside transect alone may result in biased data.

During counting, in addition to the total number of individuals, sex and age categories can also be recognized. Individual identification, herd or troupe identity is also useful to arrive at conclusions on populations or herd size, etc. Identification of individuals in a herd of elephants (Douglas-Hamilton 1972) or a troupe of chimpanzees (Goodall 1971) is possible by close observations on individual characteristics such as face, tusk, ear, tail, colour, special marks on body surface, etc).

12.4.2. Minimum Number Alive Method

This is one of the most commonly used methods for assessing rodent populations in the study area. If rodents are trapped using suitable traps baited with a preferred food for few weeks, it can be assumed that there is equal chance for all individuals of them to be trapped. If all individuals trapped are marked and released, new individuals trapped every day can be counted. This is done until the last few trapping days, when no more new individuals are available for trapping in the area as shown in the Table 7. The assumption is that all individuals of the species in the area are trapped and counted in the group of 'new capture'. Thus, in the example given, one can make an inference that there are at least 300 individuals of the study species in the area.

Table 7. Estimation of the abundance of the Norwegian rat using the minimum number alive technique

Trapping session	Capture	New capture	Marked and released
Week 1	75	75	0
Week 2	85	60	25
Week 3	90	45	45
Week 4	90	40	50
Week 5	85	45	40
Week 6	85	30	55
Week 7	70	5	65
Week 8	75	0	75
Week 9	68	0	68
Total New		300	

(The assumption is that there is equal chance for all individuals of the species in the study grids to be trapped. As there were no more new ones trapped during the last two trapping sessions, all individuals of the species in the area were trapped and counted in the group of 'new capture').

12.4.3. Capture–Mark–Recapture (CMR) Technique

This is another commonly used method to estimate populations of insects and small mammals. A 100 x 100 m^2 grid can be marked in the study area for this method. This grid can be divided into 100 equal-sized units (each of 10 x 10 m^2) to set suitably-baited traps. There should be a of minimum two trapping sessions for this method. During the first trapping session, few individuals of the study species may be trapped (P1). All of them should be marked and their essential body measurements should be collected, and then, they should be released in the same site of collection. Sufficient time should be allotted between the first and second trapping sessions for a proper mingling of the trapped ones with other members in the population. During the second trapping also, a limited number of individuals of the study species (P2) would be trapped, out of which, few (P3) may be the ones trapped in the first trapping session. From this data, the approximate population of the study species in the grid can be estimated using the following formula:

$$\frac{P1}{N} = \frac{P3}{P2} \text{ therefore, } N = \frac{P1\ P2}{P3}$$

Where N= total population estimate of the species in the study area.

Steps

1. Number of marked and released individuals in the first trapping or netting = P1 = 50
2. Number of trapped individuals in the second trapping or netting = P2 = 60
3. Number of marked individuals released after the first trapping and recaptured during the second trapping = P3 = 10
4. Population estimate = $\frac{50 \times 60}{10} = 300$

Better means to present populations are density, abundance and diversity indices. Density is the number of individuals of the species per unit area (e.g., per ha or per m^2 or per km^2). Abundance is the number of individuals of the species in the total area. Usually, biodiversity indices used are Shannon-Weaver index (H') and Simpson's Diversity index (D). Similarity index (Simpson's Similarity Index i.e., SI) denotes the similarity of different study areas or habitats with respect to the number of species common to the different study areas or habitats.

12.5. POPULATION INDICES

Population estimates alone may not have much relevance in ecology and biodiversity studies, unless it is compared and assessed in relation to factors such as the extent of area, habitat types and cumulative interactions between various organisms. These factors are important for long-term survival of the species and ecosystems. These are termed as population indices. Any student of biology with special interest in ecology and biodiversity conservation should be able to evaluate these factors in their study areas and concerned with their study species from time to time to make proper assessments on the status of the populations, population densities and population fluctuations to suggest appropriate management actions.

The total number of individuals of a species of all sex and group in an area is the population (p) of that species at that time period. It is important to note that population of a species in an area cannot be constant for a long time. There are several biotic and abiotic factors affecting the levels of populations of any of the species in the area, and hence, population estimates of an area made at any time may change in response to such factors. For example, large numbers of migratory birds are found during early December in an area of the southern hemisphere, where they migrate during the winter season. Population assessments of avian fauna of the area made in October or even in the beginning of November may not reveal any indication of the presence of such birds in the area. Their populations in the northern hemisphere can be high until the beginning of winter, when they start migratory flights to the southern hemisphere, thousands of kilometres away. Some of them make out such a long distance during a non-stop flight, one of the marvellous ethological and physiological adaptations of the wonderful world of animals. What is more interesting is their return flight to the original habitat immediately at the end of winter season, where life would be more comfortable during the spring.

Migration is a widespread phenomenon among the animal world. When basic environmental conditions are unsatisfactory, a species of animal would migrate to another part of their habitat or even to another region or continent in search of better living conditions. There is temporal and spatial migration of animals in search of favourable environmental and ecological conditions. Population levels of the species would be high (p+m; where 'm' is the number of individuals of the species migrated to the area) in the area to where they have migrated. Emigration is the behavioural process by which a species of animal leaves an area of unfavourable environmental conditions, where the population levels will be comparatively low (p – e; where 'e' is the number of individuals of the species emigrated from the area) at least for a short duration, depending on how long the emigrants can keep themselves away from their original area or habitat.

Natality, mortality, migration and emigration are biotic factors that directly affect the populations of any species from time to time. In majority of the organisms, breeding is seasonal. This has direct relevance with environmental parameters. For example, herbivores give birth during the onset of monsoon in a tropical forest habitat. This helps newborn fawns with sufficient quantity of soft palatable fodder in their immediate surroundings. If a census is made during or after the fawning season, the population figure will be more (p+f; where 'f' is the number of newborn) than the one made immediately proceeding that season.

Mortality affects the levels of populations adversely (p – d; where 'd' is the number of individuals died). Death of an individual may occur either naturally or by unnatural means. Natural death may have a certain rate, which is more or less constant over a long time period in any area, where biotic and abiotic factors are suitable for the species. However, even though it is natural, predatory effect is different from a natural death of herbivores. Predation has an indirect benefit to maintain a healthy ecosystem. The prey, which succumbs to predation, is mostly old and comparatively unhealthy among the individuals of the group. In addition to the direct effect of such predatory behaviours on the level of population, it would also help to maintain healthy populations of herbivores by eliminating the unhealthy once from further breeding activities.

From this account, it is clear that the population levels of a species assessed in an area may not be constant for a long time. It changes from time to time depending upon various factors, and hence, the population levels assessed will not have a long-standing effect. This makes it clear that detailed estimates of wildlife populations need to be made during different seasons in any of the study areas. It is also important to make frequent assessments based on seasons in conservation areas to reveal the effects of management activities.

12.5.1. Population Density (D)

The population levels alone may not be effective to assess the trends of population growth. Each habitat has its own carrying capacity with respect to the populations of any of the species it can support. Populations that may grow beyond this capacity are to be controlled under management programmes. Such levels can be revealed if the populations are estimated in terms of the extent of the area. Thus, the population density of a species is an expression of its numerical strength in its community. This can be estimated by dividing the total number of individuals of the species present in the area by the extent of the area.

Population density (D) = Total number of individuals / unit area

Density should be represented in unit area, which can be in m^2, ha^2 or km^2, where the population count was made.

Based on this estimate, it is easy to compare populations of a species in various types of habitats during different seasons. It is also possible to analyze populations of any species in different areas so as to establish habitat suitability of the species concerned.

12.5.2. Relative Density (RD)

Ecological status of an area is generally assessed on the basis of the number of species present in the area including endemic species among them, and the threat in the area. None of the natural areas is homogenous with respect to its animal or plant populations. Heterogeneity is essential for long-term survival of ecosystems. When the population and the population density of a species are assessed, only the number of that species is taken into account in terms of the extent of the area. On the other hand, when the relative density is estimated, individuals of all species of animals or plants in the area of study are taken into account. Thus, relative density is the numerical strength of a species in relation to the total individuals of all species in the study area.

$$\text{Relative Density (RD)} = \frac{\text{Number of individuals of the species}}{\text{Number of individuals of all species}} \times 100$$

When we assess the relative density of all species of the Class Mammalia in an area, the relative importance of each of the species in the area can be revealed. More specific habitat requirements of a species may be met with in any one type of habitat or in an area, which can be revealed by the relative density of it in the study area. This information will also be useful for managers while planning management policies.

12.5.3. Frequency (F)

When field observations are made on animals, all species may not be visible at the same frequency. There are more chances to see direct and indirect evidences of those species, which are more numerous in the area. There may also be more visibility of larger organisms as compared to smaller organisms. However, when we make close observations in small unit areas, smaller organisms may be more visible rather than larger organisms. For example, insects may be more frequent in a unit observation area of 1 x 1 m^2 or 2 x 2 m^2 when closely observed. At the same time, they need not be visible when a general reconnaissance survey is made in a larger area. To eliminate this problem, the frequency is calculated based on the occurrence of them among various quadrates. Thus, frequency is related to the occurrence of a species in certain quadrates in respect of the total number of quadrates studied.

$$\text{Frequency (F)} = \frac{\text{Number of quadrates in which the species was recorded}}{\text{Total number of quadrate studied}}$$

12.5.4. Relative Frequency (RF)

Relative Frequency (RF) is the representation of occurrence of any species in relation to the occurrence of all species in the area. Relative frequency can be calculated as:

$$\text{Relative Frequency (RF)} = \frac{\text{Number of occurrence of the species}}{\text{Number of occurrence of all species}} \times 100$$

12.6. ESTIMATING BIODIVERSITY

12.6.1. Habitat and Ecosystem Diversity

Habitat is the space used by an organism or a group of organisms coexisting in a landscape with specific climatic conditions, a pre-requisite for the existence of the species and of the group of species. The habitat of a species is different from that of the other, but in any special habitat, several species coexist in a cohesive manner. In the strict sense, a given habitat in one locality may have a composition of several species, all of which need not be present in the same characteristic habitat in another locality. Their number and proportion may change from area to area and from region to region. Nevertheless, what is important is that each habitat has more or less similar trophic components. An ecosystem is formed by integrated interactions of a number of habitats, thereby components of one habitat may also represent in the nearby area. Thus, they make intensive interactions between each other. Only few organisms are strictly specific to any one of the habitat types, and most of the organisms are seen in a wide range of closely-related habitats. Ecosystem classification and assessment are mainly made by analyzing the vegetation communities present (e.g., tropical moist forest, tropical evergreen forests, tropical savannah ecosystem, Afro-alpine ecosystem, etc.). It is also to be remembered that the formation of any type of vegetation in an area is also based on the soil type, climate and microclimate of the area; each one of these is inter-linked in its function in the formation of the ecosystem.

The key units in the ecological hierarchy are populations, rather than the individuals of which they are composed of or the species to which the populations belong. There is no internationally-accepted standard classification of ecosystems and related phenomena for global application. Ecosystem diversity may be assessed primarily on a local basis and in global terms. A comparison of the diversity between different areas is also possible. The number of taxa within a local site (habitat) is termed alpha (α) diversity, whereas the differences in species composition between habitats are referred to as beta (β) diversity. Gamma (γ) diversity is the diversity at regional spatial scale. The units of alpha and gamma diversity are measures of the number of species. As beta diversity designates the degree of species changes along a given habitat or physiographic gradient, it cannot be expressed in number of species. But, it is a rate of proportion, and hence, expressed in terms of the SI. An area with diverse habitats will be richer in biodiversity rather than an area with less different habitat types. Diverse habitats are the response of the presence of a variety of taxa (taxonomic units; e.g., Family, Order, Phylum). Heterogeneity is a major criterion for the assessment of the quality of habitats and ecosystems. A number of biotic factors including agents of seed dispersal are key elements in the occurrence of heterogeneity of natural ecosystems. Areas with endemic species are important habitats of the ecosystem because such habitats contribute for the uniqueness of the ecosystem (e.g., Afro-alpine, mountain ranges, oceanic islands).

12.6.2. Species Richness

The quality of a habitat is inferred from the number of species present in that habitat. The toxic richness further enhances the quality of habitats and ecosystems. Any of the habitats in the tropical forests is rich with thousands of species, whereas the habitats in the temperate forests may contain only few hundred species. As studies on biodiversity in most of the natural habitats are incomplete, our understanding on the number of species present in each habitat type is fragmentary, particularly with lower taxa.

Species are not distributed uniformly over the planet Earth. There are spatial patterns of species occurring together or far apart based on historical, causal and functional reasons. The historical events of the past (e.g., tectonic movement, continental drift, crater formation, etc.) may act as limiting factors in the distribution of species across the barriers, and may be isolated on either side of the barrier. This enhances the chances of evolving as independent species on either sides of the barrier (allopatry). In a biodiversity-rich habitat with more number of species, a community of interacting groups of species is formed where each species has a niche or role in the overall maintenance of the community or habitat (sympatry).

12.6.2.1. What to Measure and Where to Measure?

As various physical and biological factors affect the distribution patterns of organisms in any natural habitat from time to time, assessment of it requires proper planning on a long-term basis. Richness and abundance of a species assessed in an area in one particular season may not be correlated with the abundance of it in the same area during another season. The dynamics of organisms reveals specific patterns in respect of the diverse species components in the area. Huston (1994) has given a critical account on what to measure and where to measure it. When diverse organisms, such as from bacteria and other microorganisms to larger vegetative components on land such as trees or from micro-organisms such as flagellates to larger marine vertebrates such as whales, are included in samples in an ecological assessment, the sampling procedures would influence the richness and abundance. For example, all species in a community, if taken in a single sample, the sample may not reveal accurate picture of various lower organisms, when much of the higher forms of diversity would benefit better attention. Therefore, organisms to be included in a sample must be carefully selected, such as the number of species within a particular functional type (e.g. niche) or those with the number of functional types, to reveal better diversity patterns in the study area. Such categorization of organisms into functional types would reveal a picture of different ecological processes that involve a variety of interactions controlling the species composition and abundance of the species responsible for the stability of the habitat. For example, in any habitat type, there may be a number of herbivores and carnivores among the animal diversity. Further, there may be a number of herbivorous mammals as well as carnivorous mammals. In case of avian communities, it is easy to categorize as frugivorous, granivorous, carnivorous and insectivorous. Even though, the basic unit of the ecosystem is the individual species, analyzing each group of species with similar ecological functions in that ecosystem can better reveal the mechanisms that influence ecosystem organization. In an ecosystem, different habitat types with diverse species composition supporting and benefiting from each other are responsible for maintaining species-rich composition. This fundamental ecological concept should form the basis of selection of samples for wildlife ecology and biodiversity studies. Further, spatial and temporal levels of distribution of species are also critical in analyzing the species richness and abundance in an area.

Detailed wildlife ecology studies, showing spatial and temporal distribution patterns of organisms have revealed that the number of species increases when the extent of sampled area increases as revealed from the species–area curve. This shows that for a better understanding of the species richness in an area, samples involving a maximum possible extent of the area should be incorporated, where the organisms are sampled and counted. The principle behind this biological fact is that when the extent of the area is small, within that limited area of physical and biological characteristics, the organisms compete with each other and strive with difficulties. On the other hand, in an extended area, where resources are more, the level of competition between species and members is less, and hence, it is favourable for better species richness and abundance. Further, in smaller areas, more homogeneous factors can prevail whereas in larger areas, heterogeneous factors can have better chance. In response to the increase in sample size and total area, the habitats included in the samples also increase, enabling to reveal maximum count of species and individuals. When data for species–area curve are plotted for a wider spatial and temporal range, the number of species may level off and remain constant over a range of increasing areas. This would be followed by another rise before levelling off again and so on; based on the number of habitat types involved in the sampled area.

Temporal heterogeneity is also to be considered while sampling wildlife. The timing of the sampling and the duration of sampling are important factors that may affect the data on richness and abundance. For example, when the specific objective of the sampling is to find out the presence and abundance of migratory species of birds in the area, the season and the duration of the entire migratory season in particular are important.

12.6.2.2. Diversity Indices

i. Species Richness (Richness Index)

The Richness Index (Margalef's Index) can be used to evaluate the richness of various groups of organisms (butterflies, birds, mammals, etc.) in a habitat.

$$\text{Richness Index (R)} = \frac{S-1}{\ln (N)}$$

Where,

S = The number of species of that taxonomic group observed

N = Total individuals of the species observed

ii. Diversity Index

Diversity Indexes (Table 8) can be calculated using the Shannon-Weaver Index (H') (Shannon and Weaver 1949).

$$\text{Shannon-Weaver Index (H')} = -\Sigma\, pi \log pi$$

Where,

Pi = The proportion of the total sample (i.e., of the total number of individuals or total biomass) composed of species *i*.

Simpson's Index (Simpson 1949) is another index commonly used to estimate species diversity in a given habitat. It is often expressed as a diversity index, Simpson's D, of the form $D = 1 - \lambda\ (1 - \Sigma p_i^2)$, where,

Pi = The proportion of the total samples composed of *i* species.

iii. Similarity Index (Simpson's Similarity Index)

Simpson's Similarity Index (SI) can be used to assess the similarity between different habitats with reference to the composition of species and organisms (Table 9).

$$\text{Similarity Index (SI)} = \frac{2C}{a + b}$$

Where,

a = Number of species observed in habitat I

b = Number of species observed in habitat II

C = Number of species common to both the habitats

Table 8. Computation of biodiversity indices (D and H') and species Richness Index (RI)

Species (S)	Number of individuals (I)	Proportion of individuals in each species ($p_i = 1/\Sigma I$)	p_i^2	$p_i \ln p_i$	Species richness (RI)
A	30	0.36	0.130	− 0.368	1.176
B	26	0.32	0.102	− 0.365	1.227
C	18	0.22	0.048	− 0.333	1.384
D	5	0.06	0.004	− 0.169	2.485
E	3	0.04	0.002	− 0.129	3.640
$\Sigma S = 5$	$\Sigma I = 82$	$\Sigma p_i = 1$	$\Sigma p_i^2 = 0.286$ $D = 1 - \Sigma p_i^2 = 0.714$	$\Sigma p_i \ln p_i = -1.364$ $H' = -\Sigma p_i \ln p_i = 1.364$	$(\Sigma S - 1/\ln I)$

Inferences: *D and H': Simpson's Index is an index of dominance, and hence, inversely related to evenness and richness. Diversity indices measure the degree of uncertainty. In a habitat, where only few* *individual species are present, it is easy for the investigator to be more specific about the identity of the observed species. On the other hand, in an area where the number of species is high, it is difficult for the investigator to be sure of the identity of the species observed, and hence, is uncertain. In other words, if diversity is low, the certainty of picking up a particular species at random is high, and if the diversity is high, the certainty of picking up a particular species is low. When the species richness index (RI) moves towards 0, it indicates high richness of a group of organisms (e.g., birds, mammals) in the given habitat or the richness of a species (based on its population) in the habitat with respect to other species in the same habitat, or richness of a species in one of the habitats with respect to its richness in another habitat.*

Table 9. Computation of SI (In this table only two representative habitats are given. Any number of habitats can be compared for their species diversity to analyze similarity of species composition)

Species in Habitat I	Species in habitat II	Common in habitats I & II	Similarity Index (SI)=2C/a+b
A	C	C	
B	E	E	
C	F	F	
D	H	H	
E	I	I	
F	J		
G	K		
H	L		
I	M		
	N		
	O		
$\Sigma a = 9$	$\Sigma b = 11$	C = 5	SI = 2 x 5/9 + 11 10/20 = 0.5

Inference: *From the above table, it is inferred that similarity between the two habitat types based on the species composition and presence of species is 0.5. Thus, half the species present in these two habitats are the same. Hence, it can be confirmed that these two habitats have 50%*

similarity based on the species composition. When there is 50% similarity in species composition, one can also infer that these two habitats have 50% common features and characteristics enabling to have common species to the level of 50%.

(Formula for SI for three habitats: SI = 3C/a + b + c, where C is the number of common species in habitats I, II and III; and a = Total species in habitat I, b = Total species in habitat II; c = Total species in habitat III. Formula for SI for four habitats: SI = 4C/a + b + c + d, where C is the number of common species in habitats I, II, III and IV; a, b, c and d are the total species in the respective habitats I, II, III and IV).

The number of species in a sample is known as species richness. When it is expressed in terms of the extent of the area, it is species density. The relative distribution is usually referred to as 'evenness'. When the distribution is uneven, as seen in most of the cases of natural distribution of organisms in various habitat types, one or more species may tend to be dominant over others. Thus, the number of species and the evenness of relative abundance are the factors used to assess species diversity in the area. Comparative studies of species diversity of different types of habitats using appropriate samples can reveal information on the factors and the mechanisms that regulate diversity.

The distribution of each of the species in a community or in an ecosystem is governed by species-specific factors in addition to the common biological and physical factors encountered. In such a habitat, a comprehensive study on the biological diversity dealing with all organisms would be a major task. It is practically impossible for any one at any time to study all biological diversities in any area. Probably, a large group of scientific workers dealing with different groups of organisms can form a better forum for such studies. Even though, any such analysis would at first concentrate on how many species are there in the community, it should analyze the proportion of individuals in each of these species. When the number of species and the proportion of individuals of each species are revealed, the more interesting question arises; i.e., what makes these observed species alone to be in that community and why not few more species? Probably, the investigators may miss some of the species present in the community, as there is every possibility for missing several species of organisms, particularly the lower ones such as invertebrates and lower plants. This makes the need for repeated studies in the area. Such studies are not repetition of what is made earlier, but mandatory in ecology. It may also reveal the details of fluctuations in populations, seasonally or even annually. Other major questions are: what are the factors governing the pattern of distribution of the species in the community studied?, and how does this community differ from others? Finding out answers to all these questions is not a simple task; it requires manpower, fund and time.

12.7. TRAPPING WILD ANIMALS

Wild animals are trapped using appropriately sized traps, and using suitable baits. Netting is also common, particularly for birds. Pit-fall trapping is used for frogs, shrews, etc. Elephants were also trapped using this technique, but the pit should be large and deep enough to accommodate the large-sized elephant and the trapped ones should not be able to get out of the pit by itself. However, trapping elephants is legally prohibited. Wild animals such as deer are driven towards specific area (trap), using funnel-shaped temporary driving fences. These days, most of the wildlife trappings is done for translocation and management purposes only.

Questions for Review/Exercise

1. Have you understood the following?

   a. Ecological monitoring b. Species diversity

   c. Line-transect method d. Biodiversity indices

   e. Simpson's similarity index g. Absolute and partial counts

   h. Capture-Mark-Recapture (CMR) technique

2. What do you know about species-area curve? What is its importance in wildlife population ecology? Can you prepare a similar model showing the minimum extent of area to be assessed to estimate a study species in your locality?
3. What are the importance of aerial and satellite monitoring in wildlife studies?
4. Have you seen pugmarks/hoof marks/foot marks of any of the wild animals? How can you identify a wild animal by observing such marks?
5. What are the methods you have learned to estimate populations of wild animals? Can you describe any one of these methods in detail?
6. What are your perceptions of direct and indirect evidences for wildlife in a forest area?

7. See Figs 81 and 82. What are the differences you can notice by observing these figures? Find out visible signs by which we can differentiate an occupied burrow from that of an unoccupied burrow.
8. See Fig 85. Have you seen any such marks of wild animals left in their natural habitats in any of your observation areas? What are the inferences you can gather from such observations?
9. Faeces are important study resources for wildlife ecologists. Can you substantiate this view? Explain with examples.
10. See Figs 87 and 88. What are the evidences you can gather from such observations in your study area?
11. How can we make approximate assessment of the population of a species of wildlife in a study area based on faeces counts?
12. What are the observable and measurable parameters in natural ecosystems for wildlife population estimates?
13. What are your perceptions on population indices (e.g., density, relative frequency)?
14. What are the occasions when wildlife managers have to trap wild animals? Have you witnessed trapping of any wild animal?

Read the three abstracts of published research papers given below and try to understand the contents carefully. What are your perceptions on the data presented in these three abstracts? What have you learned from these? Have you realized differences in the study subjects and methods followed? Can you plan any similar research programme for your wildlife ecology project? Have you used any of these methods during your wildlife field ecology programme?

Abstract 12.1

Population density, biomass and habitat association of rodents and insectivores in Pawe area, north-western Ethiopia

*Tilahun Chekol, Afework Bekele and M Balakrishnan**

Department of Biology, Addis Ababa University, P.O. Box 1176,
Addis Ababa, Ethiopia
*Author for correspondence

Abstract: Abundance of rodents and insectivores was estimated in Pawe area of Ethiopia in natural forest, plantation, bushland, grassland, and maize farms. Grids were randomly selected in each of the habitats, and small mammals were live-trapped during wet and dry seasons. A total of 606 individuals (including recaptures) comprising 10 species of rodents and two species of insectivores were trapped. The rodents trapped were *Mastomys natalensis* (37.9%), *Arvicanthis dembeensis* (28.8%), *Stenocephalemys albipes* (9.2%), *Mastomys erythroleucus* (7.5%), *Arvicanthis niloticus* (4.5%), *Acomys cahirinus* (3.6%), *Tatera robusta* (2.6%), *Lemniscomys striatus* (1.7%), *Mus musculus* (0.9%) and *Rattus rattus* (0.3%). The two species of insectivores trapped were *Crocidura flavescens* (2.2%) and *Crocidura fumosa* (0.8%). *Mastomys natalensis* was the most abundant and widely-distributed species, while *R. rattus* was the least abundant with limited distribution. Most of the rodents and insectivores were associated with the bushland habitat. More rodents were trapped during the wet season than during the dry season. The average trap success was 17.7%. The population density ranged from 50 ha^{-1} in the plantation to 311 ha^{-1} in the bushland. Maize farm had the highest biomass of rodents and insectivores (23.9 kg ha^{-1}), followed by grassland (18.7 kg ha^{-1}) and bushland 18.0 kg ha^{-1}).

Keywords: Abundance, biomass, density, Pawe, small mammals, species composition

Reference: *Tropical Ecology* **53** 15–24, 2012

Read the abstract 12.1 and answer the following questions:

1. What are the major findings presented in this abstract?
2. Have you seen any of the species of rodents mentioned in this abstract? Can you decribe the morphological features of any of the rodents present in your locality? Can you differentiate *Rattus ratus* from *Mus musculus*?
3. How can you differentiate rodents from insectivores?
4. What is meant by trap success?
5. How can we estimate biomass of a species of animal in the study area?

Abstract 12.2

Population Status, Structure and Activity Patterns of the Bohor Reedbuck *Redunca reduncа* in the North of the Bale Mountains National Park, Ethiopia

*Bezawok Afework, Afework Bekele and Mundanthra Balakrishnan**

Department of Biology, Addis Ababa University, PO Box 1176,
Addis Ababa, Ethiopia
*Author for correspondence

Abstract: The Bohor reedbucks *Reduncа reduncа* (Pallas 1767) were studied in the Bale Mountains National Park (BMNP), Ethiopia to assess their population ecology and activity patterns. The total count method was used to assess population size. Activity patterns of a herd were studied with the recording intervals of five minutes. There were 406 Bohor reedbucks in the study area of which 96.8% (n = 393) was confined to the Gaysay-Adelay area. The population was female-biased. Excluding juveniles, 57.1% (n = 232) was females and 33.5% (n = 136) was males. More adults (71.7%, n = 291) were observed than sub-adults (18.9%, n = 77) and juveniles (9.4%, n = 38). The herd size of Bohor reedbucks varied from 3 to 5. Herds were seen in aggregations of up to 59 individuals during the dry season in swamp grassland habitat. The maximum number observed in such aggregations during the wet season was 38. The diurnal activity pattern showed feeding peaks early in the morning and late in the afternoon, with resting during the mid-day. Two feeding peaks were observed in the nocturnal activity phase: one during dusk and the other around midnight. In contradiction to the general concept of occurrence of smaller herds during the dry season, the Bohor reedbucks aggregate in swamps during the dry season. They were also active during night and day time.

Keywords: Activity patterns, Bohor reedbuck, population status

Reference: *African Journal of Ecology* **48** 502–510, 2009

Read the abstract 12.2 and answer the following questions:

1. What is your perception on total count method and how far we can expect to get the exact population size of wild animals?
2. What is meant by activity patterns of animals? How can we study activity patterns of wild animals?

3. What is your inference on the status of a wild animal population, which is female biased?
4. Do you expect any seasonal difference in the populations of wild animals?
5. What is meant by diurnal activity? How is it different from nocturnal activity?

Abstract 12.3

Amphibian Abundance and Diversity in Meru National Park, Kenya

D Victor Wasonga Wasonga[1,2] *Afework Bekele*[1], *Stefan Lotters*[2] *and Mundanthra Balakrishnan*[1]

1Department of Biology, Addis Ababa University, PO Box 1176, Addis Ababa, Ethiopia,
2National Museums of Kenya, Division of Herpetology, PO Box 40658 Nairobi, Kenya, and
3Institute for Biodiversity and Ecosystem Dynamics, Amsterdam University, PO Box 94062, 1090 GB Amsterdam, The Netherlands

Abstract: The diversity and abundance of amphibians were investigated in Meru National Park, Kenya, using transect sampling, drift-fence, pitfall trapping and opportunistic collection. A total of 430 individuals under seven genera (*Amietophrynus, Hemisus, Hyperolius, Phrynobatrachus, Phrynomantis, Ptychadena, Xenopus*) comprising eleven species were sampled in three different habitats (apart from this, two additional species are known from Meru National Park): *Acacia*-wooded grassland; *Combretum*-wooded grassland; *Acacia-Commiphora* bushland. The sex ratio for almost all species was balanced (chi-square, v2; P > 0.5) and was not affected by habitat type (ANOVA: F ¼ 8.3026, P ¼ 0.6914). Shannon-Weaver Index (2.227) and Simpson's Index (8.244) were relatively high, and most of the eleven species sampled appeared to have a relatively even distribution (Shannon's Evenness Index, E ¼ 0.927). However, *Hemisus marmoratus* and *Phrynomantis bifasciatus* were exclusively recorded in *Acacia*-wooded grassland and in low abundances. There was a positive linear relationship of body weight against snout-vent length for two randomly selected anurans (*Hyperolius glandicolor, Phrynobatrachus natalensis*) among all three vegetation communities.

Keywords: Africa, community ecology, foraging habitat, frogs, savannah

Reference: African Journal of Ecology **45** 55–61, 2009

Read the abstract 12.3 and answer the following:

1. What are your perceptions on the diversity and abundance of a group of animals in a forest habitat?
2. What are the methods used to study populations of amphibians? Can you try these methods, and collect and identify the amphibians of your study area?
3. See different statistical methods used in the above study and find out the relevance of each of these methods with respect to the parameters studied.

❑❑❑

Modern Techniques in Wildlife Ecology and Habitat Analysis 13

Camera-trapping, biotelemetry and radio-tracking, remote sensing and geographic information system, ecological and conservation genetics, DNA finger-printing

---

Recent technological advancements have helped wildlife ecologists to rely on modern tools in their efforts to understand the way of life of wild animals, and the processes and changes in natural habitats with precision. Camera-trapping helps ecologists to locate the presence of species of wild animals in remote areas, to confirm the presence of species of wild animals suspected to be extinct and to reveal the ideas on their abundance in such areas. Radio-tracking helps to find out the movement patterns of wide-ranging wild animals, which can not be followed by researchers throughout day and night and in areas of difficult terrain, with more accuracy. Remote sensing (RS) and Geographic Information System (GIS) are techniques with several applications in a variety of scientific fields. These techniques can be used to analyze the changes in land-use/land-cover patterns and habitat availability and suitability for various wild animals from time to time with the help of satellite imageries.

13.1. CAMERA-TRAPPING

A camera-trap is an automated camera set to photograph organisms moving through their natural habitats. The presence of nocturnal and rare animals in remote and difficult terrains can be confirmed as they can be photographed using this technique. A photo is taken when a motion or infrared sensor detects the presence of a moving object in front of the camera. Infrared heat and motion detectors are linked to the camera that records the date and time of every picture taken (Rowcliffe et al 2008).

Cameras designed for camera-trapping have two types of monitors; passive and active. Passive system is of single unit that uses heat and motion detectors to trigger the camera. Active system uses an infrared beam between a transmitter and a receiver and the camera is triggered when the beam is broken by the animal moving in front of it.

Camera-trapping allows confirming the presence of a species of animal in a location. With repeated efforts, it can also indicate the absence of a species in a specific area. It can also reveal the presence of an invasive species and on the survival and probable extinction of a rare species in a specific study area. If two cameras are placed one on either side of an animal track in a forest area suspected to have the species for which researchers are looking for, both flanks of the animal can be photo-captured simultaneously to allow better confirmation at the individual level. Further, if the same study area is subjected for camera-trapping studies at regular intervals, information on survival, migration and immigration of specific species can also be gathered (Box 13.1 – Figs 89 to 91).

Multiple georeferenced locations for a species allow us to document the distribution patterns of the study species and to make inferences on the habitat requirements for the species to sustain. Karant and Nichols (1998, 2002) have integrated camera-trapping and Capture–Mark–Release (CMR) algorithms to estimate density and abundance of a number of felids such as puma, jaguar, leopard and ocelots. Individuals are identified by their natural variations in pelage patterns using photographs. Repeated analyses of the photographic records in sequence gathered from one site can reveal the frequency of detection of an individual or a number of individuals of a species as well as different species of animals in the area of study. Such data can reveal information on temporal and spacial distribution of each of the species located through camera-trapping. The detection rate over the entire period of camera-trapping can be calculated as the total number of occurrences of a given species divided by the total time the camera was running. From this, a general measure on the abundance and density (based on the extent of the area where cameras are fixed) of the species can be estimated. Individuals of the same species can be identified from captured photographs using the differences in the pelage pattern of body site such as stripes, dots, signs of wounds and other marks on the body surface such as unique antler configuration in ungulates and pelage patterns in tigers, leopards and other spotted or striped animals.

Box 13.1

Camera-trapping

Helle V Goldman

Norwegian Polar Institute, NO-9296 Tromsø, Norway
(E–mail: goldman@npolar.no)

Devised at the beginning of the 20th century, camera-trapping has become an indispensible tool for research and conservation among scientists and wildlife managers. The technique is useful in general surveys of wildlife or targeted studies of particular species including rare or elusive animals. Camera-trapping can establish the presence of a species in an area, while the absence of a species can be hypothesized with sufficient camera-trapping effort and no positive result. Abundance and density can be estimated on the basis of a well designed camera-trapping effort and appropriate data analysis.

Camera-trapping is non-invasive and, if carried out with care, has little or no effect on animal behaviour. For some species, individual animals can be identified in captured images, permitting capture–recapture analysis, a method of estimating population size based on the proportion of individuals that are 'captured' on multiple occasions. In addition to yielding data concerning the activity patterns, habitat preferences and spatial–temporal spacing of animals, camera-trapping can also capture other interesting behavioural information such as sociality. Camera-trapping photographs are also invaluable for publicity, fund-raising and educational purposes.

A camera-trap is comprised of a still and/or a video digital camera connected to, or integrated with, an infrared sensor, of which there are two types: active and passive. Active sensors consist of two units between which a thin beam of light energy pulses. The camera is triggered when the beam is broken. Passive sensors consist of a single unit that is sensitive to motion and heat within the sensor's wedge-shaped field of sensitivity.

Fig 89. The Zanzibar servaline genet (*Genetta servalina archeri*) eluded scientific discovery until the mid-1990s; the first images of this endemic subspecies were captured during camera-trapping effort in 2003 (Photo: Jon Winther-Hansen and Helle V Goldman)

Determining how many traps to deploy is a trade-off between data quality and the available human resources and camera-traps. Using more traps is costlier but provides greater statistical power. Decisions also have to be taken about the duration of the trapping effort. Longer runs improve the likelihood of detecting species and obtaining multiple photographs of individual animals for capture-recapture studies. When linked to an extra battery pack, today's camera-traps record images for as long as a year after being set up. However, some studies may require keeping the survey short enough to allow for the analytic assumption that the population is 'closed'; i.e., it has not been affected by births, deaths, emigration or immigration.

Unless capture–recapture analysis is planned, a crucial part of the study design is to work out how far apart camera-traps must be placed to be spatially independent; i.e., to ensure that same individuals are not photographed by different traps within the study area, which would confound abundance estimates. Methods for estimating the area sampled by camera-traps are debated in the scientific literature. One approach is to separate camera-traps by the distance of the estimated home range of the targeted species.

Setting up traps along existing trails or near natural drinking points improves the chances of photographing animals that habitually make use of such features. However, some species avoid trails used by people. Set-ups with bait or scent lures will draw some species, but will repel others. Eschewing attractants and placing camera-traps randomly within the study area will reduce biases, but may also reduce the likelihood of photographing species of particular interest.

Camera-traps can be set to be active during particular periods of the day or night to target certain species. When equipped with conventional flash lighting, camera-traps take colour pictures at night, although these flashes disturb animals. Invisible infrared flash technology results in monochromatic images, but does not disturb animals.

Fig 90. Severed goat heads attract a Zanzibar servaline genet to a camera-trap deployed in Jozani National Park, Zanzibar. One component of the two-part active sensor system is visible (Photo: Jon Winther-Hansen and Helle V Goldman)

Camera-trap components can be secured to trees, posts or rocks or concealed inside objects such as artificial rocks. Vegetation that obstructs photographs or that could cause false triggers should be removed. Sun-warmed leaves can trigger heat or motion-sensitive passive sensors (other causes of false triggers include snow and rain).

Fig 91. Checking the settings of a passive sensor during a camera-trapping survey of carnivores in Jozani National Park, Zanzibar, in 2003; mounted on the same tree trunk is a conventional 35 mm camera, a component that has now been replaced by digital technology (Photo: Jon Winther-Hansen and Helle V Goldman)

Traps can be set up to take advantage of natural 'funnels' in rocks or vegetation that encourage animals to move into the sensor's field of sensitivity. Camera-traps with sensors aimed close to ground level document the greatest range of species. Traps aimed higher will capture taller animals and miss smaller ones. Arboreal animals are captured by camera-traps placed in trees.

Setting a delay between the trigger and capturing of an image limits repeated photographs of a single animal during a single visit. Units should be positioned so that photographed animals can be unambiguously assigned to species. If the intention is to identify individuals by their natural markings or features such as tags, each trap should comprise paired cameras that simultaneously photograph each side of the animal.

The equipment must be protected from the elements as much as possible. Sand, moisture and high humidity can damage delicate electronics. Condensation can form inside equipment when it is moved from an air-conditioned environment into the field. Only some brands of traps perform reliably at frigid temperatures. Today's commercial products are marketed as weather resistant, but user reviews should be examined to help ensure a good fit between the camera-trap and the conditions under which it will be used.

To minimize the risk of animals damaging or moving the equipment, care should be taken to avoid contaminating it with attractants. With a two-person team, a practical division of labour is for one person to handle the electronic equipment while the other handles bait or scent lures. Equipment can be wiped down with alcohol or other preparations, or handled with gloves, to reduce human scent. In some circumstances, human interference is a risk. Locking equipment in place with cables deters theft. Posting a notice explaining the purpose of the study may also avert tampering.

Video-trapping is an important development of camera-trapping with great potential for capturing observational information about wildlife. Some camera-traps on the market can be programmed to capture video or still images, and some take both at the same time. Like still images, videos are also valuable in garnering media and public attention.

As technology evolves, still and video camera-trapping will continue to increase the scale of data collection across space and time and will become more convenient and cost-effective, particularly with the wireless transmission of images and metadata from camera-traps to distant computers and hand-held devices. While still useful in determining the presence of a species and estimating density and population sizes, camera-trapping has progressed further. In combination with data management software and appropriate analytical models, camera-trapping can provide data on species interactions, socio-demographics, movement patterns and other aspects of behaviour, while minimally disturbing wildlife.

Additional Readings

Head JS, Boesch C, Robbins MM, Rabanal LI, Makaga L and Kuhl HS 2013 Effective sociodemographic population assessment of elusive species in ecology and conservation management, *Ecology and Evolution* **3** 2903–2916

Kays RW and Slauson KM 2008 Remote cameras, in *Noninvasive Survey Methods for Carnivores* (eds) RA Long, P MacKay, WJ Zielinski and JC Ray, Island Press, Washington DC, Island Press, 110–140 pages

Kays R, Tilak R, Kranstauber B, Jansen PA, Carbone C, Rowcliffe M, Fountain T, Eggert J and He Z 2011 Camera traps as sensor networks for monitoring animal communities, *International Journal of Research and Reviews in Wireless Sensor Networks* **1** 19–29

Kelly MJ and Holub EL 2008 Camera trapping of carnivores: trap success among camera types and across species, and habitat selection by species, on Salt Pond Mountain, Giles County, Virginia, *Northeastern Naturalist* **15** 249–262

13.2. BIOTELEMETRY AND RADIO-TRACKING

Biotelemetry is the technique of obtaining information about an animal through tracking radio signals from a device fixed on the animal. Wildlife telemetry studies have facilitated acquisition of detailed information on many aspects of wildlife ecology such as habitat association, home range, survivorship and mortality, migration timing and routes of migration of a variety of species of animals. The secretive nature and the difficult habitat terrains make conventional observations of many species of wildlife difficult, and hence, necessitate biotelemetry studies. In the recent past, studies using biotelemetry technique has revolutionized wildlife research due to its potential for achieving information that can not be obtained using other techniques. Tracking systems are composed of two major radio-signalling and tracking units *viz.*, the transmitter and the receiver (Figs 92, 93).

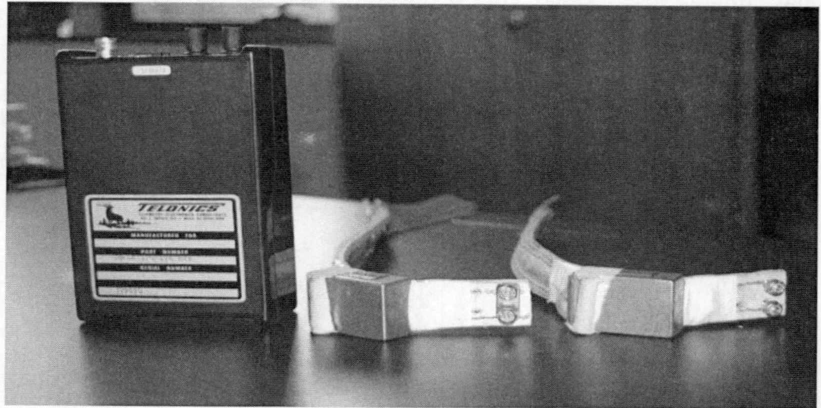

Fig 92. A telemetry receiver set (left) and two radio transmitter fixed collars (right) (Photo: M Balakrishnan)

A transmitter unit consists of a radio transmitter, a source of power (lithium and silver batteries or solar cells) and a transmitting antenna. Solar cells are not recommended for studies on animals in the areas not having good and constant access to sunlight. Identification of the location of individuals with radio-transmitters is possible by using transmitters adjusted to different radio-frequencies as in the case of analog radio stations using FM/AM radios. Transmitters can be of single stage or of double stages. Single-stage transmitters are of low weight (as low as 0.5 g) and used in tracking small animals with localized and short-range movement patterns. Double-stage transmitters are fitted with oscillators and amplifiers, and are powered by a minimum of 2.4 volts. These are larger in size and weight, and more complex and powerful, and can be used for studies on larger and wide-ranging animals. Care must be taken to reduce the weight of the transmitting unit (not more than 2%–3% of the weight of the animal) as it should not hinder the normal movement patterns of the animal on which it is attached.

Similar to a television set, the receiver unit has an antenna, a signal receiver with reception indicator, speaker and/or display and a power source. The antenna serves to increase the gain of the receiver and assists the researcher in determining the direction of the signal. Portable antennas serve better for constant field observations as the researcher can carry the antenna with him along with the receiver set, and can be pointed to all directions in search of signals while in a moving vehicle or even when walking through the habitat. Larger antennas are required to be mounted either on stationary receiving stations or on vehicles. If mounted on stationary receiving stations, ideal positioning is essential for the accuracy of information. High buildings and obstructive terrains may obstruct radio signals, and hence, positioning of the antenna should be done carefully. This unit detects and identifies signals from the transmitter. Based on the differences in the radio-frequencies, the receiver detects the location of different individuals that carry the respective transmitters. Three distinct types of radio-trackings such as very high-frequency tracking, satellite tracking, and global positioning system tracking are currently in use based on the objectives of the research.

Fig 93. A wildlife ecology student learning to track radio-collared animals (Photo: M Balakrishnan)

13.2.1. Very High Frequency (VHF) Radio-tracking

Very high frequency radio-tracking is the conventional technique in which the animal wearing the transmitter can be tracked by a person on the ground or in the air with an antenna and a receiver. This is the most useful and versatile type of radio-tracking with which location and other types of data can be gathered. Homing and triangulation are the two means of tracking radio-collared animals using VHF tracking system. Homing involves moving towards the direction of strong signals emitted from the collar until the animal is located. The receiver gain is reduced repeatedly to come closer and closer to the animal. Triangulation is the process by which the animal is tracked by using two signal bearings after using antenna from different locations (preferably at angles of about 90° to one another). The signal of the transmitter indicates the direction of the animal, but not the distance. The researcher should quickly move from the first tracking location to the second and listen to draw the second directional line towards the animal. The point where the two directional lines meet on the map is where the animal is located. If signals are received continuously from the same location for longer times, the animal may be sleeping or resting at that location. It can also be due to the death of the animal. If so, the investigator has to check on ground and reveal the truth. VHF telemetry can be used to track the location and movement patterns of both terrestrial and marine animals. Temperature and light sensors integrated with VHF transmitters can be used to monitor the body temperature of the animal to assess its health status, and thus, can even infer whether the animal is alive or dead. Light sensors help to calculate the duration of time the animal was under cover or in burrows. As VHF tracking requires the receiver to be close to the location of the animal, researchers have to be in the field most of the time to track the animal constantly for longer periods. Further, precision and accuracy of location distances obtained using VHF tracking are highly variable depending on the terrain of the study area, vegetation cover, weather condition, behaviour of the animal tracked and the intensity of efforts of the researcher.

13.2.2. Satellite Radio-tracking

Satellite radio-tracking uses a high-powered transmitter (ARGOS Platform Terminal Transmitters – PTTs) attached on an animal to send ultra-high frequency pulses to ARGOS satellite. This can transmit diverse data sets such as temperature, activity count or dive count, length of last dive and time spent out of water. Signals received through satellites are used for calculating location, and results are relayed to the computer of the researcher through internet connectivity. Satellite tracking is especially useful for monitoring long-range movements such as migration. Locating the movement patterns of migrating animals such as polar bear, caribou, whale and birds became easier and possible after establishing satellite communication technology. However, the accuracy of locations obtained by satellite tracking is low, and hence, studies on habitat selection using data from such sources are inappropriate.

13.2.3. Global Positioning System (GPS) Radio-tracking

Global Positioning System is a satellite-based radio-navigation system. Unlike VHF and satellite systems, GPS tracking is based on a radio receiver (rather than a transmitter) in an animal's collar. The receiver picks up signals from a special set of satellites, and uses an attached computer to calculate and store the locations of the animal periodically. Data stored in GPS collars can reach the researcher in three different ways. The first is a drop-off mechanism in which the collar is dropped from the animal and picked by the researcher. The second is sending data to satellites that would relay the message to the researcher, and the third is sending data according to the programmed schedule to the researcher in the study area. Thus, GPS tracking may not require frequent field visits. The GPS tracking is highly accurate and specially suited to studies where intensive and frequent data (e.g., several locations every day) are needed.

13.2.4. Telemetry technique

Telemetry equipment is to be made as per the requirements based on the study animal and the expected duration of the study. Larger animals with wider movement patterns would require more strong signals to relay, which can be received from a distance. It would require larger transmitting units. Similarly, based on the duration of the study, the transmitter must be strong enough to withstand that duration (3–4 years) as the case may be. Depending on the size and longevity, the cost of the equipment would vary.

Regardless of the type of tracking system, biotelemetry involves with capturing/darting, tagging/collaring/ implanting, tracking and retrieval. The animal to be studied must be captured or immobilized using a darter gun to fix the transmitter. This is the most critical and perhaps dangerous (when dealing with animals such as elephant, rhinoceros, lion and tiger) aspect of the biotelemetry procedure. Only an experienced person can immobilize larger animals such as elephant, tiger or leopard for radio-collaring. Effective and optimum dose of the immobilizing drug, the duration since immobilizing and collaring, and full recovery of the animal from immobilization are critical in radio-collaring. There were few cases of casualty of elephants and even tigers in India during the process of such investigations in the past, and hence, special care is required while dealing with this type of investigations. There was also human causality during the process of elephant collaring in Kenya in the recent past. Hence, in addition to expertise, utmost care is required in this process.

The transmitter unit can be fixed on a strong collar, and can be used in most of the larger mammals as a neck collar. Neck collar is also suitable for larger birds. However, in smaller animals, it is advisable to implant the transmitter subcutaneously, with the antenna of the transmitter unit projecting to the exterior, pointing towards the backside of the animal. The process of fixing the transmitter on the body of the immobilized animal should be completed in the shortest span of time, as the animal may not be retained under anaesthesia for long. The body temperature of the animal may rise at this time and it is essential to cool down by pouring water over the body including head. Immediately after fixing the transmitter, an antidote is to be administered to revive the animal from immobilization. This is also risky as far as the individual responsible for this duty is concerned. Larger animals such as elephants or tigers, when revived from the effect of anaesthesia, may attack human beings, if found around.

The strength and size of the collar are critical in the progress of the study. If the collar on which the transmitter is fixed is not strong, it may break away when the animal rubs its skin due to irritation in the beginning. If it is loose, it may come out from the neck. If it is tight, particularly if the animal is young, it may become very tight when the animal grows and ultimately the animal may die. Due to all these problems, wildlife ecologists should seriously consider the equipment type, model, capacity and weight in relation to the species of the animal, type of study, type and frequency of information required and the expected period of study well in advance, for selecting the best equipment with essential capacity to minimize the cost of biotelemetry equipment and to get satisfactory results.

Once the transmitter is fixed and activated, and the animal is revived from immobilization, the pattern of movements of the animal can be traced from the locations of signals received through the receiver system, as long as the unit works. Such signals can be analyzed using appropriate computer software tools, and information on movement patterns and major behavioural acts such as moving, resting and habitat use can be inferred.

13.3. REMOTE SENSING AND GEOGRAPHIC INFORMATION SYSTEM

Remote Sensing (RS) and Geographic Information System (GIS) are closely integrated multidisciplinary fields of science, which deals with deriving information about characteristics of objects or phenomena on the surface of the Earth from time to time without making direct contact (Colwell 1983, Wang et al 2010). Such data can be acquired with a combination of observations using multispectral sensors, opto-mechanical multispectral scanners and radar mounted on air-born or space-born platforms (aircraft or satellite) positioned at an altitude of few thousand metres to several hundred kilometres from the Earth's surface. The sensors record the reflected or emitted energy from the features on the surface in different wavelengths, which is recorded in film or in the digital form on magnetic tapes. The amount of energy reflected or emitted varies from object to object and wavelength to wavelength. In RS, this reflected electromagnetic energy from the Earth's surface is interpreted as various features of the Earth such as land-cover, soil type, water-flood area and vegetation types (Box 13.2, Table 10, Figs 94 to 96).

Remote sensing has developed as a modern tool to collect data from a wide range of areas with temporal and spectral changes and patterns, particularly to evaluate changes in the land cover, urban sprawl and natural habitats in relation to human-related impacts. Currently, remotely sensed data are considered as essential in natural resource management and related activities (Roy 1993).

Box 13.2

Remote Sensing and Geographic Information System in Wildlife Habitat Analysis

PS Roy and S Nandy

Indian Institute of Remote Sensing, Indian Space Research Organization, Dehradun, India
(E-mail: psroy13@gmail.com; subrato.nandy@gmail.com)

Habitat of an organism is an area with a combination of resources and biotic and abiotic factors that promote the occupancy of individuals of a given species (or population) and allows them to survive and reproduce. The concepts of habitat, habitat preferences and habitat association of species are important parameters in wildlife ecology and management. The quality of habitat is related to the rates of survival and reproduction of organisms living in the area. To study the habitat occupied by an organism requires understanding of its evolutionary history, ecological requirements and interactions, climatic history of the area, and even the history of movements of landmasses. Detailed information on habitat requirements of major wildlife forms is essential for habitat management and conservation activities. Identifying what constitutes the habitat of a population or species is the impetus underlying activities in wildlife management.

With the rapid increase in anthropogenic pressures, natural environments have been severely disturbed all over the world. The unprecedented growth of human population and commercial exploitation of natural resources have been prime causes of the decline of wildlife all over the world. Encroachment into natural habitats for various purposes including agriculture, grazing and poaching has been the major threats to wildlife. Designing the studies that identify habitat condition requires considerable thoughtfulness about the needs and perpetual abilities of the species under investigation, the spatial and temporal scales at which the study is to be conducted, and the methods for measuring environmental features (Morrison et al 2006). Thus, mapping and monitoring wildlife habitats have become the key components used to interpret species distribution, to evaluate population dynamics and to predict abundance of organisms. Habitat evaluation is the first crucial step for planning effective wildlife conservation and management activities. Evaluation of any habitat for its suitability or unsuitability for a particular species requires information on a host of parameters of biotic and abiotic factors.

Remote sensing has long been identified as a feasible and effective technology for mapping and monitoring wildlife habitats. Geospatial technology including Remote Sensing, Geographic Information System and Global Positioning System along with a Habitat Suitability Index (HSI) model can provide an efficient, quickest, cost-effective and accurate method for habitat evaluation (Schamberger and Krohn 1982). Remote sensing has become a vital tool for habitat mapping because of its broad spatial extents that cannot be obtained using field-based methods (Kerr and Ostrovsky 2003). Aplin (2005) pointed out that field-based approaches generate measurements with high accuracy, but it is generally impractical for studies beyond local scales due to its labour and time intensity. Even though, ground surveys have been used since time immemorial to evaluate habitats using various indices, geospatial technology can supplement or circumvent tedious ground survey methods. Remote sensing data and techniques observe the target or area of interest at scales ranging from local to global (Kerr and Ostrovsky 2003). Geospatial technology has been used as an imperative tool in wildlife studies ranging from habitat evaluation, suitability analysis and wildlife corridor monitoring. Habitat evaluation is a tool for predicting the suitability of habitat for a given species. The HSI modelling is one such technique, which provides a probability that the habitat is suitable for the species and also gives a probability that the species can occur in similar habitats. In a recent study in Jhilmil Jheel Conservation Reserve, Uttarakhand (India) (Nandy et al 2012), IRS P6 LISS-4 satellite imagery was used to generate vegetation type and land-use map. The vegetation type and land-use map was one of the inputs of multicriteria analysis for identification of potential habitat of swamp deer in the study area. For wildlife habitat modelling, various input parameters are to be taken into consideration, though the inputs vary with species and also with the area under study.

Table 10. Input parameters for wildlife habitat modelling

No.	Parameters	Source
1.	Vegetation type/land-use map	Remote Sensing (RS) data
2.	Forest canopy density map	-do-
3.	Elevation	-do-
4.	Slope	-do-
5.	Aspect	-do-
6.	Road network	High resolution RS data/ Topographic map
7.	Settlement	-do-
8.	Water body	RS data
9.	Waterholes	High resolution RS data
10.	Animal population	Animal census/RS
11.	Animal density	Field data
12.	Direct and indirect evidences of wildlife	Field data

Since the 1970s, expert opinion has served as the primary information source for wildlife habitat evaluations framed within a general set of HIS. In its simplest form, HSI is an equation of an additive, multiplicative or logical form with coefficients representing the relative value of environmental features. Typically, coefficients are scaled between 0 and 1 and are estimated using best-available knowledge as surveyed by experts or published literature. Depending on the definition of habitat suitability, model, predictions can represent environmental carrying capacity (as reflected by population density), biomass per unit area or more simply, patch occupancy (Loukmas and Halbrook 2001). In conjunction with GIS and data representing the spatial distribution of model inputs, HSI equations can be used to generate maps of ranked habitat units. In habitat suitability modelling, if the value of the index is high on a particular location, then the chance of the species occurrence is greater.

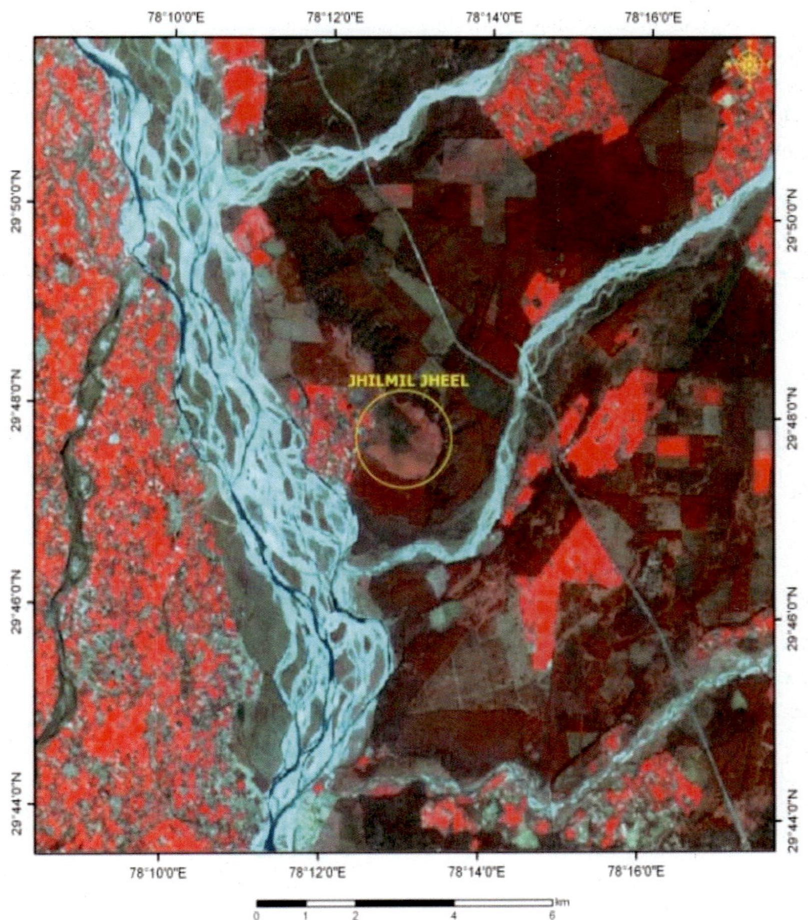

Fig 94. IRS P6 LISS-4 FCC of February 26, 2005 of Jhilmil Conservation Reserve, Uttarakhand, India

Started with habitat ranking, based on the knowledge of the expert and from published literature, HSI models use geo-statistics, logistic regression, refined logistic regression and multicriteria analysis using Analytical Hierarchy Process (AHP) and other data integration techniques to calculate an index of species occurrence (Brown et al 2000) and provide an efficient and inexpensive method for determining habitat quality characteristics.

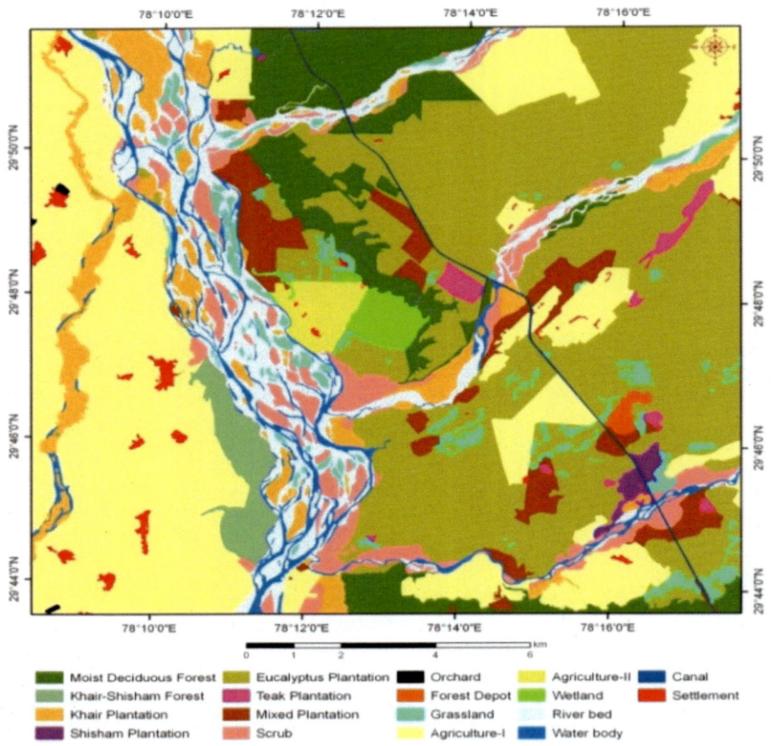

Fig 95. Vegetation type and land-use map of Jhilmil Jheel Conservation Reserve, Uttarakhand, India

Quantitative habitat models and predictive distribution maps are now important tools for conservation and management of wildlife habitats. Many new techniques such as artificial neural networks, genetic programming and machine learning have been connected with GIS to produce ecological knowledge and models. Use of different modelling approaches such as GARP, MaxEnt and BioMapper has shown their potential in wildlife research in the recent years (Baldwin 2009). In addition to optical RS data, LiDAR data and multisensor (LiDAR, SAR/ InSAR, ETM+, Quickbird) data in synergy (Hyde et al 2006) are used for wildlife habitat analysis. Geospatial technology has proven its potential in wildlife habitat analysis over the past four decades, and will continue to do so in the future with the advent of new RS data in combination with new techniques.

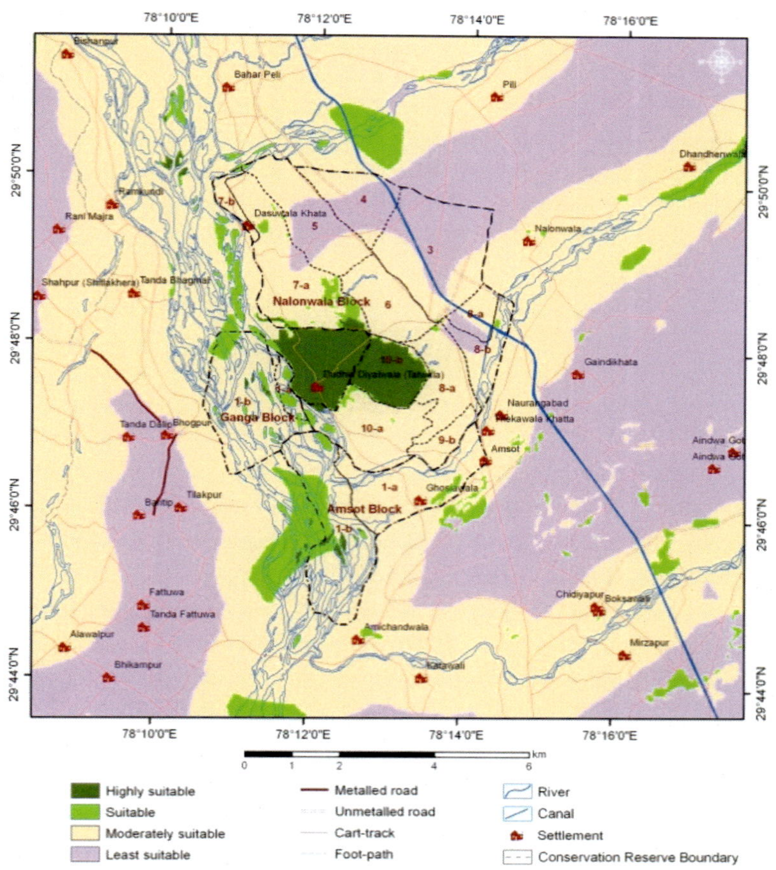

Fig 96. Potential habitat map of swamp deer in Jhilmil Jheel Conservation Reserve, Uttarakhand (Nandy et al 2012)

Reference and Additional Readings

Aplin P 2005 Remote sensing: ecology, *Progress in Physical Geography* **29** 104–113

Baldwin RA 2009 Use of Maximum Entropy Modeling in wildlife research, *Entropy* **11** 854–866

Brown M, Lewis HG and Gunn SR 2000 Linear spectral mixture models and support vector machines for remote sensing, *IEEE Transactions in Geoscience and Remote Sensing* **38** 2346–2360

Hill GJE and Kelly GD 1987 Habitat mapping by Landsat for aerial census of kangaroos, *Remote Sensing and Environment* **21** 52–60

Hyde P, Dubayah R, Walker W, Blair JB, Hofton M and Hunsaker C 2006 Mapping forest structure for wildlife habitat analysis using multi-sensor (LiDAR, SAR/InSAR, ETM+, Quickbird) synergy, *Remote Sensing of Environment* **102** 63–73

Imam E, Kushwaha SPS and Singh A 2009 Evaluation of suitable tiger habitat in Chandoli National Park, India using multiple logistic regressions, *Ecological Modeling* **220** 3621–3629

Kerr JT and Ostrovsky M 2003 From space to species: ecological applications for remote sensing, *Trends in Ecology and Evolution* **18** 299–305

Kobler A and Adamic M 2000 Identifying brown bear habitat by a combined GIS and machine learning method, *Ecological Modeling* **135** 291–300

Kushwaha SPS and Roy PS 2002 Geospatial technology for wildlife habitat evaluation, *Tropical Ecology* **43** 137–150

Loukmas JJ and Halbrook RS 2001 A test of the mink habitat suitability index model for riverine systems, *Wildlife Society Bulletin* **29** 821–826

Martinuzzi S, Vierling LA, Gould WA, Falkowski MJ, Evans JS, Hudak AT and Vierling KT 2009 Mapping snags and understory shrubs for a LiDAR-based assessment of wildlife habitat suitability, *Remote Sensing of Environment* **113** 2533–2546

Morrison ML, Marot BG and Mannan RW 2006 *Wildlife-habitat relationships: Concepts and application* (Third edition), Washington DC: Island Press

Nandy S, Kushwaha SPS and Gaur P 2012 Identification of swamp deer (*Cervus duvauceli duvauceli* Cuvier) potential habitat in Jhilmil Jheel Conservation Reserve, Uttarakhand, India using multi-criteria analysis, *Environmental Management* **49** 902–914

Nandy S, Kushwaha SPS and Mukhopadhyay S 2007 Monitoring the Chilla–Motichur wildlife corridor using geospatial tools, *Journal for Nature Conservation* **15** 237–244

Porwal MC, Roy PS and Chellamuthu V 1996 Wildlife habitat analysis for sambar (*Cervus unicolor*) in Kanha National Park using remote sensing, *International Journal of Remote Sensing* **17** 2683–2697

Roy PS, Ravan SA, Rajadnya N, Das KK, Jain A and Singh S 1995 Habitat suitability analysis of *Nemorhaedus goral* – a remote sensing and geographic information system approach, *Current Science* **69** 685–691

Schamberger M and Krohn WB 1982 Status of the habitat evaluation procedures, *Transactions of the North American Wildlife and Natural Resources* **47** 154–164

Store R and Jokimaki J 2003 A GIS-based multi-scale approach to habitat suitability modeling, *Ecological Modelling* **169** 1–15

Advantages of RS are feasibility for vast area coverage and synoptic view of the area. Data from large extents of inaccessible habitats can be more accurately gathered using RS. It can provide a wide variety of information in the field of wildlife ecology such as habitat categories, land forms, vegetation cover and waterholes, which can be combined to analyze the habitat use of wild animals, changes in forage quality, wild animal migration patterns and timing, human influence and related changes in the habitats from time to time, and even for specific periods of longer duration. For example, the habitat type and extent of coverage in a study area in 1970, 1980, 1990, 2000, 2010, etc. can be revealed from the satellite imageries of the respective years, from which information on the extent of changes of coverage of each of the habitat types in the area during such intervals can be estimated.

Geographic Information System is defined as a computer-based system for the capture, storage, retrieval, modelling, analysis, representation and display of geo-referenced spatial data to solve complex problems regarding planning and management of natural resources (Andrew 2002). Spatial datasets can be generated from RS through direct imports of images and topographic maps using GIS technology and software. It is a powerful tool to integrate and analyze raster and vector datasets to get meaningful results for decision making. Functions of GIS include data entry, data display, data management, information retrieval and analysis. Geographic Information System is being used extensively in natural resource management including management of wildlife habitats, analysis of recreation resources, floodplains, wetlands, agricultural lands and forests, and changes in these resources from time to time. It is also possible to identify to what extent any of the previous habitat type was changed to another land-use pattern (Roy 1993). As temporal satellite imageries show specific areas under any specific cover pattern, an analysis of imageries of two specific periods (e.g., 1970 and 2000) will reveal not only total changes, but also specific changes from which category and what extent was changed to which category, during that period of 30 years. Softwares such as ArcGIS and ERDAS Imagine are integral parts of GIS for image processing, classification and interpretation, and both vector and raster-based data. Organizing data for habitat analysis has become much easier due to these softwares.

It is possible to quantify and analyze wildlife attributes, corridor-use by wild animals and their status, identification of new sites for PAs, human-related impacts in wildlife habitats, habitat suitability analysis and land-use/land-cover changes using RS and GIS technologies more effectively than conventional methods of analyses. More significantly, RS and GIS techniques can substantially reduce the need for conventional detailed ground surveys, due to the fact that ground surveys in wildlife habitat evaluation are time consuming, costly and inadequate for inaccessible areas (Berwick and Saharia 1995). This technique can also be integrated with biotelemetry to study distribution and movement patterns of specific wildlife forms in relation to their habitat requirements and to generate wildlife distribution maps at regular intervals. For example, McAllister et al (1994) have shown global distribution of coral-reef fish, and Said et al (1997) have analyzed population trend and distribution patterns of wildlife and livestock in the Mara ecosystem. Further, it is also possible to predict wildlife distribution and population densities by analyzing the history of landscape, topography, soil, rainfall, temperature and other habitat factors that may influence species distribution. Remote sensing and GIS have been effectively used in such studies.

13.3.1. Land-use/land-cover Pattern Changes

Remote sensing and GIS technologies have developed as excellent tools to visualize a variety of changes on the surface of the planet Earth. Modern tools such as satellite data and GIS are efficient to check human-induced changes in the land-use/land-cover patterns and to implement mitigative measures (Raman and Gopal 2009). By using such modern tools, it is possible to conceptualize and interpret selected indices and degradation risk of natural habitats, which can be used in administration, conservation and management practices.

Habitat fragmentation and loss of forest cover have made tremendous adverse impacts on the landscape patterns in various regions in the tropics during the recent past. Consequences of such changes are now being reflected by the depletion of biodiversity, which also has implications on climate change and global warming. Remote sensing provides information on the spatial distribution of various land-use /land-cover changes and its impacts over the years (Roy 1993, Roy and Ravan 1994). Successive studies on the trends of land-use or land-cover changes using RS techniques and GIS-based mapping help to understand the severity of changes and its effects efficiently. Multi-temporal landscape pattern analysis provides an opportunity to characterize and quantify the complexity of spatial and temporal patterns and processes of land-use or land-cover changes (Gillanders et al 2008). Analysis of landscape pattern uses indices to measure landscape configuration, shape complexity and composition derived from thematic maps of multitemporal satellite data. The extent of land-cover and its relative abundance, and the spatial distribution or locations of patches are quantified by indices of composition and configuration, respectively. The integration of biological GIS-based softwares such as FRGSTAT3.2 and Patch Analyst (grid) 4.0 calculate multitudes of landscape indices such as the measure of landscape pattern to support natural and human-induced processes necessary to implement measures to prevent land degradation. Human-related changes of landscape patterns can be quantified by common landscape indices of composition and configuration. The number and size of patches can be measured to quantify the magnitude of fragmentation within the landscape (De and Tiwari 2008).

A recent study on the land-use/land-cover pattern changes, spatio-temporal landscape modification, forest degradation risk and landscape disturbance index using GIS and RS techniques around Holeta-Berga watershed in Ethiopia (Melakneh et al 2010) revealed dynamics and pattern of changes for a period of 33 years (1973–2006). The mean patch size index values of forest cover and grassland were decreased from 10 ha to 1.11 ha and from 17.4 ha to 0.87 ha, respectively. In relation to this, the landscape configuration change indicators such as the interspersion-juxtaposition index increased in crop land from 65.16% to 83.58%, and decreased in grassland from 57.685 to 34.74%. It was revealed that an extent of 981.05 ha of the study area was under high degradation risk and another extent of 8,378.58 ha was under medium to high disturbance scales. Most of the land-use/land-cover conversions and landscape pattern changes were the results of unscientifically-planned resource utilization and management programmes. Hence, it is important to understand the spatio-temporal changes of the landscape pattern and land-use/land-cover changes to implement reliable and sustainable land-use and development programmes in relation to wildlife management.

13.4. ECOLOGICAL AND CONSERVATION GENETICS

Molecular genetics has become a new tool in wildlife ecology and conservation biology, especially in the context of conservation of endemic and endangered wildlife, and those with very low populations in fragmented habitats. Even though, the theory of Island Biogeography (Diamond 1984) has established the high probability of extinction processes in isolated and fragmented habitats, the genetic basis of it could be revealed only in the recent past (Frankham et al 2002). Genetic variation, the raw material for natural selection, is constantly being created by mutations and eroded through natural selection by genetic drift. This is a fundamental concept in ecology and conservation genetics. Genetic diversity is essential for wildlife populations to adapt to environmental changes including climate change and global warming. Large populations of naturally-outbreeding species (e.g., African lion) usually have extensive genetic diversity (Newman et al 1985), but it is less in species of low populations and inbreeding populations, including those of endemics (e.g., Asiatic lion) (O'Brien 1994). Genetic diversity is reflected in the differences among individuals for characters such as eye colour, skin colour, colour banding patterns, colour of flowers, and in protein and DNA sequences of organisms. Genetic diversity represents slightly different DNA sequences, and variants may result in differences in the sequence of amino acid in the protein coded for by the locus. Such variations may result in functional, biochemical and morphological dissimilarities that cause differences in reproductive rate and survival or behaviour of individuals. Genetic diversity is described using polymorphism, average heterozygosity and allelic diversity (Frankham et al 2002).

Genetic diversity can be characterized through analysis of heterozygosity (H_o), Hardy-Weinberg expected heterozygosity (H_e), allelic diversity (A) and effective number of alleles (n_e) averaged over the loci. Observed heterozygosity can be calculated for each locus as the total number of heterozygotes divided by the sample size. The number of alleles per locus is also to be averaged and the effective number of alleles can be calculated for each locus (for details: Frankham et al 2002).

Random changes in allele frequency are related to population size. In small populations, there are more chances to change allele frequencies. Such a change of allele frequency is called genetic drift, which results due to random sampling of gametes. In extreme cases, such genetic drifts can lead to the extinction of alleles and the loss of polymorphism. Isolation of populations in fragmented habitats would lead to such extreme processes, and hence, conservation initiatives are to save such populations from extinction. Depletion of population can occur through processes such as colonization of a new habitat by a small number of individuals (a process known as the founder effect), or through habitat fragmentation, where previously wide-spread population has depleted as a result of sudden reduction in population (known as bottleneck). The result of founder effect as well as of bottleneck is that the remaining population may have a different allele frequency to that of the original population. As a result, the level of heterozygosity will erode through generations.

As resources for endangered and endemic species and for those in isolated and fragmented habitats are limited, it is important to understand the minimum population size needed to retain genetically-viable and healthy populations. To avoid inbreeding depression, the effective population size (N_e) of more than 50 individuals is recommended. Unfortunately, the current population size of most of the endangered and threatened categories of species is too small to avoid genetic deterioration. Hence, to retain the evolutionary potential of such species, their population size should be retained at a minimum of 500 in case of larger animals and around 5,000 in case of others. The Hardy-Weinberg equilibrium is fundamental to the interpretation of genetic characteristics among populations. Currently, sophisticated methods and softwares are available to analyze deviations from this equilibrium.

O'Brien (1994) reported very low genetic diversity in endangered Asiatic lions in Gir forests in India, the only place where the Asiatic lions occur naturally, with a small population of around 250 individuals. The genetic diversity of the Asiatic lion is far less than that of the non-endangered African lion. The Gir lions have almost identical DNA fingerprints, similar to that of identical human twins. The low population and related inbreeding for an extensive period have bottlenecked the population of the Asiatic lion.

Gotelli et al (1994) have characterized the genetic diversity in the endemic Ethiopian wolf, which is one of the most endangered among canids of the world. They exist in six isolated areas in Ethiopia with a total population of around 500 individuals. Habitat destruction, livestock grazing, disease spread by domestic dogs and hybridization with domestic dogs are the main threats that the Ethiopian wolves face. Populations of Ethiopian wolves have lower genetic diversity than their non-endangered relatives such as the grey wolf and coyote. This shows that the Ethiopian wolf has less evolutionary potential as compared to non-endangered canids, and hence, Ethiopian wolf is at the risk of extinction.

13.5. DNA FINGERPRINTING

It was Sir Alec Jeffreys, who first developed DNA fingerprinting and profiling technique in 1985 as a great challenge in the advancements in crime detection. Prior to this, only fingerprints were analyzed to distinguish individuals from one another as the fingerprints of two individuals never match with each other. DNA fingerprinting is a technique to study molecular genetics of organisms or of available samples, through identification and evaluation of the genetic material (DNA), including nuclear DNA, mitochondrial DNA (mtDNA) and chloroplast DNA (in plants). As it has been heavily used in forensic science, it has been emphasized in detecting criminals and to confirm doubts of paternity. However, in the field of systematic biology, it has prime role in species identification, particularly to identify and confirm variants and subspecies of known species. In addition to its use in systematic biology, it can be used in wildlife ecology as a perfect test for identification of skin, flesh, horn and tusk that are poached by wildlife criminals from different parts of the world, which are also in international export, import and trade. For example, if poached flesh is confiscated from a market, the original species and subspecies can be detected using this method by mapping DNA profile of the species of interest in data base. The shop-owner may state that the meat is of a domestic animal, but it can be from a wild species, even of a threatened whale, which is not allowed to kill. DNA fingerprint is also in use to track illegal ivory trade and poaching.

DNA fingerprinting can be done using any of the samples of the organism or the subject in doubt. The sample can be from blood, vein, heel prick, buccal swab, dried blood samples, semen, saliva, urine, fresh faeces, flesh, hair, bone, teeth and skin. In case of human beings, personal items such as toothbrush and shaving razor, and stored samples such as banked sperm or biopsy tissue samples can also be used. DNA from tissue samples can be extracted in the laboratory and then Polymerase Chain Reaction (PCR) and/or facilities of DNA sequencing are used to identify specific regions of DNA as DNA markers using specialized probes that enter the DNA sample and bind to specific DNA sequences. Binding of these different probes or primers to specific markers slowly paints a unique genetic picture of the individual from which the sample was drawn. If 4–5 DNA region markers match, it confirms that the samples are from the same individual for the purpose of crime detection, whereas when there are dissimilarities, the probability is that the samples are from different individuals. This is the criteria, which can be used to differentiate species identity in wildlife ecology, through DNA sequencing and application of other genetic markers.

Different methods used for DNA fingerprinting are as follows:

1. Restriction Fragment Length Polymorphism (RFLP) is a technique to analyze the length of the strands of the DNA molecules after digestion by restriction enzymes. This method is possible only if there are larger DNA samples.
2. The Polymerase Chain Reaction (PCR) analysis amplifies isolated regions on the strands of the DNA under examination. This method is advantageous over RFLP as it can be done using smaller samples.
3. Amplified Fragment Length Polymorphism (AFLP) is used to amplify a segment of DNA by comparing RFLP and PCR methods.
4. Microsatellites or Short Tandem Repeat (STR) technique is repeating sequences of 2–6 base pairs of DNA. Currently, this is the most widely used form of DNA fingerprinting. This system is based on the features of PCR, as it utilizes specific areas that have short sequential repeat of DNA. As markers reveal high level of inter and intra-specific polymorphism, its applications in wildlife ecology, conservation and management are high.

There are several case studies using these techniques revealing the genetic variations of diverse well-documented organisms. However, most DNA analysis for species identification of animal tissues test the presence and/or absence of one species per assay, hence requiring several tests for a complete analysis of those species potentially present in the sample. However, Buntjer and Lenstra (1998) have demonstrated that PCR with fluorescently-labelled Mammalian-wide Interspersed Repeat (MIR) primers can generate fingerprints that are suitable for rapid identification of known and unknown species on an automatic sequencing apparatus and with computer-assisted data processing. Cronin et al (2005) have quantified the genetic variation of caribou (*Rangifer tarandus* Linnaeus, 1758) at 18 microsatellite DNA loci and the cytochrome-b gene of mitochondrial DNA (mtDNA) in 11 herds of three North-American subspecies – Alaskan barren ground caribou (*R. t. granti*), Canadian barren ground caribou (*R. t. groenlandicus*) and woodland caribou (*R. t. caribou*). The subspecies do not have monophyletic mtDNA, but they have different frequencies of mtDNA genotypes. Microsatellite allele frequencies also differentiate between the woodland and barren ground subspecies. Within the subspecies, there was relatively low differentiation of microsatellite allele frequencies and mtDNA genotypes among herds of *R. t. granti* and *R. t. groenlandicus*, and relatively high differentiation of microsatellite alleles and mtDNA genotypes among herds of them in four geographically separate areas in Canada. The extent of differentiation of mtDNA genotype frequencies and microsatellite allele frequencies within and among each subspecies reflects past and the present gene flow among herds of caribou. Moreover, microsatellite can also give a clue on inbreeding depressions within certain groups of specimens or population. Overall, microsatellite marker is a powerful tool with several applications in the fields of wildlife ecology and conservation genetics.

Questions for Review/Exercise

1. Have you understood the following techniques?
   a. Camera-trapping b. Biotelemetry
   c. Satellite radio-tracking d. DNA fingerprinting
   e. Very high frequency radio-tracking
2. Have you seen a camera-trap? What is it? Why should we go for camera trapping rather than direct observational studies on wildlife?
3. What is meant by radio-tracking? Briefly discuss the methods involved in it. Find out any of the animals in your locality, region or county on which biotelemetry studies were carried out. Discuss the findings of such studies in a class seminar.
4. What are your perceptions of Remote Sensing and Geographic Information System in wildlife ecology?
5. What are your perceptions of conservation genetics in wildlife ecology? What are the inferences we can make using information on conservation genetics?
6. What is known as DNA fingerprinting? What is its role in systematic studies of organisms? How is it important in wildlife ecology?
7. What is your perception on modern molecular techniques in conservation biology? Substantiate your understanding with any of the examples you have learned (Refer to Chapter 2, Boxed item 2.3).
8. What have you learned from Boxed item 13.1? Did you get a good idea of camera-trapping technique? Are you confident that you can also find out facilities and carry out similar studies in

remote areas to find out the presence of small and shy animals, which may escape our observations during field studies?

9. See Figs 92 and 93. Check various components of biotelemetry equipment and the procedure to record the movement patterns of a wild animal with a radio collar.
10. What have you learned from Boxed item 13.2? Did you understand this modern method to study wildlife habitats?

Abstract 13.1

Land-use and Landscape Pattern Changes in Holeta-Berga Watershed, Ethiopia

Melakneh Gelet, KV Suryabhagavan* and M Balakrishnan***

* Department of Earth Sciences, Addis Ababa University, Addis Ababa, P.O. Box 1176, Ethiopia.

** Department of Biology, Addis Ababa University, Addis Ababa, P.O. Box 1176, Ethiopia.

** Author for correspondence: Email: balak212@yahoo.com

Abstract: A study on the land-use/land-cover pattern changes, spatio-temporal landscape modification, forest degradation risk and landscape disturbance index was made around the Holeta-Berga watershed in Ethiopia. The dynamics and pattern of changes for a period of 33 years (1973–2006) were analyzed using GIS and Remote Sensing techniques. The total area of built-up and crop land increased from 4,382.7 ha (4.35%) to 12,198.02 ha (12.3%) and from 33,144.3 ha (32.93%) to 62,916. 02 ha (62.51%) respectively, during this period. On the other hand, the extent of forest cover decreased at the rate of 546.32 ha/yr from 22,549.6 ha (22.4%) to 4,521 ha (4.49%) and the extent of grassland decreased from 37,416.5 ha (37.17%) to 17,437.2 ha (17.33%) during the same period. The landscape of the study area showed transformation on composition. The mean patch size index of forest cover and grassland decreased from 10 ha to 1.11 ha and from 17.4 ha to 0.87 ha, respectively. In relation to this, the landscape configuration change indicators such as the interspersion-juxtaposition index increased in crop land from 65.16% to 83.58%, and decreased in grassland from 57.685% to 34.74%. Shape complexity of the study site also increased in each of the types of land-use/land-cover classes in terms of area weighted mean fractal dimension, except grasslands. It is revealed that an extent of 981.05 ha of the study area is under high degradation risk and another 8,378.58 ha of area is under medium to high disturbance scales. Most of the land-use/land-cover conversions and landscape pattern changes are the results of unscientifically planned resource utilization and management programmes. It is important to understand the spatio-temporal changes of the landscape pattern and land-use/land-cover changes to implement reliable and sustainable land-use and development programmes.

Keywords: Land-cover changes, Forest degradation risk, GIS, Landscape disturbance index, Remote sensing, Spatio-temporal landscape modification

Reference: International Journal of Ecology and Environmental Sciences **36** 1–6, 2010

Read the abstract 13.1 and answer the following:

1. What are the ecological and environmental parameters of the study area presented in the above abstract?
2. What are the methods used to analyze the above parameters? Do you have the facilities in your institution to make similar studies?
3. What are your perceptions of the dynamics of ecosystems and its changes on a long-term basis?
4. What are your perceptions on the fast changes of ecosystem characteristics? Do you expect human involvement in such fast changes?
5. What are the major points of conclusions presented in the above abstract?

Manuscript 13.1

Remote Sensing and Geographic Information System-based African Civet Habitat Mapping in Andracha, Ethiopia

*Dagnachew Melese,[a] Karuturi Venkata Suryabhagavan,[b] Melakneh Gelet,[b] and Mundanthra Balakrishnan**

[a]Addis Ababa University, Department of Zoological Sciences, P.O. Box No. 1176, Addis Ababa, Ethiopia
[b]Addis Ababa University, Remote Sensing and GIS Stream, School of Earth Sciences, P.O. Box No. 1176, Addis Ababa, Ethiopia

*Author for correspondence: Email: balak212@yahoo.com

Abstract: Remote sensing and geographic information systems have enormous applications in ecological studies, particularly in the habitat analysis of wild animals. The present study aimed to evaluate the habitats of the African civet in Andracha, Ethiopia, to generate geo-referenced ecological data on the habitats of this species. Habitat evaluation and habitat changes during 1986 to 2012 were analyzed using LANDSAT imageries. In the year 1986, 1,017.56 km^2 (99.75%) of the study area was covered with forest, but in 2012, only 949.61 km^2 (93.09%) had forest cover. There has been a reduction of 5.97% forest cover at a rate of 0.22% per year since 1986. The study area has been classified into three suitability categories based on forest cover, water, settlement/agriculture, road and slope. Currently, 611.07 km^2 (59.9%) of the area has dense forest cover, which is less preferred by the African civet, 336.75 km^2 (33.01%) is moderately interfered with human activities and is suitable for civets, and 70.56 km^2 (6.91%) is degraded forest around human habitats, which is highly preferred by civets. Habitat suitability analysis is useful to predict potential habitats of African civets in Ethiopia and elsewhere in Africa.

Keywords: *Civettictis civetta*, Geospatial technology, Habitat suitability, Land-use/land-cover changes.

Methods

Field surveys were conducted during September 2012 to February 2013 to identify the distribution and habitat use of African civets in the study area, to collect data on the presence of civets in the area by locating civetry-locations where African civets repeatedly defecate, and to locate naturally scent-marked sites – environmental objects on which civets rub their perineal glandular area, where the glandular secretion adheres to the objects and can be seen. Eighty five GPS points of such locations were recorded and used as input for satellite data analysis and also used to obtain general information about the characteristics of the preferred habitat types of African civets. During this field survey, the assessment of the presence of the African civet was carried out by collecting relevant data points using Garmin GPS (henceforth, referred to as 'presence data'). Information about civets was also gathered from literature and from local people. All GPS points of presence data were converted to UTM WGS 84 zone 37 N projections similar to that of the satellite image, and superimposed on the study area map to portray the distribution map of African civets in the area. This study used prior knowledge, ancillary, and field data to classify LU/LC types and extents in the study area. To examine the properties and differences of the training samples, the digital spectral signature of TM of 1986 and ETM+ of 2012 were evaluated and examined using training areas of the original scenes. A maximum-likelihood classifier algorithm of supervised classification was applied using Earth Resources Data Analysis System, and quantitative area data of the overall LU/LC changes during 1986 to 2012 were compiled. The rate of change in each LU/LC categories from 1986 to 2012 based on LANDSAT temporal data was calculated as a measure of the habitat change of African civets. The rate of land-cover changes in the study area was computed as a measure of the impacts of human activities affecting the habitat range of the civets.

Parameters such as food sources, breeding sites, and signs of movement of African civets were used to model suitable habitats of the species. These parameters were grouped as proximity factors, which have influence on the access of food, reproduction, water, and effects of human activities in the habitat. Natural factors (physiographic datasets) that have a major influence on the mobility of human hunters and predators of African civets were proximity to roads, settlements, and water sources (proximity datasets). Slope was used as a major input for the representation of natural factors for mobility of the African civet.

A flowchart of the methods followed is given in Fig. 2 of the referenced Manuscript 13.2. All vector parameters were rasterized and determined to the fixed spatial extent of cell size 30 × 30 m. Proximity datasets were subjected to distance analysis and rescaled by the re-classification algorithm of ArcGIS 10.1 in the spatial analyst module. The natural physiographic dataset such as slope was rasterized and reclassified to scale data as the proximity dataset. Even though the influence level of each parameter of habitat suitability modelling varies, all the selected factors acted upon the habitat modelling process. The level of influence of each factor to the suitability model was quantified before running the analysis. Saaty's technique was applied to derive the weight factor map (level of ranking) through pair-wise comparison between criteria using IDRISI software. This technique integrates the factors and weights given in the analytical hierarchy processes.

Results

Presence Data and Habitat Characteristics of the African Civet

Four different sites (Tugiry, Beshifa, Modi, and Chegecha) in the study area were found to be suitable habitats for the African civet (Figs. 1 and 3 of the referenced Manuscript 13.2). A total of 85 locations (civetries and scent-marked sites) were located as presence data in the study sites with respective GPS points. These data were found in relation to forests, rivers, stream courses, settlements, roads, and tracks in and around natural habitats. The sample sites were about 3–5 km away from the main town (Gecha) in the study area. These sites had habitat homogeneity that can support the basic ecological requirements of African civets. They are characterized with habitats of open mixed forest and high density natural forest at a gentle to moderate slope with inter-dispersed rivers, streams, and ponds. Natural forest covers (riverine, deciduous, and afromontane) were dominant in the study area. The forest habitats comprise big and small trees and densely populated lower strata with woody climbers. The ground was mostly covered with thick litter and forest floor herbs and shrubs. Various forest trees and shrubs including fig tree (*Ficus vasta*), broom cluster fig (*Ficus sur*), spike thorn (*Maytenus arbutifolia*), and snake bean tree (*Syzygium guineense*) were sources of fruits for civets in the study area. There were settlements of small hamlets in a more dispersed way in the forest, where small farm operations were also seen. Different human activities such as deforestation, extension of croplands, built-up areas, and cattle grazing were observed on the edge of the natural forest. Road construction and tracks are being expanded in the study area through natural forests to link adjacent Kebeles (villages). Local people have revealed that the African civet is the most frequently seen animal around their locality, which they consider as a pest, as civets encroach into their farmlands and damage crops. Based on the presence data, spatial distribution of the African civet is presented in Fig. 3 of the referenced Manuscript 13.2. Most of the presence data are concentrated around 3 to 5 km away from Gecha and only sparsely distributed in remote areas, where dense natural forest is dominant.

[DOI: 10.1117/1.JRS.8 .083502]

Reference: Journal of Applied Remote Sensing **8** 1–12 published online Dec. 19, 2014.

(Tables, figures and references are not included)

Read sections of the research paper presented in Publication 13.2, and answer the following:

1. What are the methods used to analyze various habitats in the study area?
2. What are the parameters analyzed to evaluate habitat suitability for African civets during this investigation?
3. Based on the information given in the publication, what are the habitat qualities suitable for African civets?
4. Briefly write on parametric and natural (physiographic) factors used to model suitable habitats of African civets.
5. What is the major conclusion of the above investigation, as given in the abstract?

❑ ❑ ❑

References

Ancona KA and Loughry WJ 2009 Time budgets of wild nine-banded armadillos, *Southeastern Naturalist* **8** 587–598

Anilkumar S 2009 *Mutualism between Vertebrates and Cullenia exarillata in Silent Valley Forest Ecosystems, Kerala*, Ph.D. dissertation, Thiruvananthapuram: University of Kerala

Ashby KR and Santiapillai C 1983 The ecology of free-living water buffalo (*Bubalus bubalis* L.) in Sri Lanka with particular reference to Rahuwa National Park, *Tiger Paper* **10** 20–26

Azar C, Lindgren K, Larson E and Mollersten K 2006 Carbon capture and storage from fossil fuels and biomass – costs and potential role in stabilizing the atmosphere, *Climatic Change* **74** 47–79

Baillie JEM, Hilton-Taylor C and Stuart SN (eds) 2004 *2004 IUCN Red List of Threatened Species – A Global Species Assessment,* Gland: International Union for Conservation of Nature and Natural Resources

Balakrishnan M 1984 The larger mammals and their endangered habitats in the Silent Valley forests of South India, *Biological Conservation* **29** 277–286

Balakrishnan M 1987 Sebum-storing flank gland hairs of the musk shrew, *Suncus murinus viridescens*, *Journal of Zoology London* **213** 213–220

Balakrishnan M 2002 *Report of the Project on Mammal – Habitat Associations: A Case Study in Silent Valley National Park, Kerala,* Thiruvananthapuram: Department of Forest and Wildlife, Government of Kerala

Balakrishnan M 1994 Conservation of threatened African wildlife: problems and prospects, in *Tropical Ecosystems: A Synthesis of Tropical Ecology and Conservation* (eds) M Balakrishnan, R Borgstrom and Stein W Bie, New Delhi: Oxford & IBH Publishing Co. Pvt. Ltd., 193–232 pages

Balakrishnan M and Easa PS 1986 Habitat preferences of larger mammals in the Parambikulam Wildlife Sanctuary, Kerala, India, *Biological Conservation* **37** 191–200

Balakrishnan M and Ndhlovu DE 1992 Wildlife utilization and local people: a case-study in Upper Lupande Game Management Area, Zambia, *Environmental Conservation* **19** 135–144

Balakrishnan M and Sreedevi MB 2007a Captive breeding of the Small Indian Civet *Viverricula indica* (É. Geoffroy Saint-Hilaire, 1803), *Small Carnivore Conservation* **36** 5–8

Balakrishnan M and Sreedevi MB 2007b Husbandry and management of the Small Indian Civet *Viverricula indica* (É. Geoffroy Saint-Hilaire, 1803), *Small Carnivore Conservation* **36** 9–13

Barbraud C and Weimerskirch H 2001 Emperor penguins and climate change, *Nature* **411** 183−185

Bauer JJ and Ellenburg HR 1989 *The Status of Blackbuck (Antelope cervicapra) in Nepal*, Rome: Food and Agricultural Organization

Beg AR 1975 *Wildlife Habitats of Pakistan*, Peshawar: Pakistan Forest Research Institute

Bekele T, Afework B and Balakakrishnan M 2008 Scent marking by the African civet *Civettictis civetta* in the Menagesha–Suba State Forest, Ethiopia. *Small Carnivore Conservation* **38** 29–33

Berwick SH 1976 The Gir forest: an endangered ecosystem, *American Scientist* **64** 28–40

Berwick SH and Saharia VB (eds) 1995 The Development of International Principles and Practices of Wildlife Research and Management – Asian and American Approaches, Delhi: Oxford University Press

Blake DK and Loveridge J 1975 The role of commercial crocodile farming in crocodile conservation, *Biological Conservation* **8** 261–272

Blaustein AR, Han BA, Relea RA, Johnson PTJ, Buck JC, Gervasi SS and Kats LB 2011 The complexity of amphibian population declines: understanding the role of co-factors in driving amphibian losses, *Annels of New York Academy of Science* **1223** 108−119

Blower J 1975 Rhinos and other problems in Nepal, *Oryx* **12** 270–280

Bolen EG and Robinson WL 2003 *Wildlife Ecology and Management*, 5th edition, New Jersey: Pearson Education, Inc

Bolton M 1978 Crocodile farming in Papua New Guinea, *Oryx* **14** 360–365

Botkin DB 2007 Forecasting the effects of global warming on biodiversity, *Bioscience* **57** 227−236

Brutting C, Hensen I and Wesche K 2013 *Ex-situ* conservation affects genetic structure and diversity in arable plants, *Plant Biology* **15** 505–513

Bulte EH and Damania R 2005 An economic assessment of wildlife farming and conservation, *Conservation Biology* **19** 1222–1233

Buntjer JB and Lenstra JA 1998 Mammalian species identification by interspersed repeat PCR fingerprinting. *Journal of Industrial Microbiology & Biotechnology* **21** 121–127

Carrizo SF, Smith KG and Darwall WRT 2013 Progress towards a global assessment of the status of freshwater fishes (Pices) for the IUCN Red List: application to conservation programmes in zoos and aquariums, *International Zoo Yearbook* **47** 46–64

Caughley G 1969 *Wildlife and Recreation in the Trisuli Watershed and Other Areas in Nepal*, Rome: FAO/UNDP Project Report 6

Chapman CA, Chapman LJ, Struhsaker TT, Zanne A E, Clark CJ and Poulsen JR 2005 A long-term evaluation of fruiting phenology: importance of climate change, *Journal of Tropical Ecology* **21** 31–45

Chapman CA, Lawes MJ and Eeley HAC 2006 What hope for African primate diversity? *African Journal of Ecology* **44** 116–133

Chekol T, Bekele A and Balakrishnan M 2012 Population density, biomass and habitat association of rodents and insectivores in Pawe area, northern Ethiopia, *Tropical Ecology* **53** 15–24

Clarke AG 2009 The Frozen Ark Project: the role of zoos and aquariums in preserving the genetic material of threatened animals, *International Zoo Yearbook* **43** 222–230

Claxton AM 2011 The potential of the human – animal relationship as an environmental enrichment for the welfare of zoo-housed animals, *Applied Animal Behavioural Science* **133** 1–10

Coad L, Burgess ND, Bombard B and Besancon C 2009 *Progress Towards the Convention of Biological Diversity's 2010 and 2012 Targets for Protected Area Coverage*, Cambridge: United Nations Environment Programme/World Conservation Monitoring Centre

Cody ML 1986 Diversity, rarity, and conservation in Mediterranean climate, in *Conservation Biology: An Evolutionary-Ecological Approach* (ed) ME Soulé, Sunderland: Sinauer Associates, 122–152 pages

Colwell RN (ed) 1983 *Manual of Remote Sensing*, 2nd edition, Falls Church: ASP&RS

Conover M 2002 Resolving Human – Wildlife Conflicts: The Science of Wildlife Damage Management, New York: Lewis Publishers

Cronin MA, MacNeil MD and Patton JC 2005 Variation in mitochondrial DNA and microsatellite DNA in caribou (*Rangifer tarandus*) in North America, *Journal of Mammalogy* **86** 495–505

Crook 2002 Relative sensitivities of mammalian carnivores to habitat fragmentation, *Conservation Biology* **16** 488–502

Daniel JC (ed) 1980 *The Status of the Asian Elephant in the Indian Subcontinent.* Asian Elephant Specialist Group, Gland: International Union for Conservation of Nature and Natural Resources/Species Survival Commission

De A and Tiwari AK 2008 Estimation of patchiness: a measure of fragmentation in the Rajaji – Corbett National Park and adjoining areas, Uttarakhand, India, *International Journal of Ecology and Environmental Sciences* **34** 345–349

DeVos A 1982 *Deer Farming*, Rome: Food and Agricultural Organization

Diamond JM 1975 The island dilemma: lessons of modern biogeographic studies for the design of natural reserves, *Biological Conservation* **7** 129–146

Diamond JM 1984 'Normal' extinction of isolated populations in: *Extinctions* (ed) Nitecki MH, Chicago: University of Chicago Press, 191–246 Pages

Dinerstein E 1979a An ecological survey of the Royal Karnali Bardia Wildlife Reserve, Nepal, Part I Vegetation, modifying factors and successional relationships, *Biological Conservation* **15** 127–150

Dinerstein E 1979b An ecological survey of the Royal Karnali Bardia Wildlife Reserve, Nepal, Part II Habitat/animal interactions, *Biological Conservation* **16** 265–300

Dinerstein E 1979c An ecological survey of the Royal Karnali Bardia Wildlife Reserve, Nepal Part III Ungulate populations, *Biological Conservation* **18** 5–38

Dinerstein E and Wemmer C 1988 Fruits *Rhinoceros* eat: dispersal of *Trewia nudiflora* in lowland Nepal, *Ecology* **69** 1768–1774

Dinerstein E, Wemmer C and Mishra H 1988 Adoption in Greater one-horned rhinoceros (*Rhinoceros unicornis*), *Journal of Mammalogy* **69** 813–814

Douglas-Hamilton I 1972 *The Ecology and Behaviour of the African Elephant*, Ph.D. Dissertation, Oxford: Oxford University

Dubost G and Feer F 1981 The behavior of the male *Antelope cervicapra* L.: its development according to age and social rank, *Behaviour* **76** 62–127

Dudley N 2008 *Guidelines for Applying Protected Area Management Categories*, Gland: International Union for Conservation of Nature and Natural Resources

Dudley N, Stolton S, Belokurov A, Krueger L, Lopoukhine N, MacKinnon K, Sandwith T and Sekharan N (eds) 2010 *Natural Solutions: Protected*

Areas Helping People to Cope with Climate Change, Gland: IUCN/ WCPA/ TNC/UNDP/WCS/WB/WWF

Dulloo ME, Hunter D and Borelli T 2010 *Ex-situ* and *in-situ* conservation of agricultural biodiversity: major advances and research needs, *Notulae Botanicae Horti Agrobotanicini Cluj-Napoca* **38** 123–135

Dunbar RIM 1998 Impact of global warming on the distribution and survival of the gelada baboon: a modelling approach, *Global Change Biology* **4** 293–304

Easa PS 1989 *Certain Aspects of Ecology and Ethology of the Asian Elephant (Elephas maximus L) in Parambikulam Wildlife Sanctuary, South India.* Ph.D Dissertation, Thiruvananthapuram: University of Kerala

Estes R 1991 The Behavior Guide to African Mammals, Including Hoofed Mammals, Carnivores, Primates, Los Angeles: University of California Press

Estes JA, Duggins DO and Rathbun GB 1989 The ecology of extinctions in kelp forest communities, *Conservation Biology* **3** 251–264

Eisenberg JF 1980 Ecology and behavior of the Asian elephant, *Elephant* (Suppliment) **1** 36–56

Eisenberg JF and Lockhart M 1972 *An Ecological Reconnaissance of Wilpattu National Park, Ceylon*, Washington DC: Smithsonian Institution Press

Erdelen W 1988 Forest ecosystems and nature conservation in Sri Lanka, *Biological Conservation* **43** 115–135

FAO 1983 *Crocodile Breeding and Management in India. Project Findings and Recommendations*, Rome: Food and Agricultural Organization

Fashing PJ, Nguye N and Fashing NJ 2010 Behavior of geladas and other endemic wildlife during a desert locust outbreak at Guassa, Ethiopia: ecological and conservation implications, *Primates* **51** 193–197

Folke C, Carpenter S, Walker B, Scheffer M, Elmqvist T, Gunderson L and Holling CS 2004 Regime shifts, resilience, and biodiversity in ecosystem management, *Annual Review of Ecology, Evolution, and Systematics* **35** 557–581

Fox JL, Sinha SP, Chundawat RS and Das PK 1988 A field survey of snow leopard presence and habitat use in north-western India, *Proceedings of the International Snow Leopard Sympoasium* **5** 99–112

Frankham R, Ballou JD and Briscoe DA 2002 *Introduction to Conservation Genetics,* Cambridge: Cambridge University Press

Gadgil M and Berkes F 1991 Traditional resource management systems, *Research and Management Optimizer* **8** 127–141

Gadgil M and Guha R 1992 *This Fissured Land: An Ecological History of India*, New Delhi: Oxford University Press

Gametchu T, Birrie H, Wlode-Yohannes L and Gundersen SG (eds) 1994 *Present Status and Future Trends of Research on Endod and Schistosomiasis in Ethiopia*, Addis Ababa: Addis Ababa University Press

Gee EP 1964 *The Wildlife of India*, London: Collins

Gillanders SN, Coops NC, Wulder MA, Gergel SE and Nelson T 2008 Multi temporal remote sensing of landscape dynamics and pattern change: describing natural and anthropogenic trends, *Progress in Physical Geography* **32** 503–528

Goodall J 1971 *In the Shadow of Man*, Boston: Houghton Mifflin

Gottelli D, Sillero-Zubiri C, Appelbaum CD, Roy MS, Girman DJ, Garcia-Moreno J, Ostrander EA and Wayne RK 1994 Molecular genetics of the most endangered canid: the Ethiopian wolf *Canis simensis*, *Molecular Eco*logy **3** 301–312

Green MJB 1986 The distribution, status and conservation of the Himalayan musk deer *Moschus chrysogaster*, *Biological Conservation* **35** 347–375

Grubb P 2005 Order Perissodactyla, in *Mammal Species of the World: A Taxonomic and Geographic Reference* (eds) DE Wilson and DM Reeder, Baltimore: Johns Hopkins University Press, 637–722 pages

Hall LS, Krausman PR and Morrison M L 1997 The habitat concept and plea for standard terminology, *Wildlife Society Bulletin* **25** 173–182

Hohne N, Wartmann S, Herold A and Freibauer A 2007 The rules for land use, land use change and forestry under the Kyoto Protocol –lessons learned for the future climate negotiations, *Environmental Science and Policy* **10** 353–369

Huston MA 1994 *Biological Diversity: The Coexistence of Species on Changing Landscapes*, Cambridge: Cambridge University Press

IPCC 2007 *Climate Change 2007: Impacts, Adaptation and Vulnerability*, Cambridge: Cambridge University Press

Ishwaran N 1983 Elephant and woody-plant relationships in Gal Oya, Sri Lanka, *Biological Conservation* **26** 255–270

Ishwaran N 1984 *Ecology of the Asian Elephant* (*Elephas maximus* L.) *in Sri Lanka*, Ph.D. Dissertation, Michigan: Michigan State University

IUCN 1983 *Crocodiles as a Resource for Tropics*, Gland: International Union for Conservation of Nature and natural Resources

IUCN 1994a *Guidelines for Protected Area Management Categories,* Gland: International Union for Conservation of Nature and Natural Resources

IUCN 1994b *IUCN Redlist Categories,* Gland: International Union for Conservation of Nature and Natural Resources

IUCN 2001 *IUCN Red List Categories and Criteria: Version 3.1*. Gland: International Union for Conservation of Nature and Natural Resources

IUCN 2011 *IUCN Red List of Threatened Species*, Gland: International Union for Conservation of Nature and Natural Resources

Jackson RM and Hillard D 1986 Tracking the elusive snow leopard, *National Geography* **169** 792–809

Jarman M 1979 Impala social behaviour: territory, hierarchy, mating, and the use of space, *Zeitschrift fur Tierpsychologie* 21 1–92

Janzen DH 1985 On ecological fitting, *Oikos* **45** 308–310

Jnawali SR 1989 *Park People Conflict: Assessment of Crop Damage and Human Harassment by Rhinoceros (Rhinoceros unicornis) in Sauraha Area Adjacent to the Royal Chitwan National Park, Nepal*, MSc Thesis, Aas: Agricultural University of Norway

Johnsingh AJT 1983 Large mammalian prey–predators in Bandipur in India, *Journal of Bombay Natural History Society* **80** 1–49

Johnsigh AJT and Panwar HS 1990 Elephant conservation in India: problems and prospects, Rome: *Proceedings of the 5th International Theriological Congress*

Jones BTB and Barnes JI 2006 *Human–Wildlife Conflict Study: Namibian Case Study*, Windhoek: Worldwide Fund for Nature

Joslin P 1973 *The Asiatic Lion: A Study of Ecology and Behavior*, Ph.D. Dissertation, Edinburgh: University of Edinburgh

Karanth KU and Nichols JD 1998 Estimation of tiger densities in India using photographic captures and recaptures, *Ecology* **79** 2852–2862

Karanth KU and Nichols JD 2002. *Monitoring Tigers and Their Prey: A Manual for Researchers, Managers and Conservationists in Tropical Asia*, Bangalore: Centre for Wildlife Studies

King LE, Douglas-Hamilton I and Vollrath F 2007 African elephants run from the sound of disturbed bees, *Current Biology* **17** 832–833

King LE, Lawrence A, Douglas-Hamilton I and Vollrath F 2009 Beehive fence deters crop-raiding elephants, *African Journal of Ecology* **47** 131–137

Krishnan M 1975 *India's Wildlife in 1959–'70*, Bombay: Bombay Natural History Society

Kumar A 1985 Patterns of extinction in India, Sri Lanka, and elsewhere in south-east Asia: implications for lion-tailed macaque, wildlife management and the Indian conservation system, in *Monographs in*

Primatology, Vol.7. The Lion-Tailed Macaque: Status and Conservation (ed) PG Heltne, New York: Alan R Liss Inc, 65–89 pages

Laurie WA 1982 Behavioural ecology of the greater one-horned rhinoceros (*Rhinoceros unicornis*), *Journal of Zoology London* **196** 307–341

Leakey PL and Lewin R 1995 *The Sixth Extinction: Biodiversity and its Survival*, London: Phoenix

Lewis DM and Jackson J 2005 Safari hunting and conservation on communal lands in southern Africa, in *People and Wildlife: Conflict for Coexistence* (eds) R Woodroffe, SJ Thirgood and A Rabinowitz, Cambridge: Cambridge University Press, 239–252 pages

Lewis DM and Kaweche GB 1985 The Laungwa Valley of Zambia: Preserving its future by integrated management, *Ambio* **14** 362–365

Lockwood M, Worboys GL and Kothari A 2006 *Managing Protected Areas: A Global Guide,* Trowbridge: Cromwell Press

McAllister DE, Schueler FW, Roberts CM and Hawkins JP 1994 Mapping and GIS analysis of the global distribution of coral reef fishes in an equal-area grid, in *Mapping the Diversity of Nature* (ed) RI Miller, London: Chapman and Hall, 155–175 pages

MacFie E 2003 *Human–Gorilla Conflict Resolution: Recommendations for Component Within IGCP Uganda Programming*, Nairobi: International Gorilla Conservation Program

Martin EB, Martin CB and Vigne L 1987 Conservation crisis – the rhinoceros in India, *Oryx* **21** 212–218

Martin RB 2005 *Transboundary Species Project Background Study: Elephants*, Windhoek: Transboundary Mammal Project of the Ministry of Environment and Tourism

McClean CJ, Lovett JC, Kuper W, Hannah L, Sommer JH, Barthlott W, Termansen M, Smith GF, Tokumine S and Taplin JRD 2005 African plant diversity and climate change. *Annals of the Missouri Botanical Garden* **92** 139–152

McComb K, Moss C, Durant SM, Baker L and Sayialel S 2001. Matriarchs as repositories of social knowledge in African elephants, *Science* **292** 491–494

McDougal CW 1977 *The Face of the Tiger*, London: Rivington and Andre Deutsch

McNeely JA 1988 *Economics and Biological Diversity: Developing and Using Economic Incentives to Conserve Biological Resources*, Gland: International Union for Conservation of Nature and Natural Resources

McNeely JA, Miller KR, Reid WV, Mittermeier RA and Werner PA 1990 *Conserving the World's Biological Diversity,* Gland: International Union for Conservation of Nature and Natural Resources

McNeely JA and Pitt D (eds) 1985 Culture and Conservation: The Human Dimension in Environmental Planning, London: Croom Helm Ltd

Meinshausen, M., Meinshausen N, Hare W, Raper SC, Frieler K, Knutti R, Frame DJ and Allen MR 2009 Greenhouse-gas emission targets for limiting global warming to 2°C, *Nature* **458** 1158–1162

Melekneh G, Suryabhagavan K and Balakrishnan M 2010 Land-use and landscape pattern changes in Holeta-Berga watershed, Ethiopia, *International Journal of Ecology and Environmental Sciences* **36** 117–132

Milner JM, Nilsen EB and Andreassen HP 2007 Demographic die effects of selective hunting in ungulates and carnivores, *Conservation Biology* **21** 36–47

Mishra HR 1982 *The Ecology and Behaviour of Chital (Axis axis) in the Royal Chitwan National Park, Nepal*, Ph.D. Dissertation, Edinburgh: University of Edinburgh

Mishra HR and Sherpa MN 1987 Nature conservation and human needs: conflicts or coexistence: Nepal's experiment with the Annapurna Conservation Area Project, Denver: *Proceedings of the 4th World Wilderness Congress*

Mittermeier RA 1988 Primate diversity and the tropical forest: case studies from Brazil and Madagascar and the importance of the megadiversity countries, in *Biodiversity* (eds) EO Wilson and FM Peter, Washington DC: National Academy Press 145–154 pages

Mittermeier RA and Werner TB 1990 Wealth of plants and animals unites 'megadiversity' countries, *Tropicus* **4** 4–5

Mohnot SM, Rajpurohit LS and Chhangani AK 1996 Solar eclipse interrupts ciradian rhythms of Hanuman langurs: A natural experiment, *Asian Primates* 5 19–20

Mountfort G 1969 *The Vanishing Jungle*, London: Collins

Müller-Schwarze D 1983 Scent glands in mammals and their functions, in *Advances in the Study of Mammalian Behaviour* (eds) JF Eisenberg and DG Kleiman, New York: Plenum Press, 150–197 pages

Muruthi P 2005 *Human–Wildlife Conflict: Lessons Learned from AWF's African Heartlands, Working Paper*, Nairobi: African Wildlife Foundation

Myers N, Mittermeier RA, Mittermeier CG, da Fonseca GAB and Kent J 2000 Biodiversity hotspots for conservation priorities, *Nature* **403** 853–858.

Mykytowycz R 1970 The role of the skin glands in mammalian communication, in *Advances in Chemoreception, 1 Communication by Chemical Signals* (eds) JW Johnston Jr, DG Moulton and A Turk, New York: Appleton-Century-Crofts, 327–360 pages

Naidu MK 1981 Potentiality of game farming for rural development, *Tiger Paper* **8** 2–4

Nair PV, Ramachandran KK, Vijayan VS, Easa PS and Balakrishnan PV 1985 *An Ecological Study in Periyar Tiger Reserve with Special Reference to Wildlife*, Peechi: Kerala Forest Research Institute, Report No. 24

Narendran TC and Balakrishnan M 2008 *Systematics and Biodiversity Conservation,* Jodhpur: Agrobios (India)

Newman A, Bush M, Wildt DE, Dam DV, Frankenhuis MT, Simmons L, Phillips L and O'Brien SJ 1985 Biochemical genetic variation in eight endangered or threatened felid species, *Journal of Mammalogy* **66** 256–267

Nyhus PJ and Tilson R 2004 Characterizing human–tiger conflict in Sumatra, Indonesia: implicatins for conservation, *Oryx* **38** 68–74

O'Brien SJ 1994 Genetic and phylogenetic analyses of endangered species, *Annual Review of Genetics* **28** 467–48

Oza GM 1986 Threats to unique wildlife through Indian habitat destruction, *Environmental Conservation* **13** 131–136

Pacheco LF and Simonetti JA 2000 Genetic structure of a Mimosoid tree deprived of its seed disperser, the spider monkey, *Conservation Biology* **14** 1766–1775

Panwar HS 1979 A note on tiger census technique based on pugmark tracing, *Tiger Paper* **6** 16–18

Panwar HS and Sale JB 1986 The national wildlife action plan, *Indian Forester* **112** 841–844

Parmesan C and Matthews J 2005 *Biological Impacts of Climate Change*, Sinauer Associates Inc. www.sinauer.com/groom. Accessed on October 04, 2010

Parmesan C, Ryrholm N, Stefanescus C and Hill JK 1999 Poleward shifts in the geographical ranges of butterfly species associated with regional warming, *Nature* **399** 579–583

Parr JF and Sullivan LA 2005 Soil carbon sequestration in phytoliths, *Soil Biology and Biochemistry* **37** 117–124

Poulsen JR, Clark CJ and Bolker BM. 2012 Experimental manipulation of seed shadows of an Afrotropical tree determines drivers of recruitment, *Ecology* **93** 500–510

Prakash I 1977 *The Amazing Life in the Indian Desert*, Jodhpur: Central Arid Zone Research Institute

Prasad S, Nair PV, Sharatchandra HC and Gadgil M 1978 On factors governing the distribution of wild mammals in Karnataka, *Journal of Bombay Natural History Society* **75** 718–743

Prater SH 1980 *The Book of Indian Animals*. 2nd edition, Bombay: Bombay Natural History Society

Putman RJ 1984 Facts from faeces, *Mammal Review* **14** 79–97

Rahman H and Parthasarathy MD 1967 A population survey of the bonnet monkey (*Macaca radiata* Geoffroy) in Bangalore, South India, *Journal of Bombay Natural History Society* **64** 251–255

Rahman H and Parthasarathy MD 1968 The expressive movements of the bonnet macaque, *Primates* **9** 259–272

Rahman H and Parthasarathy MD 1969a The home range, roosting places and the day ranges of the bonnet macaques, *Journal of Zoology London* **157** 267–276

Rahman H and Parthasarathy MD 1969b Studies on the social behavior of bonnet monkeys, *Primates* **10** 149–162

Rahmani AR 2012 Threatened Birds of India – Their Conservation Requirements. Mumbai: Indian Bird Conservation Network, Bombay Natural History Society, Royal Society for the Protection of Birds, Oxford University Press

Rahmani AR and Manakadan R 1990 The past and present distribution of the Great Indian Bustard *Ardeotis nigriceps* Vigors, *Journal of Bombay Natural History Society* **87** 175–194

Rahmani AR and Soni RG 1997 Avifaunal changes in the Indian Thar Desert, *Journal of Arid Environments* **36** 687–703

Rajpurohit LS, Chhangani AK, Rajpurohit RS and Mohnot SM 2003 Observation of a sudden resident male replacement in a unimale bisexual troop of Hanuman langur, *Semnopithecus entellus* around Jodhpur, India. *Folia Primatologica* **74** 85–87

Rajpurohit LS and Mohnot SM 1988 Field observation on nilgai (*Boselaphus tragocamelus)*, Jodhpur, *Tiger Paper* **15** 24–29

Rajpurohit LS, Mohnot SM, Srivastava A and Chhangani AK 2001 Demography of Jodhpur langurs, 1970–2000, *Advances in Ethology* **36** 245–246

Ralls K 1977 Mammalian scent marking, *Science* **171** 443–449

Raman R and Gopal B 2009 Recent land use/land cover changes in the floodplain of the highly regulated River Yamuna upstream of Delhi, *International Journal of Ecology and Environmental Sciences* **35** 237–254

Rao JVR and Prasad NLNS 1982 *Management and Husbandry of Black Buck*, Bangkok: Food and Agricultural Organization

Rao KS, Maikhuri RK, Nautiyal S and Saxena KG 2002 Crop damage and livestock depredation by wildlife: a case study from Nanda Devi Biosphere Reserve, India. *Journal of Environmental Management* **66** 317–327

Rasmussen HB, Kahindi O, Vollrath F and Douglas-Hamilton I 2005 Estimating elephant densities from wells and droppings in dried out riverbeds, *African Journal of Ecology* **43** 312–319

Reid GM, Macbeath TC and Csatadi K 2013 Global challenges in freshwater-fish conservation related to public aquarium industry, *International Zoo Yearbook* **47** 6–15

Renwick A, Ball A and Pretty JN 2002 Biological and policy constrains on the adoption of carbon farming in temperate regions, *Philosophical Transactions of the Royal Society, London A* **360** 1721–1740

Rice CG 1986 Observations on predators and prey at Eravikulam National Park, Kerala, *Journal of Bombay Natural History Society* **83** 283–305

Rice CG 1988a Habitat, population dynamics, and conservation of the Nilgiri tahr, *Hemitragus hylocrius, Biological Conservation* **44** 137–156

Rice CG 1988b Reproductive biology of Nilgiri tahr, *Hemitragus hylocrius* (Mammalia: Bovidae), *Journal of Zoology London* **214** 269–284

Roberts JT 1977 *The Mammals of Pakistan*, London: Ernest Benn

Rodgers WA and Panwar HS 1988 *Planning a Wildlife Protected Area Network in India.* Volumes 1 & 2, New Delhi: Ministry of Environment and Forests and Wildlife, Government of India

Rodgers WA, Panwar HS and Mathur VB 2002 *Wildlife Protected Area Network in India: A Review* (Executive summary), Dehradun: Wildlife Institute of India

Roonwal ML 1977 Urbanisation in primates and comparison with man, *Journal of Scientific and Industrial Research* **36** 179–187

Rowcliffe JR, Field J, Turvey ST and Carbone C 2008 Estimating animal density using camera traps without the need for individual recognition, *Journal of Applied Ecology* **45** 1228–1236

Rowell G 1989 Annapurna: Sanctuary for the Himalaya, *National Geography* **176** 391–404

Roy PS 1993 Remote sensing for forest ecosystem analysis and management, in *Environmental Problems and Prospects in India* (ed) M Balakrishnan, New Delhi: Oxford & IBH Publishing Co Pvt Ltd, 335–364 pages

Roy PS and Raven SA 1994 Habitat management of biodiversity maintenance using aerospace sensing, in *Tropical Ecosystems: A Synthesis of Tropical Ecology and Conservation* (eds) M Balakrishnan, R Borgstrom and SW Bie, New Delhi: Oxford & IBH Publishing Co Pvt Ltd, 193–232 pages

Saha D 1987 A study of the Sundarbans Tiger Reserve, West Bengal, India, *Environmental Ecology* **5** 430–434

Said MY, Ottichilo WK, Sinange RK and Aligula HM 1997 Population and Distribution Trends of Wildlife and Livestock in the Mara Ecosystem and the Surrounding Areas, Nairobi: Kenya Wildlife Service

Santiapillai C, Chambers MR and Ishwaran N 1984 Aspects of ecology of the Asian elephant, *Elephas maximus* L. in the Ruhuna National Park, Sri Lanka, *Biological Conservation* **29** 47–61

Schaaf CD 1978 *Population Size and Structure and Habitat Relations of the Barasingha (Cervus duvauceli) in Sukla Phanta Wildlife Reserve*, Ph.D. Dissertation, Michigan: Michigan State University

Schaller GB 1967 *The Deer and the Tiger*, Chicago: University of Chicago Press

Schaller GB 1976a Mountain mammals in Pakistan, *Oryx* **13** 351–356

Schaller GB 1976b Pakistan: status of wildlife research needs, in *Ecological Guidelines for the Use of Natural Resources in the Middle-east and South-west Asia*, Morges: International Union for Conservation of Nature and Natural Resources, 133–143 pages

Scheffer M, Carpenter S, Foley JA, Folke C and Walker B 2001 Catastrophic shifts in ecosystems, *Nature* **413** 591–596

Seidensticker J 1976 On the ecological separation between tigers and leopards, *Biotropica* **8** 225–234

Sekhar NU 1998 Crop and livestock depredation caused by wild animals in protected areas: the case of Sariska Tiger Reserve, Rajasthan, India, *Environmental Conservation* **25** 160–171

Shannan CE and Weaver W 1949 *The Mathematical Theory of Communication*, Urbana: University of Illinois Press

Sharatchandra HC and Gadgil M 1975 A year of Bandipur, *Journal of Bombay Natural History Society* **72** 623–647

Shedha, MD and Treadwell BD 1995 Remote sensing applications in habitat mapping, in *Wildlife Research and Mangement – Asian and American Approaches* (eds) SH Berwick and VB Saharia, New Delhi: Oxford University Press, 55–66 pages

Sheikh PA, Corn ML, Leggett JA and Folger P 2007 *Global Climate Change and Wildlife*, Congressional Research Service Report for Congress, Code RS22597

Sillero-Zubiri C and Switzer D 2001 *Crop Raiding Primates: Searching for Alternative, Humane Ways to Resolve Conflict with Farmers in Africa*, Oxford: Oxford University

Simpson EH 1949 Measurements of diversity, *Nature* **163** 688

Singh M, Pillai S and Akram N 1987 Infant development and mother–infant interactions: An individual selection hypothesis, in *Recent Trends in Ethology in India* (eds) M Balakrishnan and KM Alexander, Bangalore: The Ethological Society of India, Bangalore,138–144 pages

Singh M, Prakash P and Pirta RS 1980 Changing patterns of behavior during the early stages of bonnet monkey *Macaca radiata*, *Proceedings of the Indian Academy of Sciences (Animal Science)* **89** 101–108

Smith JLD 1984 *Dispersal, Communication, and Conservation Strategies for the Tiger (Panthera tigris) in Royal Chitwan National Park, Nepal*, Ph.D. Dissertation, Minnesota: University of Minnesota

Smith JLD, McDougal C and Miquelle D 1989 Scent marking in free-ranging tigers, *Panthera tigris, Animal Behaviour* **38** 1–10

Smith JLD, McDougal C and Sunquits ME 1987 Female land tenure system in tigers, in *Tigers of the World: The Biology, Biopolitics, Management and Conservation of an Endangered Species* (eds) RL Tilson and US Seal, New Jersey: Park Ridge, 97–109 pages

Smith RL 1974 *Ecology and Field Biology* 2nd edition, New York: Harper and Row

Soulé ME 1985 What is conservation biology? *BioSciences* **35** 727–734

Soulé ME (ed) 1986 *Conservation Biology: The Science of Scarcity and Diversity*, Sunderland: Sinauer Associates

Soulé ME and Kohm KA (eds) 1989 *Research Priorities for Conservation Biology,* Washington DC: Island Press

Soulé ME and Wilcox BA 1980 *Conservation Biology: An Evolutionary-Ecological Approach*, Sunderland: Sinauer Associates

Sreekumar PG and Balakrishnan M 2001 Diversity and habitat preference of butterflies in Neyyar Wildlife Sanctuary, South India, *Entomon* **26** 11–22

Sreekumar PG and Balakrishnan M 2002 Seed dispersal by sloth bear (*Melursus ursinus*) in South India. *Biotropica* **34** 474–477

Stander P 1990 A suggested management strategy for stock-raiding lions in Namibia, *South African Journal of Wildlife Research* **20** 37–43

Stander P 2006 *Population Ecology and Demography of Kunene Lions: Towards Resolving Human–Lion Conflicts with Applied Research and Proactive Management*, Windhoek: Predator Conservation Trust

Stirling I, Lunn NJ and Iacozza J 1999 Long-term trends in the population ecology of polar bears in western Hudson Bay in relation to climate change, *Arctic* **52** 294–306

Stoddart LA, Smith AD and Box TW 1975 *Range Management*, New York: McGraw-Hill Book Company 532 page

Sugiyama Y 1965 On the social change of Hanuman langur (*Presbytis entellus*) in their natural condition, *Primates* **6** 382–418

Sugiyama Y 1976 Characteristics of the ecology of the Himalayan langur, *Journal of Human Evolution* **5** 249–277

Sugiyama Y, Yoshiba K and Parthasarathy MD 1965 Home range, mating season, male group and inter-troop relations in Hanuman langur (*Presbytis entellus*), *Primates* **6** 73–106

Sukumar R *1985 Ecology of the Asian Elephant (Elephas maximus) and its Interaction with Man in South India*, Ph.D. Dissertation, Bangalore: Indian Institute of Science

Sukumar R 2003 *The Living Elephants: Evolutionary Ecology, Behaviour and Conservation*, New York: Oxford University Press

Sunquist MV 1981 *The Social Organization of Tigers (Panthera tigris) in Royal Chitwan National Park*, Washington DC: Smithsonian Contribution in Zoology No. 336

Taber RD and Saharia VV 1995 Comparative approaches to wildlife management: the setting for the workshop, in *The Development of International Principles and Practices of Wildlife Management Research and Management: Asian and American Approaches*, (eds) SH Berwick and VB Saharia, Delhi: Oxford University Press, 4–9 pages

Tamang KM 1982 *Dynamics of Tiger Prey Population in Royal Chitwan National Park*, Ph.D. Dissertation, Michigan: Michigan State University

Teer JG 1995 Rangeland habitat: its evaluation and management for wildlife, in *The Development of International Principles and Practices of Wildlife Research and Management: Asian and American Approaches*, (eds) SH Berwick and VD Saharia, Delhi: Oxford University Press, 175–194 pages

Theile S 2003 *Fading Footprints: The Killiing and Trade of Snow Leopards*, Cambridge: TRAFFIC International

Thiessen DD, Own K and Lindzey G 1971 Mechanisms of territorial marking in the male and female Mongolian gerbil (*Meriones unguiculatus*), *Journal of Comparative Physiology and Psychology* **77** 38–47

Thomas CD, Bodsworth EJ, Wilson RJ, Simmons AD, Davies ZG, Musche M and Conradt L 2001 Ecological and evolutionary processes at expanding range margins, *Nature* **411** 577–581

Thomas R and Balakrishnan M 1999 Depletion of biodiversity in an *Acacia auriculiformis* introduced habitat, *International Journal of Ecology and Environmental Sciences* **25** 127–141

Thuiller W 2008 Predicting global change impacts on plant species distributions: future challenges, perspectives in plant ecology, *Evolutionary System* **9** 137–152

UNEP 1992 *Convention on Biological Diversity*, Nairobi: United Nations Environment Programme

UNEP 1995 *Global Biodiversity Assessment*, Cambridge: Cambridge University Press

Vancuylenberg BWB 1977 Feeding behavior of the Asiatic elephant in south-east Sri Lanka in relation to conservation, *Biological Conservation* **12** 33–54

Vijayan VS and Balakrishnan M 1976 *Impacts of the proposed hydro-electric project in Silent Valley on Wildlife – report of the first phase of the study*, Peechi: Kerala Forest Research Institute

Vollrath F and Douglas-Hamilton I 2002 African bees to control African elephants, *Naturwiss* **89** 508–511

Wang K, Franklin ES, Guo X and Cattet M 2010 Remote sensing of ecology, biodiversity and conservation: a review from the perspective of remote sensing specialists, *Sensors* **10** 9647–9667

Wasonga DV, Afework B, Lotters S and Balakrishnan M 2006 Habitat association of amphibians in East African bush and grasslands: an example from Meru National Park, Kenya, *Tropical Zoology* **19** 145–157

Webb G, Whitehead P and Manolis S 1987 Crocodile management in the northern territory of Australia, in *Wildlife Management: Crocodile and*

Alligators (eds) G Webb, S Manolis and P Whitehead, Chipping Norton: Beatty and Sons, 107–124 pages

Wegge P 1976 *Himalayan Shikar Reserves: Surveys and Management Proposals*, FAO Field Document No. 5, Kathmandu: Tribhuvan Univeristy Press

Wegge P 1979 Aspects of the population ecology of blue sheep in Nepal, *Journal of Asian Ecology* **1** 10–20

Wegge P 1988 *Assessment of Khunjerab National Park and Environs, Pakistan*, Glands: International Union for Conservation of Nature and Natural Resources

Whitaker R 1989 Crocodile farming as a tribal industry, *Herpetologica* **2** 1–7

Whittaker RJ 1998 *Island Biogeography: Ecology, Evolution and Conservation*, Oxford: Oxford University Press

Wilson P 1981 Ecology and habitat utilization of blue sheep *Pseudois nayaur* in Nepal, *Biological Conservation* **21** 55–74

Wondmagene D, Afework B, Balakrishnan M and Gurja B 2011 Collection of African civet *Civettictis civetta* perineal gland secretion from naturally scent-marked sites, *Small Carnivore Conservation* **44** 14–18

Workneh G, Afework B, Gurja B and Balakrishnan M 2004 Microhabitat choice and diet of small mammals in Maynugus irrigation field, Ethiopia, *African Journal of Ecology* **42** 315–321

Workneh G, Afework B and Balakrishnan M 2006 Population structure of small mammals in Maynugus irrigation field, northern Ethiopia, *International Journal of Ecology and Environmental Sciences* **31** 337–342

WRI/IUCN/UNEP 1992 *Global Biodiversity Strategy: Guidelines for Action to Save, Study and Use Earth's Biotic Wealth Sustainability and Equitability*, Washington DC: World Resource Institute

Yoshio M and Ishii M 2001 A phenological study on northward invasion of the great mormon, *Papilio memnon* L (in Japanese with English summary), *Japan Journal of Ecology* **51** 125–130

Subject Index

Author Index